COMMERCIAL POULTRY NUTRITION
SECOND EDITION

BY

STEVEN LEESON, Ph.D.
Professor of Animal Nutrition

AND

JOHN D. SUMMERS, Ph.D.
Professor Emeritus

Department of Animal & Poultry Science
University of Guelph
Guelph, Ontario, Canada

PUBLISHED BY
UNIVERSITY BOOKS
P. O. Box 1326
Guelph, Ontario, Canada
N1H 6N8

Commercial Poultry Nutrition, Second Edition
University Books,
P. O. Box 1326,
GUELPH, Ontario.
*** ***

Canadian cataloguing in publication data.
1. Poultry - Feeding and Nutrition **2.** Birds. **3.** Nutrition
I Leeson, Steven, 1948, **II** Summers, John D. 1929.

ISBN 0-9695600-2-8

ACKNOWLEDGEMENTS

We have updated and expanded the material presented in the first edition of this book, and once again, are very appreciative of the help and assistance of many individuals and organizations. We would like to extend a special thanks to our corporate sponsors as listed on the following page. Their contribution has allowed University Books to maintain a reasonable price on this new publication. We would also like to acknowledge the support of Finfeeds International.

Much of the material in this book is based on our ideas and concepts that in many instances have developed from work carried out in the Department of Animal and Poultry Science at the University of Guelph. We are, therefore, indebted to the University for providing us with an environment to develop these ideas, and to the Ontario Ministry of Agriculture Food and Rural Affairs for their generous support of our research programs over many years.

We would also like to give special thanks to Linda Caston and Diane Spratt for their invaluable assistance in proof-reading and to Papple Graphics for their input regarding artwork and layout. As in the past, our thanks to Wendy Bauer for typing so many drafts of this book and to Helen Klemencic for her invaluable assistance.

S. LEESON and J.D. SUMMERS

Guelph, 1997

SPONSORS

HUBBARD FARMS, INC.

HY-LINE INTERNATIONAL

ALLTECH

NOVUS INTERNATIONAL INC.

BASF CANADA

IMC-AGRICO COMPANY

HEARTLAND LYSINE INC.

AMERICAN PROTEINS INC.

TABLE OF CONTENTS

CHAPTER 1. NUTRITION AND WORLD POULTRY PRODUCTION

1.1. WORLD ANIMAL PRODUCTION

World food shortages, as predicted in the early 1970's, have not materialized, and in fact most developed countries are faced with surplus production of many commodities. That famine is painfully evident in certain countries is most often a reflection of economic or political restraint, rather than a shortage of food or animal feed in the world markets. Even so, our current abundance of most feeds and foods must be balanced against our continuing increase in world population. Each year we are expected to feed an extra 70 - 80 million people, and by the year 2,000 the world population is predicted at around 7 billion.

So far, we have been able to meet increased demands for human food through a combination of increased supply and through improved production efficiency. Such improvements in efficiency of production will allow us to gradually upgrade the general nutritional status of the world population as a whole and it is evident that the poultry industry is playing a major role in this important development. In the past we have had to face criticism of the energy use in animal production and especially from the point of view of inefficiency of consuming animal vs vegetable protein. Of the total energy used by most developed countries, less than 4% is used for food production. During this food production, by far the greatest quantities of energy are used during processing and household preparation to meet the stringent demands of the consumer. Consequently, of the 4% of energy used by the agrifood business, only 18% of this is actually used in primary animal production.

Increased human consumption of vegetable proteins has failed to materialize, essentially due to excessive energy use, necessary during manufacture, which is a criticism originally aimed at animal production. The production of synthetic meat analogues is thus very energy intensive, and their limited impact over the last decade attests to problems with economic viability and/or consumer acceptance. With the economy of many third world countries improving, there appears to be increased demand for animal products and especially meat and eggs. While in the immediate future there will likely be no major increase in production of meat, current trends suggest that poultry is assuming a larger proportion of this global commodity.

1.2. POULTRY MEAT PRODUCTION

Poultry meat, and to a lesser extent eggs are ideally suited to meet the increased demands for animal products with improved efficiency of production. The success of the broiler chicken in North America is now being mirrored around the world due to four major factors:

1. Ease of establishing integrated operations
2. Economically competitive price of poultry vs red meats
3. Adaptability for further processing
4. Meat composition in relation to human health

Chicken continues to be the least expensive meat in most countries, and this undoubtedly has been the main reason for its success world-wide. This competitive situation has occurred due to continued improvements in efficiency of production, that often necessitate acceptance of new ideas and innovations by producers and agribusiness organizations. On the other hand, production systems for competitive meat products have shown little change over the last two decades.

In N. America much of the success of the chicken meat industry relates to development of new consumer products, largely as a result of further processing. The most famous single product is undoubtedly the "chicken nugget", now featured by most fast-food and retail outlets. Over the last 10 years, some 30,000 non-chicken fast-food outlets in N. America have added chicken products to their menu, and during special advertising campaigns, chicken products can be the leading sales item over such conventional products as hamburgers. So-called "fast-food" stores are increasing in number in both Eastern and Western Europe, in Asia and in South America, and this will likely lead to increased demand for chicken.

In addition to developing new uses for conventional parts of the chicken and turkey carcass, the industry has also been successful in developing technology to use its own "by-products" and then finding markets for these (or vice versa). The demand for chicken wings and more recently chicken feet together with mechanically deboned meat exemplify these types of products. In addition to increasing overall poultry meat consumption, these products also lead to improved overall efficiency of production and help maintain the economic advantage seen with poultry meat.

Poultry meat is also ideally suited in terms of meeting demands for leaner meat by health conscious consumers. There has been considerable publicity over the last few years concerning the relative fat content of various meats, yet the fact remains that when comparisons are conducted on comparable products, poultry meat is by far the leanest product. Comparison of a highly trimmed steak or pork chop vs a whole broiler carcass certainly reduces the advantage usually seen with poultry. However the valid comparison is trimmed steak vs poultry breast fillet, in which case the poultry product is by far the leanest. Broiler chicken and especially turkey are therefore ideal products for segments of the food industry wishing to provide low-fat meals. Poultry meat also has the almost unique advantage of not being discriminated against due to religious or cultural beliefs, making poultry products popular with airline caterers, hotels, institutions etc. The projected production of poultry meat in 1997, relative to other meats, is shown in Table 1.1.

TABLE 1.1 Projected live animal production for 1997 ('000 tonnes)

	World	U.S.A.	Canada	Mexico	Brazil	S. America	China	Asia	Europe
Poultry	45,000	13,000	800	1,200	2,500	4,000	3,000	6,000	8,000
Pork	42,000	10,000	800	1,500	1,000	3,000	4,000	7,000	11,000
Beef	50,000	10,000	1,000	1,400	2,500	9,000	500	1,000	7,000

1.3. EGG PRODUCTION

The egg industry has not enjoyed the success of the meat industry and this situation is generally related to two main factors. Firstly, the controversy of cholesterol content of eggs and human health continues to receive undue adverse publicity although more recently these views have been questioned even by the medical profession. There is little doubt that human diets high in saturated fatty acids are harmful in terms of accentuating arterosclerosis type conditions in susceptible individuals, although the role of diet cholesterol in this issue has not been fully resolved. Certainly the most recent evidence suggests that eggs contain up to 30% less cholesterol than is stated in dated reference material often used by health practitioners and the food industry. However, regardless of absolute levels of cholesterol, the undisputable fact is that eggs do contain cholesterol and so there will likely be some concern in more affluent markets where food alternatives are plentiful. It is interesting to note that cholesterol is rarely an issue in developing countries that are striving to meet human protein needs and where the laying hen provides one of the most efficient means of converting feedstuffs, unsuited for human consumption, into the very best protein source known to humanity.

In developed countries, the egg industry faces another major problem concerning development of new products. As the poultry meat industry has flourished due to new product development, so egg consumption has stagnated or regressed for exactly the opposite reason. While a number of fast-food restaurant chains have featured egg products, they are still regarded as essentially breakfast items, and therefore have limited potential impact. Egg consumption will likely only increase, to any marked degree, through development of new food products, a trend that is now starting to emerge in N. America. Similarly the biochemical components of the egg await entrepreneurial development for both food and non-food use, and in this context manipulation of egg composition through genetic manipulation of the avian genome now seems a real possibility.

The future of the egg industry is seen by many to relate to further processing of eggs and as such, egg mass production and particularly egg solids production become the major criteria. To meet these demands, primary breeders are continually increasing egg mass production of their strains, although solids yield is often not considered in a breeding program. Considerable progress has been made in most commercial layer traits over the last 20 years although it is obvious that brown-egg birds are being changed most rapidly. Today on a world-wide basis, the distribution of white to brown egg birds is very close to 50:50. However in most these amino acids in a readily digested form. With methionine, lysine and threonine now available at competitive prices, it is possible to formulate practical diets that provide a minimum excess of amino acids and non-protein nitrogen. Unfortunately we seem unable to take this approach to its logical conclusion and formulate diets with very low levels of crude protein, that contain regular levels of essential amino acids. At some point in the reduction of crude protein we seem to lose growth rate or egg production/egg size which suggests that either we have reached the point at which non-essential amino acids become important, or that we have inadequately described the bird's amino acids needs or that the synthetic amino acids are not being used with expected efficiency. Of these factors, the need to more adequately describe amino acid needs under these specific formulation procedures, is probably most important. However, we can readily reduce crude protein supply by 15 - 20% if the use of synthetic amino acids is economical or if there is a cost associated with the disposal of manure nutrients.

In a study with broiler breeders (see Chapter 6) we have successfully used 10% rather than 16% CP by appropriate use of synthetic amino acids, and with no loss of performance, realize a 1.4 tonne annual reduction in manure nitrogen excretion per 10,000 birds. Phosphorus excretion on the other hand rarely relates to overformulation, because this nutrient is expensive and levels are normally kept to a minimum. Large quantities of phosphorus are excreted by meat birds and layers, because of the non-availability of most phosphorus derived from plant material. Much of the phosphorus in cereals and vegetable proteins is in the form of phytic acid, and this is poorly digested, especially by young birds. The only practical solution to this problem is to use synthetic phytase enzymes or possibly certain vitamin D_3 metabolites. While phytase enzymes seem to drastically reduce phosphorus excretion by birds, most of these exogenous enzymes are heat labile, and are destroyed by pelleting, expanding etc. Their routine use will probably occur where regulations dictate maximum phosphorus excretion and/or land application rates, and they will likely be sprayed on to feed following any heat processing.

In the future we may have to more closely consider the balance of other minerals, while the fate of antibiotics, anticoccidials and other feed additives as they appear in the manure, has surprisingly attracted little attention to date. In the following chapters, nutrient excretion rates in manure are considered in setting diet specifications, although they are of secondary consideration. Where economics dictate, specified crude protein levels can be modified, and then levels of "truly available" phosphorus recalculated.

The handling and disposing of manure therefore still constitutes a major problem for many poultry producers. With the public becoming more and more aware of pollution, and with stricter controls being implemented by government agencies, poultry producers will have to find a better manner in which to hold and dispose of waste materials. Much has been written about the recycling of poultry wastes as a feed for laying hens. Although the true protein level of this dried material may run as high as 8 to 10%, a significant portion of it is indigestible protein from the original feedstuffs. Regardless of the quality of the protein in dried poultry waste, the low metabolizable energy content of this material makes it an undesirable and uneconomical ingredient for use in most poultry diets. The following tables show the average composition of various poultry wastes and also the quantity of waste produced. These figures are only average values and obviously will vary depending on a number of factors, such as temperature, type of diet, type of litter, length of holding time, type of storage, etc. Poultry litter contains appreciable quantities of N, P and K, and its value as a fertilizer for vegetable and field crops can be readily calculated. Besides sources of the three major plant nutrients, poultry manure is a good source of many of the trace elements and adds a considerable quantity of organic matter to the soil. On average, birds will produce about as much fresh manure as the weight of air-dry feed consumed (Table 1.4).

TABLE 1.4 Total weight of fresh manure produced

Type of bird	Body weight (kg)	Period (weeks)	Tonnes per 1000 birds
Broiler	2.0	6	3.0
Layer	1.8	60	42.0
Turkey	5.5	12	11.0
Turkey	15.0	18	37.0

Most poultry manure is used as a source of fertilizer, and on this basis analysis for nitrogen (N), potassium (K), and phosphorus (P) is essential information (Table 1.5).

TABLE 1.5 Fertilizer value of poultry manures

	Fresh				Dry		
	% Dry Matter	% Nitrogen	% P_2O_5	% K_2O	% Nitrogen	% P_2O_5	% K_2O
Broiler litter	70	4.0	3.4	2.1	5.0	4.3	2.6
Turkey litter	70	8.0	3.8	2.0	12.0	5.7	3.0
Cage-layer manure	30	1.5	1.3	0.5	5.0	4.2	1.7
Deep-litter layer manure (1 year old)	72	2.1	2.5	1.1	2.9	3.5	1.5

TABLE 1.6 Average mineral composition of dried poultry manure (kg/tonne)

Mineral	Broiler	Caged layer	Turkey
Calcium	19.7	34.1	22.1
Magnesium	3.7	5.2	4.1
Sulphur	3.8	4.8	5.2
Manganese	0.28	0.33	0.4
Iron	1.3	0.03	1.0
Boron	0.04	0.03	0.4
Copper	0.03	0.12	0.4
Zinc	0.13	0.12	0.18
Molybdenum	0.02	0.02	0.02

TABLE 1.7 Chemical composition of peanut hull and wood shaving broiler litters

Component	Litter	
	Peanut hulls	Wood shavings
Dry matter (%)	89.1	88.9
	Percent composition of dry matter	
Crude protein	32.0	30.6
Crude fiber	15.1	14.6
Fat	2.8	2.8
Ash	17.9	19.0
N.F.E.	31.8	33.1
Lignin	9.4	10.4
Calcium	2.8	2.5
Phosphorus	2.9	2.3
	Percent of total nitrogen	
Nitrogen fractions		
True protein	45.4	44.4
Uric acid	30.5	28.8
Ammonia	13.2	15.4
Urea	2.7	2.8
Creatinine	3.5	3.6
Others	4.7	5.0

Due to the high cost of drying, alternative methods of handling manure are being studied. Recent work in our department has shown that sheep and cattle will consume diets containing relatively high levels of wet poultry manure if it is fresh, or if it has had some type of preservative added to it. Indeed, studies have suggested that ruminants can be fed poultry wastes in a liquid diet as a supplement to silage and/or grain feeding. Caution should be used when feeding poultry wastes to sheep. Sheep are far more sensitive to copper toxicity than cattle, hence one must be sure that the copper level in the manure is within the range that can be tolerated by sheep. Sheep rations containing >50 ppm of copper can cause mortality if fed for prolonged periods.

The detailed composition values in Table 1.8 can be used as a guide for feeding cattle. In the past, poultry wastes have often been fed to cattle with little or no thought to balancing nutrients. When asked how much poultry manure could be fed, the reply was often "as much as they will eat". This is the wrong approach to making efficient use of poultry wastes in a ruminant feeding program. Ruminants, like any other animal, require balanced rations if they are to perform satisfactorily. Thus, one should feed only the level of poultry manure that can be incorporated into a particular feeding program and which will not result in nutrient imbalances. Since much of the nitrogen in poultry manure is present as insoluble uric acid, it is an ideal source of slow-release nitrogen for rumen microorganisms.

TABLE 1.8 Average nutrient composition of dried caged layer manure (air dry %)

Moisture	9.6
Energy (kcal/kg)	850
Protein (N x 6.25)	27
True protein	10.6
Uric acid	5.7
Arginine	0.41
Leucine	0.60
Lysine	0.39
Methionine + cystine	0.33
Threonine	0.40
Tryptophan	0.53
Fat	1.8
Nitrogen-free extract	35.5
Fiber	14.9
Ash	26.5
Calcium	7.4
Phosphorus	2.1
Sodium	0.4
Potassium	1.3

CHAPTER 2. INGREDIENT EVALUATION AND DIET FORMULATION

Page No

Feed is the major cost of production for most classes of poultry, and so it is essential to minimize feed costs. Over the past 20 years, reduction in feed costs *per se* has been made possible by the use of least cost formulation.

2.1. LEAST COST FORMULATION

Formulation of diets requires three major inputs. Firstly, one must know the nutrient requirements of the bird to which the diet will be fed and from this knowledge establish the nutrient specifications for the diet. Such nutrients will be provided by various ingredients and therefore knowledge of the content of these same nutrients within the ingredient is the second requirement. Because a least-cost situation is desirable, the final requirement is to know the current market price of the selected ingredients. With this information, numerous computer programs are able to generate least-cost diets.

It is realized that least-cost diets may not always result in maximum profit regardless of whether the end-product is liveweight, meat, eggs, etc. In order to formulate maximum profit diets, then in addition to the information needed to formulate on the basis of simple least-cost, one also needs to know the bird's production response to options of variation in, for example, nutrient density. A key factor in formulating to maximum profit is to accurately predict animal product value at a future market age. Perhaps more importantly, it must be realized that the formulation of any single diet or series of diets, is only the starting point of a feeding program. All too often, nutritionists consider their job finished with the formulation of the diets. Whether diets are based on least-cost, maximum-profit, or any other basis, they will only be effective when used as part of a carefully designed feeding program that itself is part of a prescribed management strategy.

2.2. FEEDING PROGRAMS

In the following chapters, it is hoped to outline various feeding programs for most classes of birds. Part of these feeding programs entail diet specifications, while the remaining discussion outlines the selection criteria for the use of such diets in a feeding program. Development of such programs relies entirely on defining the goals that are expected. This statement sounds rather mundane, but all too often goals are not clearly defined and consequently poultry producers are disappointed with the results.

Production goals of feeding programs obviously vary with the species of bird, yet refinement of the programs means that goals vary within a species, within an age of a given species, and in some instances for each sex. A number of goals are universal such as feed efficiency and growth rate, but even here, there is some need for caution. For example: Is maximum growth rate the goal for all meat birds under all environmental conditions? Recent evidence suggests not, and that even a period of slower growth within a grow-out cycle may be beneficial. For laying hens, egg size is an obvious goal for selection of a feeding program, and yet requirements for weight, mass, or combinations of the two vary from country to country, and may vary seasonally within a geographical location. For laying hens, goals will also vary with age, where often there is need to maximize egg size in young birds, and then to reduce egg size as the birds get older. All these factors mean that nutritionists must have a basic understanding of the production systems, and factors influencing production economics of their various clients.

As a starting point to develop goals of feeding programs, there is one underlying theme that is important across most species namely, a knowledge of feed intake. Ultimately the nutrient intake of a bird is governed by its feed intake, and so knowledge of feed intake under various production situations is essential. Unfortunately, such information is surprisingly not always available, or is not available in sufficient detail to provide a basis for formulation. Such data often comes from small-scale experiments and extrapolation to commercial conditions requires some knowledge of bird response under these two extreme situations. Of more concern in recent years has been the estimation of feed intake of meat birds due to continual increases in the genetic potential of these birds.

2.3. INGREDIENT EVALUATION

The following tables describe the major nutrient content of a range of feed ingredients. These values are based on an ongoing data base that is revised according to fluctuations in chemical and biological analyses. In this respect, changes are made only where long-term trends are established, so as not to reflect seasonal fluctuations in growing conditions or transient changes in processing conditions. These data are presented as a comparative base for nutritionists, because it is realized that the most reliable information on ingredient composition is that which is developed locally. All values in the following tables are presented on an "as-is" basis for dry matter. In most instances, this is adjusted to 90% dry matter, the two major exceptions being corn which is assumed to be 86 - 88% dry matter, and dried whey at 95% dry matter. Most other ingredients, regardless of processing conditions, seem to stabilize at around 90% dry matter.

Available rather than total phosphorus values are given, since these are the only values of significance for the bird. Plant phosphorus is assumed to be 50% available. This may be an overestimation for the young chick and poult, because values closer to 30 - 40% are recorded with young birds. This fact is accounted for in a subsequent definition of requirements for young and older aged birds.

The vitamin A activity in feed ingredients containing pro-vitamin A is calculated at two-thirds of their average analysis values. Thus it is assumed alfalfa meal guaranteed at 220,000 U.S.P. units of vitamin A activity per kg contains only 145,200 U.S.P. units. While pro-vitamin A is well utilized by the chick at low or requirement levels, it is much less efficient than vitamin A as measured by liver storage when fed at higher levels. Levels of vitamin A used in these rations are sufficiently high to give reasonable liver storage, and the assumption of 66% availability for pro-vitamin A sources is conservative.

Choline and betaine have been considered together in Table 2.4, the figures presented being in terms of total choline plus total betaine hydrochloride when the betaine hydrochloride is divided by 1.26. This method presents choline and betaine in terms of choline equivalents. Betaine can replace the methyl donor function of choline and should be taken into account in practical ration formulation.

Theoretically, the most satisfactory measure of available energy in poultry feeds is net energy or productive energy. However, there is no practical method for measuring productive energy and the values obtained with various feedstuffs are not constant with, or independent of *(1)* age of the bird, *(2)* diet balance, *(3)* plane of nutrition (restricted or full-fed) and *(4)* productive performance of the birds. For this reason, Apparent Metabolizable Energy corrected to zero nitrogen retention is used in description of the ingredients. Such values are relatively easy to ascertain and reflect energy contribution under practical feeding situations, where synergism and antagonism between nutrients may be of importance. With the exception of details given for fat sources, only one ME value is given for each ingredient. It is known that the ME of some ingredients increases with age of bird, although such effects are quite small. Variable ME based on bird age would not likely be accepted by nutritionists during formulation. Turkeys are able to derive more energy from fibrous foods than are chickens, and those specializing in turkey nutrition often make this distinction in their data base ME values.

TABLE 2.1 Basic nutrient composition

Ingredient	Crude protein (%)	Digestible protein (%)	Metabolizable energy (kcal/kg)	Crude fat (%)	Crude fiber (%)	Calcium (%)	Available phosphorus (%)	Linoleic acid (%)
1. Yellow Corn	8.6	7.8	3329	3.8	2.5	0.01	0.13	1.9
2. Wheat	13.0	11.6	3153	1.5	2.9	0.05	0.20	0.5
3. Oats	12.0	9.9	2756	4.0	12.0	0.10	0.20	1.5
4. Barley	11.5	9.3	2795	2.0	8.0	0.10	0.20	0.9
5. Milo	9.0	7.9	3263	2.5	2.7	0.02	0.15	1.1
6. Rye	12.5	8.4	2734	1.7	2.4	0.05	0.18	0.4
7. Triticale	15.4	13.2	3110	1.0	4.5	0.05	0.19	0.4
8. Rice (rough)	7.3	5.5	2679	1.7	10.0	0.04	0.13	0.6
9. Wheat bran	15.8	11.7	1580	4.8	10.4	0.10	0.65	1.7
10. Wheat shorts	15.1	14.3	2870	4.2	3.6	0.07	0.30	1.8
11. Wheat screenings #1	14.9	11.7	3060	4.1	3.0	0.05	0.20	0.7
12. Rice bran	13.0	7.7	1900	1.7	12.0	0.06	0.90	3.4
13. Rice polishings	11.0	8.5	2750	15.0	2.4	0.06	0.18	3.3
14. Bakery by-product	10.6	9.8	3200	9.8	2.4	0.05	0.13	2.9
15. Molasses (cane)	3.0	2.1	1962	-	-	0.50	0.03	-
16. Dehydrated alfalfa meal	17.0	9.5	1647	2.0	26.0	1.40	0.10	0.3
17. Canola meal	37.5	34.0	2000	1.8	11.9	0.66	0.47	0.4
18. Full-fat canola seed	22.0	19.7	4620	40.0	6.0	0.38	0.27	8.0
19. Soybean meal (44%)	44.0	38.5	2491	0.5	7.0	0.25	0.33	0.4
20. Soybean meal (48%)	48.0	44.0	2557	0.5	3.0	0.20	0.37	0.4
21. Full-fat soybeans	37.5	33.4	3880	20.0	2.0	0.15	0.28	9.0
22. Corn gluten meal	60.0	54.4	3770	2.5	2.5	0.10	0.20	1.2
23. Corn gluten feed	22.0	14.3	1830	2.5	10.0	0.40	0.40	1.0
24. Cotton seed meal	41.0	33.2	2350	0.5	14.0	0.15	0.48	0.2
25. Peanut meal	47.0	35.7	2205	1.0	13.0	0.20	0.30	0.3
26. Peas	23.5	20.7	2550	1.3	5.5	0.10	0.20	0.9
27. Safflower meal	42.0	32.9	1630	1.1	14.5	0.37	0.63	0.5
28. Sesame meal	44.0	30.6	1984	5.0	5.0	0.20	0.75	2.0
29. Sunflower meal	46.8	35.6	2205	2.9	11.0	0.30	0.50	1.8
30. Lupins	34.5	29.8	3000	6.3	16.0	0.20	0.20	3.0
31. Flax	22.0	18.1	3600	34.0	6.1	0.25	0.17	5.2
32. Meat Meal	50.0	45.0	2500	6.0	2.5	8.00	4.00	0.6
33. Fish meal (60%)	60.0	55.4	2720	2.0	1.0	6.50	3.50	0.3
34. Poultry by-product meal	60.0	52.5	2950	8.0	2.0	3.50	2.10	2.4
35. Blood meal	80.0	71.2	2690	1.0	1.0	0.28	0.28	0.1
36. Feather meal	85.0	75.7	3016	2.5	1.5	0.20	0.75	0.1
37. Dried Whey	13.0	12.4	1918	0.5	-	0.80	0.70	0.1

TABLE 2.2 Total Amino Acid Composition

Ingredient	Methi -onine %	Cystine %	Lysine %	Hist -idine %	Trypt-ophan %	Thre-onine %	Arg-inine %	Iso-leucine %	Leu-cine %	Phenyl-alanine %	Val-ine %
1. Yellow Corn	0.2	0.1	0.2	0.2	0.1	0.4	0.4	0.5	1.0	0.5	0.4
2. Wheat	0.2	0.2	0.5	0.2	0.2	0.4	0.7	0.3	0.9	0.6	0.5
3. Oats	0.2	0.2	0.4	0.2	0.2	0.4	0.7	0.5	0.9	0.6	0.6
4. Barley	0.2	0.2	0.4	0.3	0.2	0.4	0.5	0.5	0.8	0.6	0.6
5. Milo	0.1	0.2	0.3	0.3	0.1	0.3	0.4	0.5	1.5	0.5	0.5
6. Rye	0.2	0.2	0.5	0.3	0.1	0.4	0.5	0.5	0.7	0.6	0.6
7. Triticale	0.2	0.2	0.4	0.3	0.1	0.3	0.8	0.5	1.0	0.7	0.7
8. Rice (rough)	0.1	0.1	0.2	0.2	0.1	0.3	0.6	0.3	0.7	0.3	0.5
9. Wheat bran	0.1	0.1	0.6	0.3	0.3	0.4	1.0	0.6	0.9	0.5	0.7
10. Wheat shorts	0.2	0.2	0.6	0.2	0.2	0.5	0.9	0.7	1.0	0.6	0.7
11. Wheat screenings #1	0.2	0.2	0.5	0.2	0.2	0.4	0.6	0.3	0.9	0.5	0.5
12. Rice bran	0.2	0.1	0.5	0.3	0.2	0.4	0.5	0.4	0.8	0.4	0.6
13. Rice polishings	0.2	0.1	0.5	0.2	0.1	0.3	0.6	0.3	0.7	0.4	0.7
14. Bakery by-product	0.2	0.2	0.3	0.3	0.1	0.3	0.5	0.4	0.8	0.6	0.5
15. Molasses (cane)	-	-	-	-	-	-	-	-	-	-	-
16. Dehydrated alfalfa meal	0.3	0.4	1.8	0.4	0.4	0.5	0.7	0.7	1.3	0.8	0.9
17. Canola meal	0.7	0.6	2.2	1.1	0.5	1.7	2.2	1.4	2.7	1.5	1.9
18. Full-fat canola seed	0.5	0.4	1.3	0.6	0.3	1.0	1.3	0.8	1.6	0.9	1.1
19. Soybean meal (44%)	0.6	0.6	2.9	1.1	0.6	1.7	3.2	2.5	3.4	2.2	2.4
20. Soybean meal (48%)	0.7	0.8	3.2	1.3	0.7	2.0	3.6	2.6	3.7	2.5	2.5
21. Full-fat soybeans	0.5	0.6	2.4	0.9	0.5	1.5	2.7	2.0	2.8	1.9	1.9
22. Corn gluten meal	1.6	0.9	0.9	1.4	0.3	1.7	2.2	2.4	8.1	3.2	2.6
23. Corn gluten feed	0.4	0.5	0.6	0.7	0.2	0.9	1.0	0.6	2.4	0.7	1.0
24. Cotton seed meal	0.5	0.6	1.7	1.0	0.5	1.3	4.6	1.3	2.4	2.2	1.9
25. Peanut meal	0.4	0.7	1.6	1.2	0.5	1.5	4.9	2.0	3.0	2.7	2.8
26. Peas	0.3	0.2	1.6	0.7	0.2	0.9	1.4	1.1	1.8	1.9	1.3
27. Safflower meal	0.4	0.7	1.3	0.4	0.3	0.6	2.9	0.6	1.2	1.2	1.1
28. Sesame meal	1.5	0.6	1.4	1.2	0.8	1.7	5.1	2.3	3.2	2.3	2.5
29. Sunflower meal	0.8	0.7	1.6	1.0	0.9	1.6	3.3	1.8	2.4	1.9	2.2
30. Lupins	0.3	0.6	1.7	0.9	0.4	1.2	4.5	1.4	2.4	1.3	1.4
31. Flax	0.4	0.4	0.9	0.4	0.3	0.8	2.1	1.0	1.3	1.0	1.1
32. Meat meal	0.7	0.6	3.6	0.7	0.5	1.7	3.0	1.3	3.3	1.6	2.4
33. Fish meal (60%)	1.8	1.1	5.3	1.6	0.6	2.9	4.0	4.1	5.0	2.7	3.6
34. Poultry by-product meal	1.3	2.0	3.4	1.0	0.4	2.2	3.5	2.1	4.5	1.8	3.0
35. Blood meal	1.0	1.4	6.9	4.2	1.1	3.7	3.5	1.0	10.0	6.0	7.0
36. Feather meal	0.6	5.5	1.7	0.5	0.6	4.5	6.4	4.3	6.5	4.3	7.4
37. Dried Whey	0.2	0.3	1.1	0.2	0.2	0.8	0.4	0.9	1.4	0.4	0.7

TABLE 2.3 Available amino acid composition

Ingredient	Methionine %	Cystine	Lysine %	Histidine %	Tryptophan %	Threonine %	Arginine %	Isoleucine %	Leucine %	Phenylalanine %	Valine %
1. Yellow Corn	0.18	0.08	0.16	0.18	0.07	0.33	0.36	0.44	0.8	0.42	0.33
2. Wheat	0.16	0.18	0.40	0.18	0.17	0.32	0.56	0.26	0.81	0.54	0.42
3. Oats	0.18	0.18	0.37	0.18	0.18	0.34	0.64	0.45	0.81	0.55	0.50
4. Barley	0.16	0.16	0.31	0.26	0.16	0.30	0.41	0.41	0.73	0.53	0.48
5. Milo	0.09	0.17	0.23	0.26	0.07	0.24	0.28	0.42	1.30	0.40	0.40
8. Rice (rough)	0.08	0.07	0.17	0.17	0.07	0.26	0.50	0.26	0.56	0.28	0.41
9. Wheat bran	0.08	0.07	0.42	0.24	0.24	0.28	0.79	0.48	0.72	0.41	0.55
10. Wheat shorts	0.16	0.14	0.48	0.16	0.15	0.40	0.71	0.56	0.84	0.49	0.57
12. Rice bran	0.15	0.07	0.39	0.24	0.14	0.28	0.40	0.31	0.54	0.30	0.46
13. Rice polishings	0.16	0.08	0.42	0.18	0.07	0.26	0.51	0.27	0.57	0.31	0.52
14. Bakery by-product	0.17	0.16	0.19	0.24	0.08	0.21	0.40	0.32	0.71	0.51	0.40
16. Dehydrated alfalfa meal	0.21	0.16	1.00	0.29	0.28	0.35	0.56	0.51	1.00	0.55	0.70
17. Canola meal	0.63	0.45	1.76	0.93	0.38	1.30	1.98	1.04	2.40	1.30	1.55
18. Full-fat canola seed	0.40	0.26	1.00	0.48	0.24	0.81	0.98	0.62	1.28	0.72	0.81
20. Soybean meal (48%)	0.64	0.64	2.88	1.07	0.51	1.76	3.21	2.30	3.20	2.10	2.20
21. Full-fat soybeans	0.42	0.50	2.00	0.74	0.42	1.27	2.30	1.72	2.20	1.70	1.70
22. Corn gluten meal	1.44	0.81	0.80	1.14	0.21	1.60	2.10	2.30	7.90	3.10	2.40
23. Corn gluten feed	0.33	0.35	0.42	0.56	0.14	0.65	0.87	0.48	2.12	0.63	0.83
24. Cotton seed meal	0.35	0.42	1.19	0.69	0.36	0.91	3.70	0.95	1.72	2.00	1.70
25. Peanut meal	0.33	0.55	1.28	0.96	0.38	1.20	4.00	1.80	2.70	2.30	2.40
28. Sesame meal	1.30	0.54	1.30	1.00	0.60	1.43	4.60	2.00	2.80	2.10	2.30
29. Sunflower meal	0.72	0.55	1.30	0.80	0.65	1.20	2.64	1.28	1.90	1.55	1.75
30. Lupins	0.27	0.54	1.40	0.81	0.26	1.00	4.10	1.20	2.20	1.10	1.20
31. Flax	0.33	0.30	0.72	0.32	0.34	0.65	1.76	0.72	1.10	0.76	0.95
32. Meat Meal	0.60	0.35	2.68	0.56	0.37	1.36	2.50	1.00	2.60	1.30	1.90
33. Fish meal (60%)	1.62	0.80	4.70	1.40	0.48	2.50	3.60	3.70	4.50	2.30	3.20
34. Poultry by-product meal	1.07	1.20	2.70	0.80	0.29	1.76	3.00	1.70	3.80	1.40	2.40
35. Blood meal	0.90	1.10	5.90	3.40	0.80	2.80	2.90	0.78	8.90	5.30	6.10
36. Feather meal	0.47	2.40	1.10	0.35	0.41	3.10	5.00	3.60	5.00	3.50	6.10

TABLE 2.4 Vitamin Composition

Ingredient	Vit B12 mcg/kg	Vit E mg/kg	Pantothenic acid mg/kg	Niacin mg/kg	Choline equiv. mg/kg	Riboflavin mg/kg	Thiamin mg/kg	Biotin mg/kg	Folic Acid mg/kg	Vit. A activity 10^3IU/kg	Xanthophyll mg/kg
1. Yellow Corn	-	19	5	21	660	1.3	4.4	0.07	0.02	2.40	20
2. Wheat	-	13	12	55	2200	1.1	4.8	0.11	0.40	-	3.5
3. Oats	-	18	12	15	1155	1.5	6.1	0.31	0.40	-	-
4. Barley	-	20	7	61	1210	2.2	5.5	0.20	0.51	-	-
5. Milo	-	11	11	41	660	1.1	3.7	0.29	0.02	-	1.3
6. Rye	-	9	7	1	1050	1.5	3.7	0.07	0.51	-	-
7. Triticale	-	9	8	2	950	2.5	4.2	0.10	0.60	-	-
8. Rice (rough)	-	10	8	28	990	1.1	2.8	0.90	0.24	-	-
9. Wheat bran	-	13	28	187	3674	3.1	7.9	0.11	1.78	-	-
10. Wheat shorts	-	21	17	92	3580	2.0	15.6	0.11	0.66	-	-
11. Wheat screenings #1	-	14	12	55	2090	1.1	4.4	0.12	0.40	-	-
12. Rice bran	-	55	22	286	1254	2.4	22.8	0.42	2.20	-	-
13. Rice polishings	-	85	57	528	1320	1.8	19.3	0.62	0.20	-	-
14. Bakery by-product	-	10	10	20	1100	1.0	7.2	0.12	0.30	-	-
15. Molasses (cane)	-	5	40	56	1200	4.0	2.8	0.09	0.18	-	-
16. Dehydrated alfalfa meal	11.0	120	29	44	4400	12.1	3.3	0.33	5.90	145.0	176
17. Canola meal	-	4	9	159	6450	1.2	4.0	0.28	0.55	-	-
18. Full-fat canola seed	-	60	5	92	3733	1.2	2.2	0.17	0.33	-	-
19. Soybean meal (44%)	1.1	4	14	26	2640	3.1	6.2	0.31	0.70	-	-
20. Soybean meal (48%)	1.1	2	14	20	2860	3.1	2.4	0.40	0.70	-	-
21. Full-fat soybeans	-	40	10	15	2174	2.4	1.8	0.31	0.53	3.2	-
22. Corn gluten meal	-	22	3	48	330	1.5	0.2	0.40	0.20	7.5	273
23. Corn gluten feed	-	15	16	66	1518	2.2	1.9	0.31	0.20	2.6	31
24. Cotton seed meal	-	12	7	38	3080	3.7	3.5	0.50	0.28	-	-
25. Peanut meal	-	3	52	165	1980	5.3	7.0	0.40	0.59	-	-
26. Peas	-	1	11	37	668	2.2	7.4	0.16	0.30	-	-
27. Safflower meal	-	1	44	22	2086	2.0	18.0	0.55	0.30	-	-
28. Sesame meal	-	7	6	13	1500	3.1	3.0	4.4	0.50	-	-
29. Sunflower meal	-	11	40	290	4300	3.3	2.0	1.8	0.60	-	-
30. Lupins	-	2	2	35	3100	2.8	5.3	0.04	0.40	-	-
31. Flax	-	18	4	40	3200	4.5	7.0	0.10	0.70	-	-
32. Meat Meal	70	1	4	44	1650	4.4	1.1	0.09	0.44	-	-
33. Fish meal (60%)	185	5	8	55	2860	4.8	0.7	0.22	0.24	-	-
34. Poultry by-product meal	100	5	8	45	3000	8.0	0.6	0.30	0.75	-	-
35. Blood meal	-	-	4	28	660	2.2	0.2	0.09	0.11	-	-
36. Feather meal	66.0	-	8	16	1320	2.0	0.1	0.04	0.22	-	-
37. Dried Whey	17.0	1	46	11	1980	26.4	3.7	0.39	0.90	-	-

TABLE 2.5 Mineral composition

Ingredients	Chloride (%)	Magnesium (%)	Sodium (%)	Potassium (%)	Iron (%)	Manganese (mg/kg)	Copper (mg/kg)	Zinc (mg/kg)	Selenium (mg/kg)
1. Yellow Corn	0.04	0.15	0.05	0.38	0.01	4	3	29	0.04
2. Wheat	0.08	0.16	0.09	0.52	0.01	48	7	40	0.51
3. Oats	0.10	0.17	0.06	0.37	0.01	38	5	31	0.30
4. Barley	0.18	0.12	0.02	0.56	0.01	16	7	40	0.35
5. Milo	0.08	0.17	0.04	0.34	0.01	14	9	26	0.03
6. Rye	0.37	0.12	0.02	0.26	0.01	66	7	30	0.45
7. Triticale	0.41	0.15	0.04	0.41	0.01	51	6	35	0.43
8. Rice (rough)	0.30	0.14	0.03	0.34	0.01	15	3	10	0.17
9. Wheat bran	0.30	0.15	0.06	1.24	0.02	115	12	89	0.95
10. Wheat shorts	0.10	0.26	0.07	0.85	0.01	104	9	99	0.81
11. Wheat screenings #1	0.05	0.15	0.09	0.52	0.01	48	7	40	0.55
12. Rice bran	0.07	0.85	0.10	1.35	0.02	425	14	30	0.20
13. Rice polishings	0.10	0.65	0.11	1.17	0.02	310	8	30	0.12
14. Bakery by-product	0.48	0.20	0.52	0.60	0.02	30	7	41	0.30
15. Molasses (cane)	0.65	0.40	0.30	3.50	0.02	50	20	35	0.08
16. Dehydrated alfalfa meal	0.45	0.34	0.16	2.40	0.03	50	9	41	0.06
17. Canola meal	0.05	0.51	0.09	1.47	0.02	61	7	44	0.92
18. Full-fat canola seed	0.03	0.31	0.01	0.81	0.02	35	6	26	0.52
19. Soybean meal (44%)	0.05	0.25	0.05	2.61	0.02	32	35	54	0.12
20. Soybean meal (48%)	0.05	0.27	0.05	2.54	0.01	27	36	52	0.11
21. Full-fat soybeans	0.04	0.21	0.03	1.50	0.01	20	27	41	0.09
22. Corn gluten meal	0.06	0.05	0.10	0.03	0.04	7	28	66	0.31
23. Corn gluten feed	0.20	0.29	0.95	0.60	0.05	5	47	45	0.17
24. Cotton seed meal	0.03	0.39	0.05	1.20	0.01	18	16	40	0.05
25. Peanut meal	0.55	0.04	0.07	1.10	0.03	29	6	80	0.12
26. Peas	0.06	0.12	0.03	1.10	0.01	18	16	20	0.05
27. Safflower meal	0.03	0.27	0.10	0.69	0.03	24	9	80	0.13
28. Sesame meal	0.05	0.50	0.04	1.20	0.04	48	4	27	0.06
29. Sunflower meal	0.03	0.75	0.02	1.00	0.10	15	3	100	0.25
30. Lupins	0.01	0.13	0.10	1.00	0.01	70	4	30	0.18
31. Flax	0.05	0.30	0.08	1.20	0.02	74	17	91	0.12
32. Meat Meal	0.90	1.00	0.55	1.23	0.04	18	8	98	0.40
33. Fish meal (60%)	0.55	0.21	0.47	0.32	0.06	25	8	119	1.83
34. Poultry by-product meal	0.40	0.18	0.36	0.28	0.05	20	6	79	0.92
35. Blood meal	0.26	0.14	0.33	0.21	0.03	6	8	80	0.60
36. Feather meal	0.41	0.20	0.70	0.30	0.05	15	12	7	0.72
37. Dried Whey	0.07	0.13	0.43	0.19	0.02	4	43	105	0.06

TABLE 2.6 Composition of fat sources

	Metabolizable energy (kcal/kg)		Fat	M.I.U.[5]	Fatty acid profile (%)							
Ingredient	1[1]	2[2]	%	%	12:0	14:0	16:0	18:0	16:1	18:1	18:2	18:3
38. Tallow	7400	8000	98	2		4.0	25.0	24.0	0.5	43.0	2.0	0.5
39. Poultry fat	8200	9000	98	2		1.0	20.0	4.0	5.5	41.0	25.0	1.5
40. Fish oil	8600	9000	99	1		8.0	21.0	4.0	15.0	17.2	4.4	3.0[3]
41. Vegetable oil	8800	9200	99	1		0.5	13.0	1.0	0.5	31.0	50.0	2.0
42. Coconut oil	6500	7800	99	1	50.0[4]	20.0	6.0	2.5	0.5	4.0	2.1	0.2
43. Palm oil	7200	8000	99	1		2.0	42.4	3.5	0.7	42.1	8.0	0.4
44. Vegetable soapstock	7800	8100	98	2		0.3	18.0	3.0	0.3	29.0	46.0	0.8
45. Animal-Vegetable blend	8200	8600	98	2		2.1	21.0	15.0	0.4	32.0	26.0	0.6
46. Restaurant grease	8100	8900	98	2		1.0	18.0	13.0	2.5	42.0	16.0	1.0

[1]ME for young birds up to 3 weeks of age
[2]ME for birds after 3 weeks of age
[3]Contains 25% unsaturated fatty acids $\geq$ 20:4
[4]Contains 15% saturated fatty acids $\leq$ 10:0
[5]Moisture, impurities, unsaponifiables

TABLE 2.7 Calcium, phosphorus and sodium sources

Ingredient	% Ca	% P
47. Limestone	38.0	-
48. Oyster shell	38.0	-
49. Calcium carbonate	40.0	-
50. Bone meal	26.0	13.0
51. Monocalcium phosphate	17.0	25.0
52. Dicalcium phosphate	23.0	20.0
53. Tricalcium phosphate	21.0	19.0
54. Defluorinated rock phosphate	34.0	19.0
55. Curaco phosphate	35.0	16.0
56. Phosphoric acid (75%)	-	25.0
Ingredient	**% Na**	**% Cl**
57. Plain salt	39.0	60.0
58. Iodized salt	39.0	60.0 (I, 70 mg/kg)
59. Cobalt iodized salt	39.0	60.0 (I, 70 mg/kg; Co, 40 mg/kg)
60. Sodium bicarbonate	27.0	-

TABLE 2.8 Trace mineral sources

Ingredient		% of Major Mineral	Ingredient		% of Major Mineral
61. Cobalt	oxide	71.0	65. Manganese	oxide	77.0
	chloride	24.0		chloride	27.5
	sulphate	21.0		sulphate	32.5
	carbonate	46.0		carbonate	47.0
62. Copper	oxide[1]	79.0	66. Zinc	oxide	78.0
	chloride	37.0		chloride	48.0
	sulphate	25.5		sulphate	36.0
	carbonate	55.0		carbonate	52.0
63. Iron	oxide[2]	77.0	67. Selenium	sodium selenite	46.0
	chloride[3]	34.0		sodium selenate	42.0
	sulphate[2]	32.0			
	carbonate[2]	40.0			
64. Magnesium	oxide	56.0	68. Iodine	potassium iodine	77.0
	carbonate	30.0		calcium iodate	65.0

[1]Cupric; [2]Ferrous; [3]Ferric

TABLE 2.9 Synthetic amino acids

Amino Acid	Relative Activity (%)	Crude Protein equivalent (%)
69. DL-Methionine	100	59
70. Methionine hydroxy analogue (liquid)	88	0
71. L-Lysine	100	120
72. L-Lysine HCL	79	96
73. L-Arginine	100	200
74. L-Arginine HCL	83	166
75. L-Tryptophan	100	86
76. L-Threonine	100	74
77. Glycine	100	117
78. Glutamic acid	100	177

2.4. INDIVIDUAL INGREDIENT DESCRIPTIONS

The following section is intended as a guide to ingredient selection during formulation. Ingredient choice is essentially based on economic input of the nutrients as shown in the preceding tables, although an understanding of particular advantages or problems with specific ingredients is essential in setting minimum and/or maximum constraints during formulation. It is not the intent of this section to provide an all encompassing description of every nutrient listed in the preceding tables. For such information, readers are directed to more specialized texts on individual ingredients or classes of ingredients. Rather it is intended to highlight those points which should be considered in setting ingredient constraints.

Following these general descriptions are suggestions for ingredient minimum and/or maximum constraints for poultry. Ingredient numbers correspond to those used in the preceding tables on nutrient specification. These constraints are obviously guidelines, and nutritionists are encouraged to modify these dependent upon knowledge of local conditions.

1. CORN

Corn has become the major cereal source in poultry diets, and because of its inclusion level, it usually is the major source of energy. Most corn used in animal feeds will be US #2 Grade or less (Table 2.10.).

TABLE 2.10 U.S. Corn Grades

Grade	Bushel Weight		Damaged kernels	Broken kernels and foreign matter
#	(lbs/bushel)	(kg/hl)	(%)	(%)
1	56	72.2	3	2
2	54	69.6	5	3
3	52	67.0	7	4
4	49	63.2	10	5
5	46	59.3	15	7

The energy value of corn is contributed by the starchy endosperm which is composed mainly of amylopectin, and the germ which contains most of the oil. Most corn contains 3 - 4% oil, although newer varieties are now available which contain up to 6 - 8% oil, and so contribute proportionally more energy. These high-oil corn varieties also contain 2 - 3% more protein, and proportionally more essential amino acids. The protein in corn is mainly as prolamin (zein) and as such its amino acid profile is not ideal for poultry. This balance of amino acids, and their availability, must be seriously considered when low-protein diets are formulated, because under these conditions the corn prolamin can contribute up to 50 - 60% of the diet protein. Corn is also quite high in the yellow/orange pigments, usually containing around 5 ppm xanthophylls and 0.5 ppm carotenes. These pigments ensure that corn-fed birds will have a high degree of pigments in their body fat and in egg yolks.

While #2 grade is the standard for animal feeds, lower grades are often available due to adverse growing, harvesting or storage conditions. Dependent upon the reason for lower grade, the feeding value of corn usually declines with increase in grade number. Table 2.11 shows the metabolizable energy value of corn necessarily harvested at various stages of maturity due to adverse late-season growing conditions.

TABLE 2.11 Corn maturity and energy value

Corn description	Moisture at harvest (%)	100 kernel wt at 10% moisture (g)	ME(kcal/kg) at 85% dry matter
Very immature	53	17	3014
Immature	45	22	3102
Immature	39	24	3155
Mature	31	26	3313

The energy value of corn declines by 10 - 15 kcal/kg for each 1 lb reduction in bushel weight below the standard of 56 lb/bushel. However these lower bushel weight samples show no consistent pattern with protein or levels of most amino acids, although there is an indication of loss of methionine content with the immature samples. Another potential problem with handling immature, high-moisture corn, is that the drying conditions must necessarily be more harsh, or more prolonged in order to reduce moisture level to an acceptable 15%. Excessive or prolonged heating causes caramelization of corn which then has a characteristic smell and appearance, and there is concern that lysine will be less available because of Maillard reaction with carbohydrates. Dale and co-workers recently reported on the nutritive value of broken corn and so-called "foreign material" that is always found in corn (Table 2.10). Protein level differed little in the various fractions, although broken kernels and foreign material were some 2.5 and 11.0% lower in energy respectively, compared to whole intact kernels.

Depending upon the growing season and storage conditions, molds and associated mycotoxins can be a problem. Aflatoxin contamination is common with insect damaged corn grown in hot-humid areas, and there is little that can be done to rectify the horrendous consequences of high levels of this mycotoxin. There is an indication of aluminosilicates partially alleviating the effects of more moderate levels of aflatoxin. If aflatoxin is even suspected as being a problem, corn samples should be screened prior to blending and mixing. Zearalenone is another mycotoxin that periodically occurs in corn. Because the toxin ties up vitamin D_3, skeletal and egg shell problems can occur. With moderate levels of contamination, water-soluble D_3 via the drinking water has proven beneficial.

Mold growth can be a serious problem in corn that is transported for any length of time. With corn shipped at ≥16% moisture and subjected to ≥25°C during shipping, mold growth often occurs. One solution to the problem is to add organic acids to the corn during loading for shipment. However it must be remembered that while organic acids will kill molds, and prevent re-infestation, they have no effect on any mycotoxins already produced.

If corn is to be fed in mash diets, then there seems to be an advantage to grinding to as uniform a particle size as possible, and to size of about 0.7 - 0.9 mm. This size is often referred to as "medium" ground. Birds fed fine or coarse-ground corn seem to exhibit lower digestibility values. Corn presents some problems to the manufacture of pelleted diets, and often good pellet durability in diets containing ≥30% corn can only be obtained by inclusion of pellet-binders.

CONSTRAINTS: **Max:** **poor pellet quality**
potential mycotoxin contamination
skin/egg yolk color

1a. HIGH MOISTURE CORN

With the availability of organic acids for preserving high moisture corn, some interest has been shown in the use of this product for poultry feeding. High moisture corn is very palatable to poultry, and high levels can be used in diets if it can be efficiently incorporated. Flow of the ground product, and the finished feed itself, can present a problem in handling. When incorporating high moisture corn in a ration, one must consider its moisture content and make appropriate allowances. For example, if a sample of corn contains 30% moisture, then 122 kg of this material should be used for every 100 kg of kiln-dried corn (moisture content of 15%) needed in the diet. The feeding value of high moisture corn can be considered equal to that of regular dried corn on a dry matter basis. One of the main problems that could occur with the use of high moisture corn, is mycotoxin contamination. Assuming one is starting with high moisture corn that has been properly stored in an air tight silo, even where a preservative has been added to the corn before storing, mold problems could result with the finished feed if it is not properly handled. High moisture corn diets should be mixed at least every second day and more frequently in hot weather. Also care must be taken to ensure that there is no build-up of feed in storage bins or in the feed trough. Hence, it is imperative that feed troughs be allowed to run empty at least once a day. Alternatively, organic acids can be used to prevent mold growth, although these products lead to increased corrosion of metal feeders and augers etc.

2. WHEAT

Wheat is commonly used in many countries as the major energy source in poultry diets. There is often confusion regarding the exact type of wheat being used, because wheats are described in a number of different ways. Traditionally wheats were described as being winter or spring varieties and these were traditionally grown in different regions because of prevailing climate and soil conditions. Wheats are sometimes referred to as white or red, depending upon seed coat color, and finally there is the classification of hard vs soft. In the past, most winter wheats were white and soft, while spring wheats were hard red. In terms of feeding value, the main criterion is whether wheat is soft or hard, because this will have an effect on composition, and especially protein. Because of developments in plant breeding, the color and time of planting can now be more variable. Hard wheats have a greater proportion of protein associated with the starch and so contain more protein that is also higher in lysine. The proteins in hard wheat are useful in bread making, while the soft wheats are only used for manufacture of cookies and cakes. Durum wheat, used in manufacture of pasta is a very hard wheat. The physical hardness of these wheats is due to the strong binding between starch and the more abundant protein.

As with corn, the grading system for wheat relates to bulk density and the proportion of broken grains and foreign material (Table 2.12).

TABLE 2.12 US Wheat Grades

	Hard Wheat Bushel Weight		Soft Wheat Bushel Weight		Total % foreign material and broken kernels
Grade #	lbs/bu	kg/hl	lbs/bu	kg/hl	
1	58	74.8	60	77.4	3
2	57	73.4	58	74.8	5
3	55	70.9	56	72.2	8
4	53	68.3	54	69.6	12
5	50	64.5	51	65.8	20
Feed grade	<50	<64.5	<51	<65.8	>20

The composition of wheat is usually more variable than that of other cereals. Even within the hard wheats, protein level can vary from 10 - 18%, and this may relate to varietal differences and variance in growing conditions. Most hard wheats will not have to be dried after harvest, although drying conditions and moisture content of wheat at harvest appear to have little effect on feeding value. Environmental temperature during growing seems to have a major effect on wheat nitrogen content, and although high temperature can result in 100% increase in nitrogen level, the relative proportion of both lysine and starch tend to be decreased.

Depending upon the growing region, frost damaged or sprouted wheat is sometimes available to the feed industry. Frost damage effectively stops starch synthesis, and so kernels are small and shrunken. While 100 kernel weight should be around 27g, with severe frost damage, this can be reduced to 14 - 16g. As expected the metabolizable energy level of this damaged wheat is reduced, and under these conditions there is a very good correlation between bushel weight and metabolizable energy. For non-frosted wheat however, there does not seem to be the same relationship between energy level and bushel weight.

Sometimes wheat will sprout in the field. Sprouted wheat would probably be rejected simply based on appearance, although research data suggests that metabolizable energy level is only reduced by 3 - 5%. There are no problems to feeding spouted wheat, as long as it has been dried to ≤14% moisture, and can be economical if discounted accordingly. Wheat contaminated with rust however seems to more seriously affect feeding value, and metabolizable energy value can be reduced by up to 25%.

While wheats are much higher than corn in protein content, and provide only slightly less energy, there are some potential problems from feeding much more than 30% in a diet, especially for young birds. Wheat contains about 5 - 8% of pentosans which can cause problems with digesta viscosity, leading to reduced overall diet digestibility and also wet manure. The most components are arabinoxylans which are linked to other cell wall components, and these are able to adsorb up to 10 times their weight as water. Unfortunately birds do not produce adequate quantities of xylanase enzymes, and so these polymers increase the viscosity of the digesta. The 10 - 15% reduction in ME of wheats seen with most young birds (< 10 d age) likely relates to their inability to handle these pentosans. Variability in pentosan content of wheats per se likely accounts for most of the variability of results seen in wheat feeding studies, together with our inability to predict feeding value based on simple proximate analyses. These adverse effect on digesta viscosity seem to decrease with increased storage time for wheats. Problems with digesta viscosity can be controlled to some extent by limiting the quantity of wheat used, especially for young birds, and/or by using exogenous xylanase enzymes (see section 2.8g).

Wheats also contain α-amylase inhibitors. Although these inhibitors have not been fully identified, they are thought to be albumin proteins, found predominantly in the starch. These inhibitors seem to be destroyed by the relatively mild temperatures employed during pelleting. Compared to corn, wheat is also very low in levels of available biotin. Whereas it is sometimes difficult to induce symptoms of biotin deficiency in birds fed corn diets devoid of synthetic biotin, problems soon develop if wheat is fed. While newly hatched chicks have liver biotin levels of around 3,000 ng/g, this number declines to 600 ng/g within 14 d in the wheat fed bird. Adding just 50 μg biotin/kg diet almost doubles the liver biotin reserve, while adding 300 μg/kg brings levels back to that seen in the day-old chick. There is also concern that wheat causes a higher incidence of necrotic enteritis in broiler chicks. It seems as though wheat provides a more suitable medium for the proliferation of certain pathogenic bacteria. The problem is most severe when wheat is finely ground, and incidence of necrotic enteritis can be tempered by grinding wheat through a roller mill. Fine grinding of wheat can also cause beak impaction in young birds. The proteins in wheat tend to be "sticky", and so adhere to the beak and mouth lining of the bird. Severe beak impaction with feed tends to reduce feeding activity, increase feed deposited in open bell drinkers, and provides a medium in the mouth region that is ideal for bacterial and fungal growth. These problems can be resolved by coarse grinding of wheat.

Using wheat in diets for meat birds does however improve pellet durability. The same proteins that enhance the baking characteristics of hard wheats, also help to bind ingredients during pelleting. Adding $\geq$ 25% to a diet has the same effect as including a pellet binder in diets that are difficult to pellet.

One advantage of wheat, is that it can be fed as whole grain to birds after 10 - 14 d of age. Offering whole wheat and a balancer feed with adequate minerals and vitamins provides a very economical way for farmers to utilize home-grown wheat. In recent studies we offered broilers a conventional three-diet program, or after 7 d of age, a choice between whole wheat and crumbled broiler starter through to 49 d. From 7 - 21 d, male broilers voluntarily consumed about 15% of their ration as wheat, while from 21 - 35 d and 35 - 49 d this increased to 34% and 41% respectively. Table 2.13 shows performance data of these birds.

TABLE 2.13 Broiler performance with free-choice wheat

Diet	Body Wt 49d (g)	Feed:Gain	Protein intake (g/kg B.wt)	Energy intake (kcal/kg B.wt)	Carcass Wt (g)
Control	3030	1.93	370	6044	2230[b]
Free-choice wheat	2920	1.99	364	6106	2135[a]

Adapted from Leeson and Caston 1993

Body weight was only slightly depressed, although carcass weight was significantly reduced and breast yield was reduced by about 10%. The free-choice wheat system did however show a saving of 10% in feed cost per kg liveweight gain although feed cost per kg of breast meat was not different. Another advantage claimed for feeding whole wheat to broilers, is natural control over coccidiosis. Whole wheat feeding stimulates gizzard and gastric motility and the enhanced activity in this acidic environment is thought to reduce oocyst viability.

CONSTRAINTS: **Max: wet compacted litter**
digesta viscosity
Min: 10% for pellet quality

3. OATS

Oats are grown in cooler moist climates although they are of minor importance on a global scale, representing only about 1.5% of total cereal production. Most oats are used for animal feed, and about 85% of this quantity is used locally and there is little trade involved. The hull represents about 20% of the grain by weight, and so this dictates the high fiber - low energy characteristics of oats. The amino acid profile is however quite good, although there is some variation in protein and amino acid levels due to varietal and seasonal effects. The best predictor of the energy value of oats, is simply the crude fiber content which is negatively correlated with ME. Oat lipids are predominantly oleic and linoleic acid, although a relatively high proportion of palmitic acid leads to a "harder" fat being deposited in the bird's carcass.

As for other small grains, oats contain an appreciable quantity of β-glucans, and these cause problems with digesta and excreta viscosity. Most oats contain about 3 - 7% β-glucans and so with moderate inclusion levels of oats in a poultry diet it may be advantageous to use supplemental β-glucanase enzyme. There has been some interest in development of so-called naked oats, which are similar in composition to oat groats. Naked oats contain up to 17% CP with 0.68% lysine and 1% methionine plus cystine. The ME value is around 3200 kcal/kg, making these oats comparable to wheat in most characteristics. As with regular oats, β-glucan can still be problematic and their adverse effect can be overcome with use of exogenous enzymes, and to a lesser extent antibiotics such as neomycin. Much of the phosphorus in naked oats is as phytic acid, and so availability is very low. There have been some reports of reduced skeletal integrity in birds fed naked oats unless this reduced availability is taken into account. There are reports of broilers performing well with diets containing up to 40% naked oats, and with layers, up to 50% has been used successfully.

CONSTRAINTS: **low energy**
wet litter related to β-glucans

4. BARLEY

Barley is a medium energy-protein ingredient, falling between oats and wheat in most characteristics. Young birds are less able to digest barley, although this may be a consequence of **β**-glucan content, and so this effect may relate to variety and growing conditions. The protein content of barley is usually around 11 - 12%, although much higher levels to 14 - 16% are sometimes encountered. These high-protein barleys are often little changed in content of essential amino acids. The lysine content of barley within the range of 10 - 14% CP is described by the equation 0.13 + 0.024% x CP. The metabolizable energy level of barley is highly correlated with bushel weight, and there is a strong negative correlation with fiber.

Barley contains moderate levels of trypsin inhibitor, whose mode of action relates to sequestering of arginine, although by far the major problem with barley is content of **β**-glucans. Most barley samples will contain 4 - 9% **β**-glucan, although with dry growing conditions that include rapid maturation and early harvest, the content can increase to 12 - 15%. As previously described for wheat, the main problem of these **β**-glucans is the bird's inability to digest the structure, resulting in the formation of a more viscous digesta. This increased viscosity slows the rate of mixing with digestive enzymes and also adversely affects the transport of nutrients to the absorptive mucosal surface. The rate of diffusion to the intestinal microvilli is a function of the thickness of the unstirred boundary layer, and this increases with increased digesta viscosity. Motility of the digesta will also indirectly affect the thickness of the unstirred boundary layer, which will also affect rate of absorption of all nutrients. The adverse effect of **β**-glucans is most pronounced with dietary fats and fat soluble compounds. Adding synthetic **β**-glucanase enzymes to diets containing more than 15 - 20% barley seems to resolve many of these problems, the usual outward sign of which is wet litter. Unfortunately the description of exogenous enzymes is not standardized, as neither is the standard for units of efficacy, and so it is often difficult to compare products on the basis of the concentration of specific enzymes. Early studies show that any product should provide at least 120 units **β** glucanase/kg diet.

Enzymes seem to be less efficacious as the birds get older. Our studies show slight improvement in energy value of high **β**-glucan barley when enzymes are used in diets for adult birds, but that some enzymes actually cause reduction in energy value when used with low **β**-glucan barley. With this low **β**-glucan barley, the addition of **β**-glucanase enzymes actually caused birds to be in severe negative nitrogen balance for the 3 d duration of the balance study. For younger birds however, the efficacy of **β**-glucanase enzymes is well established, and many nutritionists consider barley plus enzymes as being equivalent in feeding value to wheat. These values can be used as a basis for economic evaluation of enzymes. While **β**-glucans are most often regarded as being problematic to birds, there seems to be one advantage to their inclusion in the diet. Feeding **β**-glucans reduces blood cholesterol in birds, and this likely positive effect is reversed by use of synthetic **β**-glucanases. The mode of action of **β**-glucans may well be simply via sequestering of fats and bile acids in the digesta.

Barley can be used in choice-feeding studies, as previously described for wheat. Due to the physical structure of the kernel however, with its sharp spinets, birds are often reluctant to consume whole barley grain. Turkeys at least seem to readily eat whole barley in a choice-feeding situation, at least after 50 d of age.

CONSTRAINTS: Max: wet litter and β-glucan content

5. MILO

This ingredient is probably known by more different names around the world than any other cereal or feedstuff - milo, sorghum and kaffir corn being the most common names. In many respects milo is almost comparable to corn in feeding value. There seem to be more varietal differences with sorghum, although on average, its energy value will be slightly less than that of corn. For those not wanting any marked degree of pigmentation of eggs or skin, milo offers the best high-energy alternative to corn.

The feeding value of milo is essentially 95 - 96% that of corn, although in many markets it is priced at less than this. The starch in milo is intimately associated with the protein, and this leads to slightly reduced digestibility, especially in the absence of any heat processing. The major concern with milo, is the content of tannins, which are a group of polyphenols having the property of combining with various proteins. Birds fed tannins therefore exhibit reduced growth rate and in some instances increased incidence and severity of skeletal disorders. Hydrolysable tannins are characterised by having a gallic acid unit combined by ester linkages to a central glucose moiety. Condensed tannins on the other hand are based on flavan-3-ols (catechin). Because tannins in milo are essentially condensed tannins, studies involving tannic acid (hydrolysable) as a source of tannin may be of questionable value. The tannins are located in the outer seed coat and the underlying testa layer. Generally, the darker the seed coat, the higher the tannin content, although the tannins in the testa layer may be more indicative of general tannin content in the milo.

So-called bird-resistant milos are usually very high in tannin, and are characterised by a darker seed coat color. These higher levels of tannin can result in up to 10% reduction of dry matter and amino acid digestibility. There is a good correlation between tannin content and AMEn, and as a generalization the following formula can be used: AMEn = 3900 - 500 (% tannin), kcal/kg. Tannins are most detrimental when fed to young birds, and especially when protein content of the diet is marginal. For example, it is usually recommended that high - tannin ($\geq$ 1%) milo not be used for turkeys under 8 weeks of age. The relationship between tannins and diet protein or amino acids is not clear. Certainly feeding more protein or higher levels of certain amino acids seems to temper any growth retardation. The fact that methionine supplementation can overcome detrimental effects of tannins on growth rate, without alleviating problems with digestibility, suggests that birds can compensate for inferior digestibility by increasing their feed intake. Tannins also seem to increase the incidence of leg problems, especially in broiler chickens. The exact mechanism is unknown, although because bone mineral content is little affected, it is assumed to relate to derangement in the development of the organic matrix, especially in the region of the growth plate. There seems no advantage to increasing supplemental levels of any minerals or vitamins when high-tannin milos are necessarily used.

Various mechanisms have been tried to reduce the level or effect of tannins in milo. Unfortunately, most of these processes involve wet chemical treatment, which although quite simple, are expensive when re-drying of the milo is considered. Water treatment (25%) with propionic acid for 10 d has been shown to improve protein and energy availability by up to 10%. Alkali treatment seems the most effective means of reducing tannin levels, and products such as potassium and sodium hydroxide have both been used. Adding non-ionic polymers, such as polyethylene glycol also seems to be beneficial, while the problem of impaired bone development can be partially corrected by adding up to 0.8% available phosphorus to the diet of young birds.

Tannin levels in milo should be assayed routinely, or at least seed coat and testa color measured as a qualitative indication of tannin content (see later section on ingredient analyses).

CONSTRAINTS: **Max: tannin content**

6. RYE

Although the nutrient content of rye is essentially similar to that of wheat and corn, its feeding value for poultry is poor due to the presence of various antinutritional factors. Rye contains a water insoluble fraction, which if extracted, improves it's feeding value. Various other treatments such as water soaking, pelleting, irradiation and the dietary supplementation of various antibiotics all help to improve the growth of chicks fed rye.

One of the most noticeable effects of feeding rye, other than reduced performance is the production of a very sticky and watery excreta. The sticky droppings are due to the pectin-like components present in rye. Structural arabinozylans, present in rye endosperm cell walls are responsible for creating viscous digesta. These viscous products reduce the rate of diffusion of other solutes in the digesta so affecting nutrient uptake from the gut. In recent years enzyme preparations have been developed that markedly reduce the antinutritional factor and eliminate the wet-sticky fecal problem with rye based diets.

CONSTRAINTS: **Max: pentosans**
low energy
wet manure

7. TRITICALE

Triticale is a synthetic small grain cereal resulting from the intergeneric cross of wheat and rye. Its higher yield per acre, as compared to rye or wheat, make it of agronomic interest in areas of the world not suitable for corn production. Numerous cultivars have been developed with protein contents varying from 11 to 20% and amino acid balance comparable to wheat and superior to that of rye. Energy content is also similar to that of wheat and superior to that of rye. Like wheat, triticale has a significant phytase content and so is a better source of available phosphorus than other cereals such as corn or milo. There are reports of increased enhancement of other dietary phosphorus with triticale supplementation. The starch content of triticale is as digestible as that of wheat and presents no wet litter or sticky manure problem. Where triticale is available, high levels can be used in poultry diets without any adverse problems. Similar to wheat and rye it contains little or no xanthophyll and with fine grinding can result in beak impaction with young birds. Also like wheat its feeding value can be increased by appropriate enzyme supplementation of the diet.

8. RICE

Almost without exception, rice is grown for human consumption, although periodically in rice growing areas, samples unfit for human consumption, or damaged samples are available for animal feeding. Rice is a relatively poor quality ingredient for poultry, containing only 7 - 8% CP and providing just 2600 - 2700 kcal ME/kg. Rice does contain quite high levels of trypsin inhibitor that will be destroyed by normal pelleting temperatures.

By far the most important component of rice, is the bran that is removed during processing, and this provides a very useful ingredient for the feed industry (ingredient #12 - 13).

CONSTRAINTS: Max: trypsin inhibitor in some samples.

9.-11. WHEAT BY-PRODUCTS

Wheat by-products result as inedible waste material from flour manufacture. There is considerable variation in the classification of various wheat by-products, and names for the same product differ in different countries. Traditionally wheat was ground through a series of grinders of decreasing sieve size, producing so called middlings, most of which became flour. The non-usable portion of this flour stream was often termed wheat middlings as opposed to wheat shorts which were a "by-product" of the extraction of bran. Today only three major animal feed ingredients are available (although again nomenclature may vary in different countries). Bran is from the outer seed coat, and is obviously high in fiber, and apart from specialized applications is little used in poultry nutrition. After extraction of flour, bran and wheat germ (also for human consumption) the remaining product is now most commonly referred to as wheat shorts. Shorts are therefore "by-products" remaining from extraction of the other three distinct products (flour, bran and germ). In N. America the term "red-dog" was sometimes used, although this is now commonly included in the shorts. Red dog represented the very finest particles of bran, endosperm and germ.

Wheat Bran: The main characteristics are high fiber, low bulk density and low metabolizable energy. Bran is however quite high in protein, and amino acid profile is comparable to that seen in whole wheat. Bran has been claimed to have a growth promoting effect for birds which is not directly related to any contribution of fiber to the diet. Such growth promotion is possibly derived from modification of the gut microflora. The energy value of bran may be improved by up to 10% by simple steam pelleting, while the availability of phosphorus is increased by up to 20% under similar conditions. Bran would only be considered where limits to growth rate are required, and where physical feed intake is not a problem. High bran diets promote excessive manure wetness, and transportation costs of bran diets are increased in proportion to the reduced bulk density of the diet.

CONSTRAINTS: Max: fiber and energy contribution.

Wheat Shorts: Because shorts are most commonly an end point for various wheat milling by-products, then composition can be quite variable. If shorts contain much more than 5% crude fiber, it is an indication of a greater proportion of bran-type residues, and so the energy value will be correspondingly reduced. Shorts are higher in protein than is whole wheat, but provide about 20% less energy, and therefore should be priced accordingly. Higher fiber-content wheat shorts are obviously worth proportionally less. Sometimes wheat shorts are combined with bran to produce wheat mill-run.

CONSTRAINTS: Max: fiber and energy contribution

Wheat screenings: Wheat screenings are a by-product of the cleaning and grading of wheat that itself is usually destined for human consumption. The product is therefore available in most countries that have significant wheat production. In addition to broken and cracked wheat kernels, screenings will also contain wild oats and buckwheat as well as weed seeds and other organic contaminants. The higher grades (1 or 2) contain significant proportions of wheat, and so their nutrient profile is very similar to that of wheat. The weed seeds, depending upon variety may be of some nutritional value. Since certain weed seeds produce a feed-refusal type reaction in layers and so only the best grades should be considered for high producing stock. The weed seeds can pose problems to arable farmers that use manure from birds fed coarsely-ground diets containing screenings, since some of the weed seeds can pass undamaged through the digestive tract. The level of screenings used in finisher diets of meat birds should also be severely limited, since breakage of the gut during processing leads to fine particles of black weed seeds adhering to the fat pads of the bird - such birds are easily recognized and often condemned due to fecal contamination. Number 1 and 2 grade screenings can be used up to 40% of the diet for broilers and layers.

CONSTRAINTS: Max: problems of weed seed contamination
downgrading of carcasses at processing

12.-13. RICE BY-PRODUCTS

Rice by-products are the result of dehulling and cleaning of brown rice, necessary for the production of white rice as a human food. Rice by-products are one of the most common cereal by-products available to the feed industry, with world production estimated at around 50 M tonnes. The by-product of preparing white rice, yields a product called rice bran, which itself is composed of about 30% by weight of rice polishing and 70% true bran. In some regions the two products are separated, being termed polishing and bran. Alternatively the mixture is sometimes called rice bran, whereas in other areas the mixture may be called rice pollards. The polishings are very high in fat content and low in fiber, while the true bran is low in fat and high in fiber. The proportions of polishing and true bran in a mixed product will therefore have a major effect on its nutritive value. In the following discussion, rice bran refers to the mixture of polishing (Ingredient #13, Table 2.1) and bran (Ingredient #12, Table 2.1). The composition of any sample of mixed rice bran can be calculated based on proportional values of these components shown in Table 2.1.

Because of a high oil content (6 - 10%) rice bran is very susceptible to oxidative rancidity. Raw bran held at moderate temperatures for 10 - 12 weeks can be expected to contain 75 - 80% of its fat as free fatty acids, which are themselves more prone to rancidity.

Rice bran should be stabilized with products such as ethoxyquin. Higher levels of ethoxyquin give greater protection against rancidity although economical levels appear to be around 250 ppm. Rice bran can also be stabilised by heat treatment. Extrusion at 130°C greatly reduces chances of rancidity, and of the development of free fatty acids.

When high levels of raw rice bran are used ($\geq$ 40%) there is often growth depression and reduction in feed efficiency, likely associated with presence of trypsin inhibitor and high levels of phytic acid. Trypsin inhibitor, which seems to be a relatively low molecular weight structure, is destroyed by moist heat, although phytic acid is immune to this process. The phosphorus content of rice bran is assumed to be only 10% available for very young birds. However, phosphorus availability may increase with age, and if this happens it could create an imbalance of calcium:phosphorus. This latter effect is suggested as the reason for improved growth response in older birds fed rice bran when extra calcium is added to the diet. Because of the potential for high fiber content, use of rice bran may be improved with addition of exogenous arabinoxylanase enzymes.

CONSTRAINTS: Max: fiber content
oil content
trypsin inhibitor
phytic acid

14. BAKERY BY-PRODUCT

As its name implies, bakery by-product meal is a by-product of the bakery and confectionary industries, and so considerable quantities are often available in large urban areas. Unfortunately the composition of bakery by-product can vary considerably, because of variation in raw ingredients used, and also the potential inclusion of inert fillers. The meal can vary considerably in fat content and also in fiber and ash depending upon fillers used to improve flow characteristics of high-fat meals. If soybean hulls are used, the meal will be high in fiber, whereas when limestone is used ash and calcium levels are elevated. The AMEn will obviously be less as the fiber and/or ash content increase:

AMEn (kcal/kg) = 4000 - [100(% fiber) + 25 (% ash)]

Quality control procedures necessarily involve monitoring fiber and ash content. Bakery by-product meal can also be very high in sodium content, because of inherently high salt content of some raw ingredients such as potato chips. Because of subsequent problems with water intake, the meal should be checked periodically for sodium content. Stringent quality control procedures are most important when there is a change in supplier.

CONSTRAINTS: Max: fiber, ash and salt contents

15. MOLASSES

Molasses is a by-product of the sugar refining industry, where either sugar beet or sugar cane are used as raw materials. Because of a high water content and concomitantly low energy value, it is only used extensively in poultry diets in areas close to sugar refineries. The molasses usually available for animal feeding is so called final or blackstrap molasses, which is the product remaining after all sugar has been extracted for human consumption. Depending upon local conditions, high-test and type A and B molasses are sometimes available. The high-test product is basically unrefined cane or beet juice that has had its sugars inverted to prevent crystallisation. Type A and B molasses are intermediate to final molasses. As expected, the energy level of molasses decreases as more and more sugar is extracted. Molasses is usually quantitated with a Brix number, measured in degrees, and these numbers relate very closely to the sucrose concentration in the product. Both cane and beet molasses both contain about 46 - 48% sugar.

Although molasses contains relatively little energy and protein, it can be used to advantage to stimulate appetite and to reduce dustiness of feed. For example, feed intake is usually increased in birds such as young Leghorn pullets, if molasses is poured directly onto feed in the feed trough. It is doubtful that molasses improves "taste" of feed under these conditions, rather it presents a novel feed texture to the bird. A major problem with molasses is a very high potassium content, at 2.5 - 3.5%, which has a laxative effect on birds. While most birds perform well on balanced diets containing up to 20% molasses, inclusion levels much above 4% will likely result in increased water intake and increased manure wetness.

CONSTRAINTS: **Max:** **low energy**
laxative due to potassium

16. DEHYDRATED ALFALFA

Dehydrated alfalfa meal can be quite high in protein (18 - 20%) although because the product is heated during drying, availability of essential amino acids such as lysine is often 10 - 20% below expected values. Alfalfa is very high in fiber content, and is most often added to poultry diets as a source of xanthophylls for pigmentation, or as a source of so-called unidentified growth factors.

Alfalfa products should contain a minimum of 200,000 IU vitamin A activity per kg, although in most cases this will only be 70% available. In order to achieve intense yellow skin color in broilers or egg yolk color of $\geq$10 on the Roche scale, diets should contain 5% alfalfa as one source of xanthophylls in the diet. Alfalfa levels much in excess of 5% have only a moderate effect on pigmentation, and so other natural or synthetic sources must be used to ensure consistently high levels of pigmentation. At high levels of inclusion (20%) problems can arise due to the saponins and phenolic acids normally present in alfalfa. If alfalfa contains any appreciable mold count, then estrogen level can be high. Fresh grass is thought by some nutritionists to contain an unidentified growth factor which is of particular significance to turkeys, although much of this "factor" is destroyed by the dehydrating process. Even so, many nutritionists still insist on adding 1 - 2% dehydrated alfalfa to turkey feeds, especially pre-starter and starter diets. The addition of small quantities of alfalfa also impart a darker color to diets which helps mask any minor fluctuations in appearance due to regular changes in formulations. The quality of alfalfa products has been improved considerably in recent years due to the use of inert gas storage, pelleting and addition of antioxidants.

CONSTRAINTS: **Max:** **high fiber**
too much color in yolks or skin

Min: color of yolks or skin
general diet appearance
unidentified factor for turkeys and breeders

17. CANOLA MEAL

Canola is a widely grown crop in western Canada and production is increasing in other parts of the world. Production increases have been influenced by the marked increase in the demand for canola oil as well as the ability of this high protein oilseed to grow in more northern climates not suitable for the production of soybeans.

While canola is derived from the original rapeseed varieties its composition has been altered through genetic selection. The level of goitrogens and erucic acid, two of the more detrimental constituents of the original rapeseed cultivars have been markedly reduced. Erucic acid levels are now negligible while goitrogen levels are down to less than 20 μg/g. These levels are low enough to be of little or no problem to poultry. Varieties containing such levels of toxins are classified as canola and are often referred to as "double zero varieties".

Canola still has enough goitrogen activity to result in significant increases in thyroid weight, although this does not appear to be a problem affecting the performance of poultry. The tannin levels in canola can also be relatively high at up to 3% for some cultivars. Again research has shown that the canola tannins have little influence on the utilization of the protein in diets containing appreciable levels of the meal.

Canola meal also contains significant quantities (1.5%) of sinapine. While this compound poses no problem to most classes of poultry, a significant percent of brown-egg birds produce eggs with a fishy and offensive odor when fed canola sinapines. One of the end products of the degradation of sinapine in the intestinal tract is trimethylamine and it is this compound which is involved in the production of fishy flavored eggs. A proportion of today's brown egg laying birds lack the ability to produce trimethylamine oxidase, and consequently the trimethylamine is deposited into the egg. Even 1% sinapine can produce off-flavored eggs. It should be pointed out that eggs produced by broiler breeders and most other medium or heavy weight birds, producing brown shelled eggs, have no problem with the utilization of sinapine-containing diets.

While canola meal has been accepted by the feed industry as a high quality feedstuff for poultry, there continues to be isolated reports of increased leg problems with broilers and turkeys, smaller egg size with layers and in some cases reports of increased liver hemorrhages when diets contain significant amounts of canola meal. There are several reports which suggest that increased leg problems resulting from feeding canola may be due to its having a different mineral balance than does soya. The addition of dietary K, Na and in some cases Cl have, under certain conditions, altered bird performance. Canola is also high in phytic acid and thus there is speculation that the high level of this compound may be tying up zinc and so affecting bone development. The smaller egg size reported with canola meal diets seems to be a direct result of lower feed intake. Canola levels should therefore be limited in diets for very young laying hens, or at least until feed intake plateaus at acceptable levels.

Within the past few years there are reports which suggest that the high level of sulphur in canola meal may be responsible for some of the leg problems and reduced feed intake noted with canola meal diets. Canola meal contains 1.4% sulphur while soybean meal contains around .44%. Of the sulphur in soya, 75% is contributed by the sulphur amino acids compared to around 20% for a similar comparison for canola meal. High levels of dietary sulphur have been reported to complex intestinal calcium and lead to increased calcium excretion. This could explain the reports suggesting low availability of calcium in canola meal and possibly a factor in enhanced leg problems reported with canola diets. While lower weight gain has at times been reported with canola diets it is usually noted that feed:gain ratios are little affected. This suggests that the reduction in gain was not the result of poorer diet quality but rather a direct effect on appetite, resulting in reduced feed intake.

Recent work demonstrates quite clearly that a soybean meal diet containing the same level of sulphur as that in canola diets results in similar weight gain and feed intake (Table 2.14).

TABLE 2.14 Interaction of sulphur and calcium in canola and soybean meal diets.

Protein source	Suppl. S. (%)	Total S. (%)	Calcium level (%)	Weight Gain (g)
Canola Meal	-	.46	.37	424
	.26	.72	.37	371
	-	.46	1.32	560
	.26	.72	1.32	481
Soybean Meal	-	.14	.37	525
	.13	.27	.37	519
	.26	.40	.37	479
	.39	.53	.37	373
	-	.14	1.32	635
	.13	.27	1.32	598
	.26	.40	1.32	559
	.39	.53	1.32	451

In this study the unsupplemented canola diet contained .46% sulphur while the soya diet contained .14%. Adding sulphur to the soybean meal diet resulted in a decrease in weight gain. The level of sulphur in the unsupplemented canola diet (.46%) lies part way between the levels found in the .26 and .39% sulphur supplemented soybean meal diet. Weight gain for the unsupplemented canola meal diet was 424 g while the average for the two soybean meal diets was 426g. Higher dietary calcium levels partially overcame the growth depressing effect of high dietary sulphur thus demonstrating the negative effect of sulphur on calcium retention.

In view of the depressing effect on appetite and calcium retention from high dietary sulphur levels, these need to be closely monitored if substantial levels of canola are included. High levels of methionine or sulphate salts, along with ingredients with significant amounts of sulphur, like some phosphate supplements, can add considerable sulphur to a diet. Some sources of water can also be high in sulphur.

Broilers can tolerate dietary sulphur levels of up to around .5% without any effect on performance while laying hens can handle even higher levels. There are reports which suggest that part of the response to increased levels of dietary sulphur is due to their influence on dietary acid-base balance. While Mongin in his original work suggested considering K+Na-Cl in the dietary acid-base balance equation, S, being a strong anion, should also be considered in this equation.

Canola meal contains less lysine than does soybean meal but slightly more sulphur amino acids, per unit of dietary protein. It is also lower in energy than is soya. Levels of up to 8% canola meal are used in laying diets without any adverse effects on performance although egg size may be reduced by up to 1g. Energy content is the factor that usually limits inclusion level.

CONSTRAINTS: Max: low energy
glucosinolates
fish taint with brown-egg birds
depressed egg size with young layers

18. FULL-FAT CANOLA

The nutrient profile of canola seed makes it an ideal ingredient for higher nutrient dense diets. Periodically, grades unfit for oil extraction are available for animal feeding. Canola seed suffers from the same problems as described for canola meal, although obviously most harmful elements are diluted by the high oil content. Seeds must be ground adequately to ensure normal digestion, and this is best accomplished by mixing with ground corn prior to passing through a hammer mill. The oil provides considerable energy, and is an excellent source of linoleic acid. The ground seed is not too oily, and so provides a practical way of adding considerable quantities of fat to a diet. Due to early frost damage, some samples of canola contain oil that is contaminated with chlorophyll - while unacceptable to the human food industry, this contaminant does not seem to pose any major problems to poultry.

CONSTRAINTS: Max: see canola meal
rancidity

19-20 SOYBEAN MEAL

Soybean meal has become the world-wide standard against which other protein sources are compared. Its amino acid profile is excellent for most types of poultry, and when combined with corn or sorghum, methionine is usually the only limiting amino acid.

The protein level in soybean meal can be variable, and this may be a reflection of seed variety and/or processing conditions involved in fat extraction. Traditionally the higher protein meals are produced from de-hulled beans, whereas the lower protein (44% CP) meals invariably contain the seed hulls, and are higher in fiber and lower in metabolizable energy. There is some variation in seed type used and this can affect protein and fat content which are negatively correlated. Whereas fat content of the seed is dictated early in seed development, protein is deposited almost to the end of maturity, therefore growing and harvesting conditions tend to have more effect on protein content of the seed. For soybean processors, about 65% of the value of soybeans is attributed to their protein content, and 35% to the oil.

During processing, soybeans are dehulled (about 4% by weight) and then cracked prior to conditioning at 70°C. The hot cracked beans are then flaked to about 0.25 mm thickness so as to enhance oil extraction by a solvent, which is usually hexane. Hexane must be removed from the meal because it is a highly combustible material and also a potent carcinogen. Problems occurring during processing that result in residual hexane in the meal are usually noticed by severe and sudden liver damage in birds. Soybean meals tend to be very dusty, and in mash diets, soya is responsible for some of the dust found in controlled environment houses. Soybean meal is also notorious for its poor flow characteristics and for bridging in storage bins. Addition of products such as bentonite clays, even at levels as low as 2.5 kg/tonne, can greatly improve flowability of soybean meal.

Soybeans contain a number of natural toxins for poultry, the most problematic being trypsin inhibitor. As with most types of beans, the trypsin inhibitors will disrupt protein digestion, and their presence is characterised by compensatory hypertrophy of the pancreas. Apart from reduced growth rate and egg production, presence of inhibitors is diagnosed by a 50 - 100% increase in size of the pancreas. Fortunately, the heat treatment employed during processing is usually adequate to destroy trypsin inhibitors and other less important toxins such as haemaglutinins (lectins). In developing countries, trypsin inhibitor levels are sometimes controlled by simply fermenting or germinating beans, where after 48 hrs of treatment, protein digestibility is almost equivalent to that seen in conventionally heated beans. Trypsin inhibitor levels are usually "assayed" indirectly by measuring urease activity in processed soybean meal. Urease is of little consequence to the bird, although the heat-sensitivity characteristics of urease are similar to those of trypsin inhibitors, and urease levels are easy to measure. Residual urease in soybean meal has therefore become the standard in quality control programs. Urease is assessed in terms of change in pH during the assay, where acceptance values range between 0.05 and 0.2. Higher values mean there is still residual urease (trypsin inhibitor) and so the test is useful to indicate undercooked meal. However, while low values mean the toxins have been destroyed, there is no indication of potential overcooking, which can destroy lysine and reduce ME value. For this reason other tests are sometimes used. A fairly easy test to accomplish is protein solubility in potassium hydroxide. Dale and co-workers at the University of Georgia have shown a good correlation between the amount of protein soluble in 2% KOH, and chick growth determined in a bioassay. Heating tends to make the protein less soluble, and so high values suggest undercooking, while low values mean overcooking. Values of $\geq$ 85% solubility indicate underprocessing and $\leq$ 70% mean the sample has been overprocessed. The assay is influenced by particle size of soybean meal and time of reaction, and so these must be standardized within a laboratory.

As soybean meal is heated, its color changes, and again this can be used in quality control programs. Simply measuring color in a Hunterlab Color spectrophotometer can indicate degree of cooking. Degrees of "lightness" "redness" and "yellowness" can be measured and each change with cooking temperature and time. Again it is important to control particle size during this assay.

Over the last few years there has been growing concern about some of the less digestible carbohydrates in soybean meal. The α-galactoside family of oligosaccharides cause a reduction in metabolizable energy with reduced fiber digestion and quicker digesta transit time. Birds do not have a α-1:6 galacto oxidase enzyme in the intestinal mucosa. Apart from reduced digestibility, there is concern about the consistency of excreta and its involvement in foot pad lesions in both young turkeys and broiler breeders. Soybean meal usually contains about 6% sucrose, 1% raffinose and 5% stachyose, none of which are well digested by the bird. Adding raffinose and stachyose to isolated soybean protein to mimic levels seen in soybean meal, causes a significant reduction in metabolizable energy. These problems limit the diet inclusion level of soybean meal, especially in turkey prestarters. The solution to the problem relates to change in soybean processing conditions or use of exogenous feed enzymes. Extracting soybeans with ethanol, rather than hexane, removes the oligosaccharides, while there are now some galactosidase enzyme products available which are designed specifically to aid digestion of vegetable proteins.

CONSTRAINTS: Max: litter condition
foot pad lesions in poults
trypsin inhibitor
oligosaccharides

21. FULL-FAT SOYBEANS

Soybeans provide an excellent source of both energy and protein for poultry. As with any ingredient, their usage rate depends upon economics, although in the case of soybeans such economics relate to the relative price of soybean meal and of supplemental fats. Soybeans contain only about 18% crude protein, but of course are a concentrated source of energy because they also contain about 20% oil.

As with the manufacture of soybean meal, soybeans must be heat processed in some way to destroy the trypsin inhibitors and also to improve overall protein digestibility. Feeding raw soybeans, or improperly processed soybeans will cause poor growth rate or reduced egg production and egg size. If processing conditions are suspect the bird's pancreas should be examined, because if trypsin inhibitors are still present, pancreas size can be expected to increase by 50 - 100%. While processed beans should be periodically tested for trypsin inhibitor or urease levels, a simple on-going test is to taste the beans. Under-heated beans have a characteristic "nutty" taste, while over-heated beans have a much darker color and a burnt taste. The problem with overheating is potential destruction of lysine and other heat-sensitive amino acids.

Heat-treated soybeans can be easily ground in a hammer mill, even though they are high in fat, and the ground product is a relatively free-flowing material. Because of the high oil content, ground beans should not be stored for any length of time due to potential for oxidative rancidity. However it is important that beans be well ground because it is necessary to release fat from the plant cells in order to aid digestion. Coarsely ground beans have lower fat digestibility than do more finely ground material. Heating beans by whatever means usually results in considerable "shrinkage" which is mainly due to loss of water. In many situations, shrinkage will be up to 7%, but of this less than 1% will be real loss of dry matter.

Recently there has been growing interest in processing beans through extruders or expanders. The heat necessary to destroy trypsin inhibitors and other hemaglutinins found in raw beans is dependent upon exposure time, and so high temperatures for a shorter time period are as effective as lower temperatures for longer times. Because both expanders and extruders are fast throughput, the beans have a relatively short dwell time in the conditioning chamber. Consequently slightly higher temperatures are necessary, and depending upon design, such machines are best operated at 140 - 155°C. Again the effectiveness of expanding and extrusion can be measured by tests for urease and available lysine content.

A potential problem with soybeans is the direct deposition of their very unsaturated fat into the carcass of poultry. Carcasses containing appreciable quantities of unsaturated oil tend to be more oily to the touch, and likely have reduced shelf life due to oxidative rancidity. This latter problem can be resolved to some extent by using much higher levels of vitamin E ($\geq$50 IU/kg) in the finisher diet.

CONSTRAINTS: **Max:** **oil rancidity**
carcass composition

22.-23. CORN GLUTEN PRODUCTS

Corn gluten meal and corn gluten feed are by-products of the wet milling processes of corn for manufacture of high-fructose corn syrup. Removing the energy yielding starch component and the germ means that these corn by-products represent a concentration of the original protein components. Corn gluten meal is very high in protein (60%) and as such is compared with the animal proteins during formulation. Corn gluten feed is much lower in protein (20%) because of dilution with fibrous hull material. In certain areas of the world, the products are given different or even the same name, and so protein content should always be specified. Corn gluten meal is the product most commonly used in poultry diets. Corn gluten meal is very deficient in lysine, although with appropriate use of synthetic lysine sources, the product is very attractive where high nutrient density is required. Gluten meal is also very high in xanthophyll pigments (up to 300 mg/kg) and is a very common ingredient where there is a need to pigment poultry products.

CONSTRAINTS: **Max:** **pigmentation**
lysine deficiency

Min: **pigmentation**

24. COTTONSEED MEAL

Cottonseed meal is not usually considered in diets for poultry, although for obvious economic reasons it is often used in cottonseed producing areas. A high fiber content and potential contamination with gossypol are the major causes for concern. Gossypol is a yellow polyphenolic pigment found in the cottonseed "gland". In most meals, the total gossypol content will be around 1%, although of this only about 0.1% will be free gossypol. The remaining bound gossypol is fairly inert, although unfortunately binding can have occurred with lysine during processing, making both the gossypol and the lysine unavailable to the bird. So-called "glandless" varieties of cottonseed are virtually free of gossypol.

Birds can tolerate fairly high levels of gossypol before there are general problems with performance, although at much lower levels there can be discoloration of the yolk and albumen in layer birds. Characteristically the gossypol causes a green-brown-black discoloration in the yolk depending upon gossypol levels, and length of egg storage. As storage time increases, the discoloration intensifies, especially at cool temperatures (5°C) where there is more rapid change in yolk pH. Gossypol does complex with iron, and this activity can be used to effectively detoxify the meal. Adding iron at a 1:1 ratio in relation to free gossypol greatly increases the dietary inclusion rate possible in broiler diets and also the level at which free gossypol becomes a problem with laying hens. Because most cottonseed samples contain around 0.1% free gossypol, detoxification can be accomplished by adding 0.5 kg ferrous sulphate/tonne feed. With addition of iron, broilers can withstand up to 200 ppm free gossypol, and layers up to 30 ppm free gossypol without any adverse effects.

If cottonseed meal contains any residual oil, then cyclopropenoid fatty acids may contribute to egg discoloration. These fatty acids are deposited in the vitelline membrane, and alters its permeability to iron that is normally found only in the yolk. This leached iron complexes with conalbumin in the albumen producing a characteristic pink color. Addition of iron salts does not affect this albumen discoloration, and the only preventative measure is to use cottonseed meals with very low residual fat content.

CONSTRAINTS: Max: yolk discoloration
albumen discoloration
residual fat level
gossypol content
lysine availability

25. PEANUT (GROUNDNUT) MEAL

The peanut is an underground legume, and because of warm moist conditions in the soil, is very susceptible to fungal growth with aspergillus contamination being of most concern. Grown essentially for their oil, peanuts yield a solvent extracted meal containing 0.5 - 1% fat with about 47% protein. As with soybeans, peanuts contain a trypsin inhibitor that is destroyed by the heating imposed during oil extraction. Potential aflatoxin contamination is the major problem with groundnut meal. Being a potent carcinogen, aflatoxin causes rapid destruction of the liver, even at moderate levels of inclusion. Peanut meal that is contaminated with aflatoxin can be treated by ammoniation which seems to remove up to 95% of the toxin. Alternatively products such as sodium-calcium aluminosilicates can be added to the diet containing contaminated groundnut, because these minerals bind with aflatoxin preventing its absorption.

CONSTRAINTS: Max: aflatoxin

26. PEAS

Peas are a medium energy-protein ingredient that can be used effectively in poultry diets depending upon local economic conditions. The major limitation to using peas is low levels of sulphur amino acids and moderate energy level. With high-tannin peas, there may be an advantage to some type of heat treatment, although such processing is of little value for regular pea varieties. Protein digestibility is reduced by about 6% for each 1% increase in tannin content. Peas do have some of their carbohydrate as oligosaccharides, and so enzyme systems being developed to improve the digestibility of soybean meal may be of use with peas. Peameal is a very dense material and bulk density of the final diet should be taken into consideration for diets containing $\geq$ 15% peas.

CONSTRAINTS: Max: digestibility (tannins)
bulk density

27. SAFFLOWER MEAL

Safflower is grown mainly for its oil content which, although variable, can be as high as 40%. The residual meal contains in excess of 20% fiber and is referred to as undecorticated safflower meal. It is possible to commercially remove a large portion of the hull resulting in a meal containing 42 to 44% protein with a fiber content of around 14%. This product is referred to as decorticated meal. Safflower meal is very deficient in lysine although with appropriate lysine supplementation the protein quality of safflower meal is similar to that of soybean meal. However, with the high fiber content the available energy is still relatively low and so its value does not equal that of soybean meal. Where safflower meal is available relative large quantities can be used in poultry diets if proper consideration is given to nutrient availability.

CONSTRAINTS: Max: lysine deficiency
low energy

28. SESAME MEAL

Sesame meal is very deficient in available lysine, and this is sometimes used to advantage in formulating lysine deficient diets for experimental reasons. Sesame also contains high levels of phytic acid which can cause problems with calcium metabolism leading to skeletal disorders or poor eggshell quality. If diets contain >10% sesame, then the diet should be formulated to contain an extra 0.2% calcium.

CONSTRAINT Max: lysine deficiency
phytic acid

29. SUNFLOWER MEAL

Sunflowers are grown in cooler climates, being harvested for either oil extraction or increasingly as a source of wild bird seed. The extracted meal is very low in energy and quite deficient in lysine and also in available threonine. For meat birds the high fiber-low energy characteristics are problematic, and so maximum inclusion relates to the birds ability to increase feed intake in order to maintain energy intake. Pelleting significantly improves feed intake of young birds fed 10% sunflower meal, and there is also an indication of improved metabolizable energy with some form of heat processing.

CONSTRAINTS: Max: low energy
lysine availability

30. LUPINS

Low alkaloid lupins are being increasingly used as an alternative feedstuff for poultry in certain areas of the world. These new cultivars have been reported to contain low levels of the toxic alkaloids (less than .01%) normally found in wild varieties. These low alkaloid lupin seeds are often referred to as sweet lupins and can either be white, yellow or blue seeded varieties. The high level of fiber in the seeds (12 to 18%) results in low metabolizable energy values compared to soybean meal. Mature lupin seeds contain little or no starch, the bulk of their carbohydrate being oligosaccharides (sugars) and non-starch polysaccharides. Many reports suggest that sweet lupins are comparable to soybeans with respect to protein quality although they are slightly lower in methionine and lysine. Their low oil content (6 to 10%) and absence of antinutritive factors means that they can be inexpensively processed for on-farm animal feeding. Recent studies have shown that dehulling lupins can result in a marked increase in their nutritive value. Also with the proper dietary enzyme supplement the feeding value of raw lupins is improved.

CONSTRAINTS: **Max: low energy**

31. FLAXSEED

Flax is grown essentially for its oil content, although in Europe there is still some production of the fibrous plant material for linen production. Fat-extracted flax, which is commonly called linseed meal, has traditionally been used for ruminant feeds. Over the last few years there has been interest in feeding full-fat flax seeds to poultry, because of its contribution of linolenic acid. Flax oil contains about 50% linolenic acid (18:3w3) which is the highest concentration of this fatty acid outside of marine oils. It has recently been shown that 18:3w3, and its desaturation products docohexanoic acid and eicosapentaenoic acid are important in human health, and especially those individuals at risk from chronic heart disease. Health Canada and government agencies in other countries now recognise the importance of linolenic acid in human health, and suggest the need to increase average daily intake, and especially intake in relation to that of linoleic acid.

Feeding flax seeds to poultry results in direct incorporation of linolenic acid into poultry meat and also into eggs. Feeding laying hens 10% flax results in a 10 fold increase in egg linolenic acid content, while feeding 20 and 30% results in 23 and 39 fold increase respectively. Eating two such modified eggs each day provides adults with most of their daily recommended allowance of linolenic acid. Linolenic acid enriched eggs therefore provide an attractive alternative to consumption of oily fish. Linolenic acid is essentially responsible for the characteristic smell of fish oils, and undoubtedly flax oil does have a "paint-type" smell. There is some concern about the taste and smell of linolenic acid-enriched eggs, and this area needs more careful study with controlled taste panel work. Whole flax seeds are not well digested by birds, and consequently the seeds must be ground prior to adding to finished feed. Flax seed can be mixed with cereal, and then ground in a hammer mill.

CONSTRAINTS: **Min: linolenic acid**
Max: digestibility, oxidative rancidity

32. MEAT MEAL

Most meat meals are by-products of beef and swine processing, and can be of variable composition. For each 1 tonne of meat prepared for human consumption, about 300 kg is discarded as inedible product, and of this about 200 kg ends up in meat meal. In the past meat meal referred to only soft tissue products, while meat and bone meal also contained variable quantities of bone. Today, meat meal most commonly refers to animal by-products with bone where protein level is around 50% and calcium and phosphorus at 8 and 4% respectively. Because the mineral comes essentially from bone, the calcium phosphorus ratio should be around 2:1, and deviations from this usually indicate adulteration with other mineral sources.

Variation in calcium and phosphorus content is still problematic, and the potential for overfeeding phosphorus is a major reason for upper limits on inclusion level. Meat meals usually contain small residues of fat, and the best quality meals will be stabilized with antioxidants such as ethoxyquin. Some of the variability in composition is now being resolved by so-called "blenders" that source various meat meal products and mix these to produce more consistent meat meals.

Recent evidence suggests that the metabolizable energy content of meat meal, and other animal protein by-products, is higher than the most common estimates used in the past. In bioassays, ME values determined at inclusion levels of 5 - 10% are much higher than those determined at more classical levels of 40 - 50% inclusion. The reason for the higher values is unclear, although it may relate to synergism between protein or fat sources, and these are maximised at low inclusion levels. Alternatively, with very high inclusion levels of meat meal, the high calcium levels involved may cause problems with fat utilization due to soap formation, and so energy retention will be reduced. The ME value shown in Table 2.1 reflects this higher energy estimate.

A more recent concern with meat meal is microbial content, and especially potential for contamination with salmonella. Due to increasing awareness and concern about microbial quality, surveys show that the incidence of contamination has declined, but remains at around 15%. Protein blends are at most risk, because obviously a single contaminated source can lead to spread of salmonella into various blended products. One means of reducing microbial load is to treat freshly processed meals with organic acids. In many studies it is shown that meals are virtually sterile when they emerge from the cooking chambers, and that problems most often occur with recontamination. Certainly most feed ingredients contain salmonella, however, because of the relative proportion of meat meals used in a diet, the actual chance of contamination of a single bird may, in fact, come from corn (Table 2.15).

TABLE 2.15 Relative risk due to salmonella from various ingredients

	% Salmonella Contamination	Diet %	Relative Risk Factor (%)
Corn	1	60	60
Vegetable proteins	8	30	24
Meat Meals	15	5	7.5

The relative risk to an individual bird is therefore claimed to be much higher for cereals because, even though they are not usually contaminated, their much higher inclusion level results in a greater potential risk. However this type of argument is open to the real criticism that meat meals are much more likely to contaminate the feed, trucks, equipment etc., and that salmonella numbers will likely increase after feed manufacture. Pelleted and extruded/expanded diets will have much lower microbial counts than corresponding mash diets.

CONSTRAINTS: **Max:** **microbial quality**
calcium phosphorus

33. FISH MEALS

Because of the decline in activities of most fisheries directed at human food industries, fish meals are now almost exclusively produced from smaller oily fish caught specifically for meal manufacture. Menhaden and anchovy are the main fish species used for meal manufacture, with lesser quantities of herring meal produced in Europe. Fish meal is usually an excellent source of essential amino acids, while energy level is largely dependent upon residual oil content. Because of variable oil and protein content, expected ME value can be calculated based on knowledge of their composition in the meal

$$\text{ME (kcal/kg)} = 3000 \pm (\text{Deviation \% fat} \times 8600) \pm (\text{Deviation \% CP} \times 3900).$$

where standard fat content is 2%, and CP is 60%. Therefore a 4% fat, 63% CP sample is expected to have an ME of 3289 kcal/kg, while a 1% fat, 58% CP sample will have ME closer to 2836 kcal/kg. The ash content of fish meal will be predominantly calcium and phosphorus and the latter can be considered approximately 90% available, as is any phosphorus from quality animal proteins.

All fish meals should be stabilized with antioxidants such as ethoxyquin. This is especially true for high-oil content samples, but even with only 2% residual oil, there is good evidence to show reduced oxidation (in terms of production of oxidation products and free fatty acids, as well as reduced heat production) by adding 100 ppm ethoxyquin during manufacture.

Potential problems in feeding fish meal are taint of both eggs and meat, and gizzard erosion in young birds. With inadequately heat-treated fish meal there is also the potential problem of excessive thiaminase activity. Depending upon geographical location, taint in eggs and meat can be detected by consumers when birds are fed much more than 4 - 5% fish meal. Problems of taint will be more acute with high fat samples, and of course the problems are most acute if fish oil *per se* is used. Even at levels as low as 2.5%, some brown-egg birds produce tainted eggs which may be related to the trimethylamine content of fish meal, and the genetic predisposition of certain birds failing to produce sufficient trimethylamine oxidase. Excess trimethylamine is shunted to the egg, producing a characteristic fishy taint (see Canola meal). The trimethylamine content of fish meal is around 50 - 60 mg/kg, and assuming a 2.5% inclusion level, and maximum feed intake of these brown egg layers of 125g/day, means that the bird is taking in about 0.2mg/day. Each affected egg contains around 0.8 mg, and so even at 80% egg production, it is obvious that the diet contains sources of trimethylamine other than fish meal, or that there is microbial synthesis in the intestine.

For young chicks, and especially the broiler chicken, a major concern with feeding fish meal, is gizzard erosion. A proportion of chicks fed almost any level of fish meal develop gizzard lesions, although there is a strong dose-relationship. Affected birds have signs ranging from small localized cracks in the gizzard lining, through to severe erosion and hemorrhage which ultimately leads to total destruction of the lining. The hard lining is required for physical processing of feed, and also acts as a barrier for the sensitive underlying mucosa against the degrading effects of acid and pepsin produced by the proventriculus. Because of disrupted protein degradation, the affected birds show very slow growth rate. The condition is most common when fish meal is included in the diet, although similar signs are seen with birds fed high levels of copper (250 ppm) or vitamin K deficient diets, or simply induced by starvation. Gizzard erosion was initially thought to be associated with histamine levels in fish meal. Feeding histamine to birds simulates the condition, as does feeding a semi-purified diet containing histidine that has been heat-treated. Fish meals contain histamine, and following microbial degradation during pre-cooking storage, bacteria possessing histidine decarboxylase will convert variable quantities of histidine to histamine. Histamine has the effect of stimulating excessive acid production by the proventriculus, and it is this acid environment that initiates breakdown of the gizzard lining. More recently, a product known as gizzerosine has been isolated from fish meal, and this has histamine-type properties in terms of stimulating acid secretion. Gizzerosine is formed by heating histidine and a protein during manufacture of fish meal. The most common components are lysine and histidine. Gizzerosine is almost 10 x as potent as histamine in stimulating proventricular acid production and some 300 x more potent in causing gizzard erosion. Currently the only useful screening test is to feed high levels (25 - 50%) to young chicks and score the degree of gizzard lesions (see ingredient quality control).

Because the mode of action of gizzerosine is via acid production and a change in gizzard pH, there have been attempts at adding buffers to prevent the problem. For example adding sodium bicarbonate has been reported to lessen the severity of gizzard erosion. However levels as high as 10 kg/tonne are required to change gizzard pH by only 0.3 units. Variable levels of gizzerosine in fish meals likely relate to pre-processing holding time and storage temperature, and also to the time and temperature of the cooking and oil extraction procedures.

CONSTRAINTS Max: taint of meat and eggs
gizzard erosion

34. POULTRY BY-PRODUCT MEAL

As with meat meal, poultry by-product meal is produced essentially from waste generated during poultry meat processing. Because only one species is used, PBM should be a more consistent product than meat meal, and certainly calcium and phosphorus levels will be lower. Variability in composition relates to whether or not feathers are added during processing, or kept separate to produce feather meal. PBM and feathers are best treated using different conditions, because feathers require more extreme heat in order to hydrolyze the keratin proteins. PBM with feathers may therefore mean that either the feather proteins are undercooked or that the offal proteins are overcooked. Overcooking usually results in a much darker-colored product. PBM contains more unsaturated fats than does meat meal, and so if much more than 0.5% remains in the finished product, it should be stabilized with a good quality antioxidant.

There is current interest in ensiling various poultry carcasses and/or poultry by-products prior to heat processing. Ensiling allows for more control over microbial contamination prior to processing, and allows the potential to better utilize smaller quantities of poultry carcasses on-farm or from sites more distant to the PBM processing plant. Ensiling is also being considered as a means of handling spent layers prior to production of PBM. Poultry carcasses or offal do not contain sufficient fermentable carbohydrate to allow lactic acid fermentation which will quickly reduce pH to about 4.2, and stabilize the product. These lactic acid producing mircrobes can therefore be encouraged to proliferate by adding, for example, 10% molasses or 10% dried whey to ground carcasses. These mixtures quickly stabilize at around pH 4.2 - 4.5, and can be held for 10 - 15 d prior to manufacture of PBM. Carcasses from older birds may require slightly higher levels of these carbohydrates, and because of their inherently high fat content may be mixed with products such as soybean meal in order to improve handling characteristics. Ensiled whole carcasses, as is now being produced with spent fowl, may present problems with availability of feather proteins for reasons outlined previously in terms of ideal processing conditions for tissue versus feathers. In the future, this problem may be resolved by adding feather-degrading enzymes to the ensiling mixture.

CONSTRAINTS: **Max: feather content**
rancidity

35. BLOOD MEAL

Blood meal is very high in crude protein, and while it is an excellent source of lysine, it is very deficient in isoleucine and this imbalance needs correcting if any substantial quantity is used in a diet. Blood meal is essentially the solids of the blood from processing plants, and consists mainly of hemoglobin, cell membranes, cellular electrolytes and a small quantity of lipid. Historically, the level of blood meal used in diets has been severely limited, mainly because of problems of palatability and poor growth rate and abnormal feathering. All these problems relate to inherent amino acid balance and also to low digestibility/availability induced by overheating of the blood during processing. With less harsh drying treatments, the amino acids are more stable, and there are few problems with palatability. If blood meal is overheated, it has a much darker color, tending to be black rather than reddish-brown. The amino acid balance of blood meal can be "improved" by combining it with other ingredients. For example a 50:50 mixture of blood meal and hydrolysed hair meal gives a product with a reasonable amino acid balance, and certainly a balance that is preferable to either product alone. Such a mixture may be used in least-cost formulation, whereas either ingredient is unlikely to be used independently because of amino acid balance.

CONSTRAINTS: **Max: palatability**
amino acid balance

36. FEATHER MEAL

Feather meal can be an excellent source of cystine and a good source of crude protein where this is needed to meet regulatory requirements. However its use is severely limited by deficiencies of several amino acids, including methionine, lysine and histidine. Feather meal usually contains about 4.5 - 5.0% cystine, and this should be around 60% digestible. The energy value of feather meal is quite high, being around 3300 kcal ME/kg, and Dale and co-workers at the Univ. Georgia suggests TMEn of feather meal is highly correlated with its fat content (2860 + 77x% fat, kcal/kg). Variability in quality is undoubtedly related to control over processing conditions. Feathers are partially dried and then steam-treated to induce hydrolysis, and within reason the higher the temperature and/or processing time, the better the chance of complete hydrolysis. Obviously extreme processing conditions will cause destruction of heat-labile amino acids such as lysine. Feather meal also contains an amino acid called lanthionine, which is not normally found in animal tissues. Total lanthionine levels can therefore be used in assaying meat meal type products for potential contamination with feathers. Lanthionine can occur as a breakdown product of cystine, and there are some research results which indicate a very good correlation between high lanthionine levels and poor digestibility of most other amino acids. In most feather meal samples, lanthionine levels should be at 20 - 30% of total cystine levels. A potential problem in using lanthionine assays in quality control programs, is that it is readily oxidized by performic acid, which is a common step used in preparation of samples for amino acid analysis and particularly where cystine levels are of interest.

As with other animal proteins, there is current interest in alternate methods of processing. Treating feathers with enzyme mixtures that presumably contain keratinase enzyme, has been shown to improve overall protein digestibility and bird performance. More recently it has been shown that a pre-fermentation with bacteria such as Bacillus licheniformis for 5 d at 50°C, produces a feather lysate that is comparable in feeding value to soybean meal when amino acid balance is accounted for.

CONSTRAINTS: Max: amino acid balance

37. DRIED WHEY

Only small quantities of whey are used in poultry diets, and this is often as a source of so-called unidentified growth factors in turkey prestarter diets. Whey permeate is very high in lactose, a sugar that has recently been considered as beneficial in limiting the colonization of salmonella in the intestine in very young birds.

CONSTRAINTS: Min: unidentified growth factors
salmonella control

38-46 FATS AND OILS

Fats provide a concentrated source of energy, and so relatively small changes in inclusion levels can have significant effects on diet ME. Most fats are handled as liquids, and this means heating of most fats and fat blends that contain appreciable quantities of saturated fatty acids.

Depending upon the demands for pellet durability, 3 - 4% is the maximum level of fat that can be mixed with the other diet ingredients. To this, 2 - 3% can be added as a spray-on coat to the formed pellet. Newer technology of spraying fat onto the hot pellet as it emerges from the die means that much higher inclusions will be possible since the hot pellet seems better able to absorb the fat. Under these conditions, there is concern for manufacturers who demand extreme pellet durability, since fines will already be treated with extra fat, prior to its recycling through the pellet mill. All fats and oils must be treated with an antioxidant such as ethoyxquin. Ideally, this should be added at the point of manufacture. Fats held in heated tanks at the mill must be protected from rancidity. The more unsaturated a fat, the greater the chance of rancidity. Fats also provide varying quantities of the essential nutrient linoleic acid. Unless a diet contains considerable quantities of corn, it may be deficient in linoleic acid, because all diets should contain a minimum of 1% linoleic acid. A major problem facing the industry at the moment is the increasing use of restaurant grease in feed-grade fats. These greases are obviously of variable composition in terms of fatty acid profile and content of free fatty acids. Also, dependent upon the degree of heating that they have been subjected to, these greases can contain significant quantities of undesirable break-down products.

In order to ensure adequate levels of linoleic acid, and to improve palatability and reduce dustiness of diets, all diets require a minimum of 1% added fat, regardless of other economic or nutritional considerations. There is considerable information published on factors that influence fat digestibility, but in most instances this knowledge is not used during formulation. In large part variability is due to the fact that digestibility is not a static entity for any fat, but rather its utilization is variable with such factors as bird age, fat composition and inclusion level. Unfortunately, these variables are difficult to factor into formulation programs. Other concerns about fats are potential for rancidity and their effect on carcass composition. Following are some of the factors known to influence utilization of feed grade fats.

a) Moisture, Impurities, Unsaponifiables

Feed grade fats will always contain some non-fat material, that is generally classified as M.I.U. Because these impurities impart no energy or little energy, they act as diluents. A recent survey indicated M.I.U's to range from 1 - 9%. Each 1% MIU means loss in effective value of the fat by about $3 - 4/tonne, and more importantly energy contribution will be less than expected. The major contaminants are moisture and minerals. It seems as though moisture can be quickly detected by Near Infra Red Analysis. Moisture and minerals also lead to increased peroxidation.

b) Rancidity and Oxidation

The feeding value of fats can obviously be affected by oxidative rancidity that occurs prior to, or after feed preparation. Rancidity can influence the organoleptic qualities of fat, as well as color and "texture", and can cause destruction of other fat soluble nutrients, such as vitamins, both in the diet and the body stores. Oxidation is essentially a degradation process that occurs at the double-bond in the glyceride structure. Because presence of double-bonds infers unsaturation, then naturally the more unsaturated a fat, the greater the chance of rancidity. The initial step is the formation of a fatty free radical when hydrogen leaves the α-methylenic carbon in the unsaturated group of the fat. The resultant free radical then becomes very susceptible to attack by atmospheric oxygen (or mineral oxides) to form unstable peroxide free radicles. These peroxide free radicles are themselves potent catalysts, and so the process becomes autocatalytic and rancidity can develop quickly. Breakdown products include ketones, aldehydes and short chain fatty acids which give the fat its characteristic "rancid" odor. Animal fats develop a slight rancid odor when peroxide levels reach 20 meq/kg while for vegetable oils problems start at around 80 meq/kg.

Oxidative rancidity leads to a loss in energy value, together with the potential degradation of the bird's lipid stores and reserves of fat soluble vitamins. Fortunately we have some control over these processes through the judicious use of antioxidants, such as ethoxyquin. Most antioxidants essentially function as free radical acceptors - these radicals are, however, stable and do not cause autocatalytic reactions. Their effectiveness, therefore, relies on adequate dispersion in the fat immediately after processing. As an additional safety factor, most diets will also contain an antioxidant added via the premix. The Active Oxygen Method (AOM) is most commonly used to indicate potential for rancidity. After 20h treatment with oxygen, quality fats should develop no more than 20 meq peroxides/kg.

c) *Fatty acid profile*

Fat composition will influence overall fat utilization because different components can be digested with varying efficiency. It is generally recognized that following digestion, micelle formation is an important prerequisite to absorption into the portal system. Micelles are complexes of bile salts, fatty acids, some monoglycerides and perhaps glycerol. The conjugation of bile salts with fatty acids is an essential prerequisite for transportation to and absorption through the microvilli of the small intestine. Polar unsaturated fatty acids and monoglycerides readily form this important association. However, micelles themselves have the ability to solubilize non-polar compounds such as saturated fatty acids. Fat absorption is, therefore, dependent upon there being an adequate supply of bile salts and an appropriate balance of unsaturates:saturates.

Taking into account the balance of saturated to unsaturated fatty acids, can therefore, be used to advantage in designing fat blends. This type of synergistic effect is best demonstrated using pure fatty acids (Table 2.16). In this study the metabolizable energy of the 50:50 mixture of the unsaturated oleic acid with the saturated palmitic acid, is 5% higher than than expected based on the mean value of 2710 kcal/kg. We, therefore, have a boost of 5% in available energy that likely comes from greater utilization of the palmitic acid because of the presence of the unsaturated oleic acid.

TABLE 2.16 Metabolizable energy of layer diets containing various fatty acids

	Determined ME (kcal/kg)	Expected ME (kcal/kg)
Oleic	2920	
Palmitic	2500	
50:50 mixture	2850	2710 (+5%)

(Atteh and Leeson, 1985)

This type of synergism can, however, have a confounding effect on some research results. If we want to measure the digestibility of corn, it is possible to feed just corn for a short period of time and conduct a balance study. For obvious reasons it is impossible to feed pure fats, and we have to conduct studies involving graded fat additions to a basal diet, with extrapolation of results to what would happen at 100% feeding level. In these studies we assume the difference in digestibility between any two diets is due solely to the fat added to the diet. If, because of synergism, the added fat improves digestibility of basal diet components, then this "boost" in digestibility is attributed to the fat and an erroneously high value is projected. However, it can be argued that this "boost" to fat's value occurs normally when fats are added to diets, and that these higher values more closely reflect the practical value of fat in a poultry diet. We have proposed this synergism to account for some of the so-called "extra-caloric" effect of fat often seen in reported values, where metabolizable energy can sometimes be higher than corresponding gross energy values (which theoretically cannot occur). Table 2.17 shows results from this type of study where corn oil was assayed using different types of basal diet.

TABLE 2.17 Variation in ME value of corn oil attributed to fatty acid saturation of the basal diet	
Basal diet	**Corn oil ME (kcal/kg)**
1. Predominantly unsaturated	8390[a]
2. Predominantly saturated	9380[b]
3. Practical ingredients	8510[a]

When the basal diet contains saturated fatty acids, there is an apparent increase in the ME of corn oil. This effect is possibly due to the unsaturates in corn oil aiding in utilization of the basal diet saturates. However, because of methods of diet substitution and final ingredient ME calculation, any such synergism is attributed to the test ingredient (corn oil).

ME values of fats will therefore vary with inclusion level, although this effect will be influenced by degree of fat saturation. A ratio of 3:1, unsaturates:saturates is a good compromise for optimum fat digestibility for all ages of bird. However this ratio may not be the most economical type of fat to use, because of the increased cost of unsaturates relative to saturates.

d) Level of free fatty acids and fatty acid hydrogenation

Concern is often raised about the level of free fatty acids in a fat, because it is assumed these are more prone to peroxidation. Acidulated soapstocks of various vegetable oils contain the highest levels of free fatty acids, which can reach 80 - 90% of the lipid material. For young birds there is an indication that absorption of fatty acids is highest in birds fed triglycerides rather than free fatty acids and this may relate to less efficient micelle formation or simply to less bile production. Wiseman and Salvador (1991) demonstrated this effect in studying the ME value of tallow, palm oil and soy oil that contained various levels of free fatty acids (soapstock of the respective fat). Table 2.18 shows a summary of these results, indicating energy values for the respective fats containing the highest and lowest levels of free fatty acids used.

TABLE 2.18 Effect of level of free fatty acid and bird age on fat ME value (kcal/kg)			
		10d age	**54d age**
Tallow -	13% FFA 95% FFA	7460 4920	7940 6830
Palm -	6% FFA 92% FFA	6690 3570	7800 6640
Soy -	14% FFA 68% FFA	9290 8000	9300 8480
			Adapted from Wiseman and Salvador (1991)

These data suggest that free fatty acids are more problematic when the fat is predominantly saturated and this is fed to young birds. Contrary to these results others have shown comparable results with broilers grown to market weight and fed tallows of varying free fatty acid content.

Hydrogenation of fats becomes an issue with the general use of these fats in restaurants, and the fact that restaurant grease is now a common, and sometimes the major component of feed grade fat blends. Hydrogenation results in a high level of trans oleic acid (40 - 50%) and the predominantly vegetable oils having physical characteristics similar to those of lard. There seems to be no problem in utilization of these hydrogenated fats by poultry with ME values of restaurant greases being comparable to that of vegetable oils. The long-term effect of birds eating trans fatty acids is unknown at this time.

e) *Bird age and bird type*

Young birds are less able to digest saturated fats, and this concept has been known for some time. With tallow for example, palmitic acid digestibility increases from 50 to 85% over 14 -56 d of age, which together with corresponding changes for other fatty acids means that tallow ME will increase by about 10% over this time period. The reason why young birds are less able to digest saturated fats is not well understood, although may relate to less bile salt production, less efficient recirculation of bile salt or less production of fatty acid binding protein.

f) *Soap formation*

When fats have been digested, free fatty acids have the opportunity of reacting with other nutrients. One such possible association is with minerals to form soaps that may or may not be soluble. If insoluble soaps are formed, there is the possibility that both the fatty acid and the mineral will be unavailable to the bird. There is substantial soap formation in the digesta of broiler chicks and this is most pronounced with saturated fatty acids, and with increased levels of diet minerals. Such increased soap production is associated with reduced bone ash and bone calcium content of broilers. Soap production seems to be less of a problem with older birds. This is of importance to laying hens that are fed high levels of calcium. In addition to calcium, other minerals, such as magnesium, can form soaps with saturated fatty acids. In older birds and some other animals, there is an indication that while soaps form in the upper digestive tract, they are subsequently solubilized in the lower tract due to changes in pH. Under these conditions both the fatty acid and mineral are available to the bird. Control over digesta pH may, therefore, be an important parameter for control over soap formation.

g) *Variable ME values*

It seems obvious that the use of a single value for fat ME during formulation is a compromise, considering the foregoing discussion on factors such as inclusion level, bird age, soap formation etc. Table 2.6 gives different ME values for birds younger or older than 21 d, and this in itself is a compromise. Following is an attempt to rationalize the major factors affecting ME of a given fat, although it is realised that such variables are not easily incorporated within a formulation matrix (Table 2.19).

TABLE 2.19 Factors affecting fat ME values

		Relative fat ME
i) Bird age:	28d+	100%
	7-28d	95%
	1-7d	88% (especially for saturates)
ii) Free Fatty acids:	0 - 10%	102%
	10 - 20%	100%
	20 - 30%	96%
	30%+	92% (especially for saturates)
iii) Inclusion level	1%	110%
	2%	100%
	3%	98%
	4%	96%
	5%+	94%
iv) Calcium level	<1%	100%
	>1%	96% (especially for birds <56d age)

38. TALLOW

Tallow has traditionally been the principle fat source used in poultry nutrition. However over the last 10 years, there has been less use of pure tallow and greater use of blended oils. Tallow is solid at room temperature and this presents some problems at the mill, especially when heated fats are added to very cold ingredients taken in from outside silos. Being highly saturated, it is not well digested by young chickens, although there is some evidence of better utilization by young turkeys. The digestibility of tallow can be greatly improved by the experimental addition of bile salts suggesting this to be a limiting feature of young chicks. However, the use of such salts is not economical and so inclusion of pure tallow must be severely restricted in diets for birds less than 15 - 17d of age.

CONSTRAINTS: Max: pelleting quality
rancidity
digestibility by young birds

39. POULTRY FAT

This fat source is probably the best for most types and ages of poultry in terms of its fatty acid profile. Due to its digestibility, consistent quality and residual flavor, it is in high demand by the pet food industry, and this unfortunately reduces its supply to the poultry industry. As discussed with poultry meal, there is a concern with integrated poultry operations that fat-soluble contaminants may be continually cycled (and concentrated) through the birds. This can obviously be resolved by breaking the cycle for a 1 or 2 bird crop period.

CONSTRAINTS: Max: pellet quality
rancidity

40. FISH OIL

There is current interest in the use of fish oils in diets for humans and animals, since its unusual component of long chain fatty acids is thought beneficial for human health. Feeding moderate levels of fish oils to broiler chickens has been shown to increase the eicosapentaenoic acid (long chain fatty acid) content of meat. However, with dietary levels in excess of 1%, distinct fish type odor is often present in both meat and eggs, which is due mainly to the contribution of the omega-3 fatty acids.

CONSTRAINTS: Max: organoleptics

41. VEGETABLE OILS

A large range of vegetable oils is available as an energy source, although under most situations, competition with the human food industry makes them uneconomic for animal feeds. Most vegetable oils provide around 8700 kcal ME/kg and are ideal ingredients for very young birds. If these oils are attractively priced as feed ingredients, then the reason(s) for refusal by the human food industry should be ascertained ie: contaminants.

CONSTRAINTS: Max: pellet quality
rancidity

42. COCONUT OIL

Coconut oil is a rather unusual ingredient in that it is a very saturated oil. Coconut oil is more saturated than tallow. It contains 50% of saturated fatty acids with chain length less than 12:0. In many respects, it is at the opposite end of the spectrum to fish oil in terms of fatty acid profile. There has been relatively little work conducted on the nutritional value of coconut oil, although due to its saturated fatty acid content it will be less well digested, especially by young birds.

CONSTRAINTS: Max: digestibility

43. PALM OIL

Palm oil production is now only second to soybean oil in world tonnage. Palm oil is produced from the pulpy flesh of the fruit, while smaller quantities of palm kernel oil are extracted from the small nuts held within the body of the fruit. Palm oil is highly saturated, and so will have limited usefulness for very young birds. Also, soapstocks produced from palm oil, because of their free fatty acid content, will be best suited for older birds. There is potential for using palm and coconut oils as blends with more unsaturated oils and soapstocks, so as to benefit from potential fatty acid synergism.

CONSTRAINTS: Max: digestibility

44. VEGETABLE SOAPSTOCK

As a by-product of the vegetable oil refining industry, soapstocks provide a good source of energy and essential fatty acids. Soapstocks can be quite high in free fatty acids, and so stabilization with an antioxidant is essential. Soapstocks may also be acidulated, and this may pose problems with corrosion of metallic equipment. Some impurities may be added to soapstocks as a means of pollution-free disposal by refineries, and therefore quality control becomes more critical with these products. Moisture level may also be high in some samples, and this simple test is worthwhile for economic evaluation.

CONSTRAINTS: Max: impurities
level of free fatty acids

45. ANIMAL-VEGETABLE BLENDS

Some manufacturers mix animal and vegetable based fats, to give so-called blended products. The vegetable source is usually soapstock material. The blend has the advantage of allowing for some synergism between saturated fatty acids of animal origin and unsaturates from the soapstock. Animal-vegetable blends are therefore ideally suited for most classes of poultry, without the adverse problem of unduly increasing the unsaturates in meat which can lead to increased rate of oxidative rancidity (reduced shelf life).

CONSTRAINTS: Max: general considerations of a fat

46. RESTAURANT GREASE

An increasing proportion of feed fats is now derived from cooking fats and oils, and the generic product is termed restaurant grease. Their use has increased mainly due to problems of alternate disposal. Traditionally restaurant greases were predominately tallow or lard based products and this posed some problems in collection and transportation of solid fats. In recent years due to consumer concerns about saturated fats, most major fast food and restaurant chains have changed to hydrogenated vegetable cooking fats and oils. The fats are hydrogenated to give them protection against high-temperature cooking. Today, restaurant greases contain higher levels of oleic acid, and much of this will be trans-oleate. Assuming there has not been excessive heating, and that the grease has been cleaned and contains a minimum of M.I.U, then the energy value will be comparable to that of poultry fat. Future use of non-fat "cooking fats" will lead to considerable variation in the nutrient profile of these products.

CONSTRAINTS: Max: M.I.U
composition

47-56. CALCIUM AND PHOSPHORUS SOURCES

Calcium

Constraints are not usually imposed on these ingredients, because there should be fairly stringent constraints imposed on minimum and maximum levels of calcium and phosphorus in a diet. There has been considerable controversy in the past concerning the relative potency of limestone vs oyster shell as sources of calcium, especially for the laying hen. Perhaps of more importance than the source of calcium, is particle size. Usually the larger the particle size, the longer the particle will be retained in the upper digestive tract. This means that the larger particles of calcium are released more slowly, and this may be important for the continuity of shell formation, especially in the dark period when birds are reluctant to eat. Oyster shell is a much more expensive ingredient than limestone, but it offers the advantage of being clearly visible in the diet to the egg producer and so there is less chance of omission during feed manufacture. Birds also have some opportunity at diet self-selection if oyster shell is given, and this may be of importance in maintaining optimum calcium balance on egg-forming versus non egg-forming days.

Limestone should be in as large a particle size as can be readily manipulated by the bird's beak. For laying hens, this means a fairly coarse crumble consistency. There has been some concern in recent years of the variability in solubility of limestone from various sources. This can easily be checked by measuring pH changes when limestone is added to hydrochloric acid at initial pH of 4.0. Obviously 100% solubility is desirable, yet ideally this should be achieved over a prolonged period of time which hopefully correlates with the slow release of calcium into the blood stream.

Periodically, dolomitic limestone is offered to the feed industry. Dolomitic limestone contains at least 10% magnesium, and this complexes with calcium or competes with calcium for absorption sites. The consequence of feeding dolomitic limestone is induced calcium deficiency, usually manifested by poor skeletal growth or egg shell quality. The major user of dolomitic limestone is the steel industry and so problems with this ingredient seem to mirror the economic malaise in steel production. Dolomitic limestone should never be used in poultry diets.

Phosphorus

A considerable number of inorganic phosphorus sources are used around the world. Most naturally occurring phosphate sources are unavailable to the bird unless they are heat-treated during processing. As with limestone, the solubility in HCl at pH 4 can be used as a measure of quality. Insoluble phosphate sources are unlikely to be available to the bird - however solubility is not a guarantee of subsequent availability. Solubility tests are therefore only useful in screening out insoluble sources. Tests for biological availability are much more complex, because they necessarily require a chick bioassay where growth and bone ash are measured.

The phosphorus in most phosphate sources with the exception of soft phosphate can be regarded as close to 100% available. Rock phosphate and curaco phosphate are the major exceptions because these sources may only be 60 - 65% available to the bird. Anhydrous dicalcium phosphate is about 10% less available than the hydrated form, and this seems to relate to solubility. In this context, ingredients that stimulate gastric secretion, and hence HCl production, seem to result in improved utilization of the anhydrous form. Some rock phosphates contain various contaminants that may be of concern for poultry. The most common of these is vanadium. At just 7 - 10 ppm of the diet, vanadium will cause loss in internal egg quality. At slightly higher levels (15 - 20 ppm), there is a change in the shell structure where the shell takes on a somewhat translucent appearance, and appears more brittle. Rock phosphates can also contain as much as 1.5% fluorine. Because fluorine can influence calcium metabolism, there are often regulations governing the maximum permissable levels in feed. Only de-fluorinated rock phosphates are recommended although it must be remembered that this product usually contains about 5% sodium. Most mineral sources are detrimental to the pelleting process because they create significant friction at the pellet die. With phosphates, there is a distinct advantage to using rock phosphates rather than mono-or dicalcium phosphate in terms of pelleting efficiency. Numerous studies show 10+% increase in pellet throughput and energy saving with rock phosphates, although because of shorter dwell time in the steam conditioning chamber, pellet quality will deteriorate unless changes are made to the production system.

57-60. SODIUM SOURCES

Most diets will contain some added salt, usually in the form of sodium chloride. Where iodine is not added as a separate ingredient, iodized salt must be used. In most countries the various salt forms are differentiated by color, with common salt being a natural white color, and iodized salt being red. Cobalt iodized salt is often used in diets for swine and ruminants, and this can be used without any problems for poultry. This type of salt is usually colored blue. Because high levels of sodium chloride can lead to increased water intake, then a substitution of sodium bicarbonate for a portion of this chloride salt has been shown to be beneficial. Under these conditions, up to 30% of the supplemental salt can be substituted with sodium bicarbonate without loss in performance, and such birds often produce drier manure. For inclusions of sodium bicarbonate above 30%, care must be taken to balance the chloride levels - under commercial conditions it is often difficult to add economical sources of chloride other than salt. Chloride contributed by ingredients such as choline chloride and lysine-HCl should be accommodated during formulation.

61-68. TRACE MINERAL SOURCES

Trace minerals are available in a variety of forms, and periodically problems arise due to lack of knowledge of the composition, and/or stability of mineral salts. Most research into mineral availability has been conducted with so-called reagent-grade forms of minerals, which are very pure and of known composition and purity. Unfortunately, the feed industry cannot afford the luxury of such purity, and so obviously, feed grade forms are used.

One of the most important factors to ascertain prior to formulation is the state of hydration of a mineral. Many mineral forms contain "bound" water which obviously dilutes the effective mineral concentration. For example, hydrated cupric sulphate (white crystal) contains about 40% copper, whereas the more common pentahydrate (blue) contains 26% copper. It should also be emphasized that the various processing conditions used in manufacturing will likely reduce the biological availability. A combination of these two factors can mean a substantially lower potency of trace mineral sources relative to chemical standard values (Table

2.8). For this reason, feed manufacturers are encouraged to take great care in ordering trace minerals based on valency, hydration and purity. All minerals sources should be analyzed, on an "as is" basis for the major mineral component.

Because feed manufacturers are often concerned about "space" in the diet during formulation, there is a trend towards making very concentrated mineral and vitamin premixes. In considering concentration of mineral sources, oxides appear attractive, since they invariably contain the highest mineral concentration (Table 2.8). Oxides however are potent oxidizing agents, and if stored with premixed vitamins for any length of time, can cause the destruction of vitamins that are susceptible to oxidation. Since oxides are generally less available than other mineral salts, they should not be used exclusively in mineral premixes.

61. COBALT

The major source of cobalt is cobalt sulphate or cobalt carbonate. Both products are good sources of cobalt, with the cobalt as sulphate being slightly more available than in the carbonate form. Cobalt oxide has very low availability, and should not usually be considered during formulation.

62. COPPER

Copper oxide, sulphate and carbonate are used by the feed industry. Copper oxide can be of very low biological availability, especially with poor quality samples than contain significant amounts of metallic copper. Good quality copper oxide can be considered as available as is copper sulphate. As previously mentioned, the degree of hydration of copper sulphate must be specified.

63. IRON

Ferrous salts should be used in feed manufacture. As with copper, the major contaminant can be the metal itself, and this has a very low biological availability. Ferrous carbonate and ferrous sulphate are the preferred forms of iron. Ferrous salts are prone to chemical change during storage, such that 10 - 20% of ferric salts can be produced from original ferrous forms after 3 - 6 months storage at around 25°C.

64. MAGNESIUM

Magnesium carbonate and oxide are both available in feed grade form. The oxide can take up both water and carbon dioxide when stored for any length of time, and such activity obviously reduces the relative potency of magnesium.

65. MANGANESE

The major source of manganese used in the feed industry is manganese oxide. Sulphate and carbonate sources both have higher biological availability, yet these are usually uneconomical to use. Manganese oxide has a biological availability of 50 - 70%, yet this can be greatly influenced by its major contaminant, namely manganese dioxide. Manganese dioxide is only 50% as bioavailable as is the oxide, and so an appreciable content of dioxide can lead to a marked reduction in effectiveness of manganese oxide. Oxides should not contain more than 10% dioxides, and undoubtedly the range of availability quoted in research findings, is a reflection of dioxide contamination.

66. ZINC

Zinc oxide and zinc sulphate are the most common forms of zinc used in the feed industry. Zinc is often used as a catalyst in various industrial processes, and unfortunately catalysts sometimes find their way into the feed industry and are of low biological availability. Zinc sources can be contaminated with aluminum, lead and cadmium. If good quality sources are considered, then zinc oxide and zinc sulphate appear to be of comparable biological availability.

67. SELENIUM

Selenium is most often added to feeds as sodium selenite or sodium selenate. The most common naturally occurring form of selenium is selenomethionine, and this seems to have a much lower potency than either of the salt forms. There seems to be a greater availability of selenium with low protein diets, although this may be related to the fact that when birds are growing at a slower rate, their absolute selenium requirement is reduced. Selenium availability, from whatever source, is improved when diets contain antioxidants such as ethoxyquin.

Selenite is more readily reduced to elemental selenium, and for this reason selenate is sometimes preferred. Selenium metal is less available and can form insoluble complexes with other minerals. Whichever form of selenium is used, it must be remembered that the final diet inclusions are extremely low in relation to the other minerals, and so some degree of premixing is essential prior to incorporation in diets or premixes.

68. IODINE

If iodine is added to a mineral premix, rather than supplied with salt, then potassium iodide and calcium iodate are the preferred sources. Potassium iodide is very unstable and deteriorates rapidly with moderate exposure to heat, light and/or moisture. Calcium iodate is therefore the best source of supplemental iodine.

Mineral chelates and complexes

Chelates are mixtures of mineral elements bonded to some type of carrier such as an amino acid or polysaccharide. These carriers, or ligands, have the ability to bind the metal, usually by covalent bonding through amino groups or oxygen. The formed chelate is usually a ring structure with the divalent or multivalent metal held strongly or weakly through two or more covalent bonds. Iron in hemoglobin is the classical example of a chelate. The covalent bonding is such that the chelate has no electrical charge.

Chelated, or complex minerals are usually much more expensive than inorganic minerals, and so one expects improved bird performance through either enhanced absorption or better utilization in some way. It is difficult to rationalize the cost of chelated minerals based solely on improved absorption in the intestine. Even a 50% difference in absorption can be most economically resolved by doubling the level of inorganic mineral used. However there are limits to the level of any one mineral to be used, because of potential negative effects on absorption and utilization of other minerals and other nutrients. The mineral availability from some inorganic sources can be very low. For example the manganese in some samples of manganese sulfate has been reported at just 5%, and in this instance a 20 fold increase in inclusion level, while correcting the potential manganese absorption problem, will likely have adverse effects on utilization of phosphorus, calcium and iron.

Factors affecting the uptake of haem iron is often used to support the concept of using chelated minerals. There are a number of other trace minerals, such as copper, manganese and phosphorus, that can affect absorption of inorganic iron, while uptake of haem iron will be little affected. The uptake of chelated minerals is therefore expected to be more consistent, and less effected by adverse (or enhanced) environments in the gut lumen. Bioavailability of minerals from chelates should also be consistent, because of standardization during manufacture, versus less standard conditions with some supplies of inorganic salts. There are also claims of chelated minerals being used more effectively at the cellular level , following absorption. There are few classical supporting claims for these suppositions, and so enhanced performance of meat birds and layers is discussed in terms of stimulation of various biological processes by the mineral and/or that the chelated mineral enters certain pools with greater affinity or efficiency.

Ultimately the choice of using inorganic versus chelated minerals is one of economics, which obviously relates to cost benefit. Such results may vary depending upon the levels and spectrum of trace minerals used and the bioavailability to be expected from inorganic sources that are available.

69-78 SYNTHETIC AMINO ACIDS

Synthetic sources of methionine and lysine are now used routinely in poultry diets, and tryptophan and threonine will likely be used more frequently as future prices decline. In most situations, the use of synthetic amino acids is an economic decision, and so their price tends to shadow that of soybean meal, which is the major protein (amino acid) source used world-wide. By the year 2000 lysine use in N. America is estimated at 100,000 tonnes while that for methionine will be around 75,000 tonnes, of which the poultry industries use 40 - 75%.

Lysine is usually produced as the hydrochloride salt, and consequently, the commercial products have 79% lysine activity on a weight basis. In N. America, lysine tends to be considered a commodity, and as such its use is directly related to that of other ingredients. In general, there is greater L-lysine HCl usage when soybean meal price increases, or when corn price declines. In Europe however, because of inherently higher commodity prices, L-lysine HCl tends to be used less as a commodity, and more as a means of improving performance. Care must therefore be taken in interpretation of cost benefit of lysine use in research results reported from these two regions.

Tryptophan is not thought to be a limiting amino acid in most poultry diets, and so the move to greater synthetic tryptophan use comes from the swine industry. Tryptophan will become a limiting nutrient as crude protein levels of diets are reduced, although currently its efficient use is somewhat hampered by complexity involved in diet analysis. Tryptophan levels in ingredients and feed are much more difficult to assay than are the other common amino acids, and in part, this situation leads to variability in research results aimed at quantitating response to tryptophan. This amino acid is most likely to be considered when diets contain appreciable quantities of meat or poultry by-product meal.

Methionine is available in a number of forms and also as an analogue. Over the years there has been considerable research into the potency and use of these various sources. There are basically four different sources of methionine (Table 2.20)

TABLE 2.20 Methionine sources

DL-methionine	DL-methionine Na	Methionine hydroxy analogue	Methionine hydroxy analogue-Ca
CH_3 | S | CH_2 | CH_2 | $H\text{-}C\text{-}NH_2$ | COOH	CH_3 | S | CH_2 | CH_2 | $H\text{-}C\text{-}NH_2$ | $CooNa^+$	CH_3 | S | CH_2 | CH_2 | H-C-OH | COOH	CH_3 | S | CH_2 | CH_2 | H-C-OH | $COOCa^+$
Powder	Liquid	Liquid	Powder

Dow Chemical was the first to produce powdered DL-methionine in commercial quantities in the 1940's, while Monsanto introduced the calcium salt of methionine hydroxy analogue in the 1950's. Since this time, the market has established demand for both DL-methionine and the analogue, in both powdered and liquid forms.

It has been known for some time that most essential amino aids can be replaced by the corresponding α-keto acid (analogue). With the exception of lysine and threonine, which are not involved in transamination processes, it is possible to replace amino acids with their keto acid analogues. Presumably the bird produces the corresponding amino acid by transamination involving mainly non-essential amino acids such as glutamic acid. Such transamination can occur in various tissues, and some bacteria in the intestine may also synthesize amino acids prior to absorption. The question of relative potency of products such as liquid MHA (eg. Alimet®) often arises in selection of methionine sources. Liquid MHA has a base value of 88%, based on normal chemical structure. Availability of this 88% value has then been shown to vary from 60 - 100%. It seems inconceivable that any nutrient could have such variable efficacy, and so one must look at experimental conditions, and diet formulation in assessing such results. Potency of MHA relates to variable uptake in the intestine, degradation in body tissues and or degree of elimination by the kidney. Another major variable in response to MHA under commercial conditions is ingredient methionine levels used in formulation and diet specifications for methionine and cystine. There are usually logical reasons why nutritionists use different potency values. The bottom line is cost per kg of meat/eggs produced, and the value of products such as MHA quickly establish themselves over time within an integrated operation.

Liquid products, whether MHA or methionine, are often considered over dry products, due to ease of incorporation at the feed mill.

2.5. INGREDIENT CONSTRAINTS

The following values are intended as guidelines for minimum and maximum constraints of ingredient levels used during formulation. These levels are based on consideration of the previous discussions of each ingredient, and the need to produce diets of acceptable textured quality. A minimum constraint for fat of 1% is given for most classes of bird, because all diets, regardless of nutrient specifications should contain this level of fat related to bird performance, dust level and stimulation of feed intake. Where minimum levels on other ingredients are given, it is an indication of likely benefits to accrue from their inclusion.

TABLE 2.21 Ingredient constraints - young birds 0-4 weeks (% Diet inclusion)

Ingredient	Min	Max
Yellow corn	20	60
Wheat (+ enzyme)		20(40)
Oats		10
Barley (+enzyme)		10(30)
Rice (rough)		10
Wheat bran		8
Wheat shorts		10
Wheat screenings		10
Rice bran		15
Rice polishings		15
Milo		40
Rye		0
Triticale		2
Bakery by-product		10
Molasses		1
Dehydrated alfalfa		5
Canola meal		5
Full-fat canola		5
Soybean meal (48%)	10	30
Soybean meal (44%)	10	25
Full fat soybean		15
Corn gluten meal		15
Corn gluten feed		15
Cotton seed meal		5
Peanut meal		8
Peas		5
Safflower meal		8
Sesame meal		5
Sunflower meal		8
Lupins		8
Flax		8
Meat meal		8
Fish meal (60%)		8
Blood meal		2
Feather meal		1
Dried whey		5
Tallow		0
Lard		0
Poultry fat	1	5
Fish oil	1	5
Vegetable oil	1	5
Coconut oil		0
Palm oil		2
Vegetable soapstock	1	5
Animal-vegetable blend	1	5
Restaurant grease	1	5

TABLE 2.22 Ingredient constraints - meat type birds 4-18 weeks (% Diet inclusion)

Ingredient	Min	Max
Yellow corn	20	70
Wheat (+ enzyme)		25(50)
Oats		25
Barley (+enzyme)		15(40)
Rice (rough)		20
Wheat bran		20
Wheat shorts		20
Wheat screenings		20
Rice bran		20
Rice polishings		20
Milo		50
Rye		10
Triticale		15
Bakery by-product		15
Molasses		5
Dehydrated alfalfa		10
Canola meal		8
Full-fat canola		10
Soybean meal (48%)		30
Soybean meal (44%)		25
Full fat soybean		20
Corn gluten meal		20
Corn gluten feed		20
Cotton seed meal		10
Peanut meal		10
Peas		10
Safflower meal		10
Sesame meal		10
Sunflower meal		10
Lupins		10
Flax		10
Meat meal		8
Fish meal (60%)		10
Blood meal		2
Feather meal		2
Dried whey		10
Tallow	1	5
Lard	1	5
Poultry fat	1	8
Fish oil	1	1-8
Vegetable oil	1	8
Coconut oil	1	5
Palm oil	1	5
Vegetable soapstock	1	8
Animal-vegetable blend	1	8
Restaurant grease	1	5

Table 2.23 Ingredient constraints - immature egg layers and immature breeding stock 4-18 weeks (% Diet inclusion)

Ingredient	Min	Max
Yellow corn	20	70
Wheat (+ enzyme)		20(40)
Oats		25
Barley (+enzyme)		20(40)
Rice (rough)		25
Wheat bran		15
Wheat shorts		25
Wheat screenings		25
Rice bran		20
Rice polishings		20
Milo		50
Rye		5
Triticale		15
Bakery by-product		15
Molasses		5
Dehydrated alfalfa		5
Canola meal		5
Full-fat canola		10
Soybean meal (48%)	10	30
Soybean meal (44%)	10	30
Full fat soybean		20
Corn gluten meal		25
Corn gluten feed		20
Cotton seed meal		10
Peanut meal		15
Peas		10
Safflower meal		15
Sesame meal		15
Sunflower meal		15
Lupins		10
Flax		10
Meat meal		10
Fish meal (60%)		8
Blood meal		2
Feather meal		2
Dried whey		10
Tallow	1	2
Lard	1	3
Poultry fat	1	5
Fish oil	1	5
Vegetable oil	1	5
Coconut oil	1	2
Palm oil	1	2
Vegetable soapstock	1	5
Animal-vegetable blend	1	5
Restaurant grease	1	5

TABLE 2.24 Ingredient constraints - mature egg layers and mature breeding stock (% Diet inclusion)

Ingredient	Min	Max
Yellow corn	20	60
Wheat (+ enzyme)		20(40)
Oats		20
Barley (+enzyme)		15(30)
Rice (rough)		20
Wheat bran		10
Wheat shorts		20
Wheat screenings		15
Rice bran		15
Rice polishings		15
Milo		40
Rye		0
Triticale		10
Bakery by-product		10
Molasses		5
Dehydrated alfalfa		5
Canola meal (brown egg)		8(0)
Full-fat canola (brown egg)		8(0)
Soybean meal (48%)	10	25
Soybean meal (44%)	10	20
Full fat soybean		20
Corn gluten meal		20
Corn gluten feed		20
Cotton seed meal		10
Peanut meal		10
Peas		10
Safflower meal		10
Sesame meal		10
Sunflower meal		15
Lupins		15
Flax		15
Meat meal		8
Fish meal (60%)		2-10
Blood meal		2
Feather meal		2
Dried whey		5
Tallow	1	5
Lard	1	5
Poultry fat	1	8
Fish oil	1	1-8
Vegetable oil	1	8
Coconut oil	1	5
Palm oil	1	5
Vegetable soapstock	1	8
Animal-vegetable blend	1	7
Restaurant grease	1	6

2.6. ECONOMIC VALUE OF INGREDIENTS

Least-cost feed formulation has largely replaced the need to consider the economic value of individual ingredients. With most formulation programs, the economic value of ingredients being used, or rejected, is usually available. For new ingredients, parametric analyses gives an indication of relative worth. For those individuals without access to these sophisticated programs, some means of economic analysis is essential. Because protein and energy are the major nutrient costs for most diets, economic comparison taking into account these nutrients gives an 85 - 90% estimation of overall economic worth. Also, because corn and soybean meal are usually the major energy and protein sources in poultry diets, they can be used as a basis for comparison. Therefore, if one knows the current price of corn and soybean meal, and their contributions of protein and energy, one can evaluate the relative worth of any other ingredient. Comparison of this predicted value with a market price of the new ingredient will give the potential buyer a guide of relative worth of the new ingredient.

The solution to this question is resolved by solving simultaneous equations:

- if corn contains 8.6% crude protein and provides 3300 kcal metabolizable energy (ME) per kg, then 1 tonne of corn provides 86 kg protein and 3300 kcal x 1000 kg = 3,300,000 kcal or 3300 Mcal of ME.

- if soybean meal contains 48% crude protein and provides 2550 kcal ME/kg, then 1 tonne of soybean meal provides 480 kg protein and 2550 Mcal of ME.

- If corn is selling for \$150/tonne and soybean meal for \$300/tonne, then:

a) $86\,X + 3300\,Y = 150$ (\$)

b) $480\,X + 2550\,Y = 300$ (\$)

Where X = cost 1 kg protein and Y = cost of 1000 kcal ME (1 Mcal ME).

Equation a) can be re-written as $X = \frac{150-3300Y}{86}$

and substituting this value for X into equation b;

then $480\,\frac{(150-3300Y)}{86} + 2550Y = 300$

From this equation Y = 0.03385 \$ = cost 1000 kcal energy

Substituting Y into equation a;

$86\,X + 3300\,(.03385) = 150$

From this equation X = .4453 \$ = cost 1 kg protein

These equations now give the relative cost of protein and energy in corn and soybean meal, and there is potential to judge the economic worth of other ingredients.

For example is wheat a good purchase at \$180/tonne?

Wheat provides 130 kg crude protein and 3150 Mcal/ME per tonne

Therefore its relative worth is 130 x .4453 = 57.89
plus 3150 x .03385 = 106.63
$164.52

Therefore, when corn is priced at $150/tonne and soybean meal is priced at $300/tonne, based on the contribution of energy and protein, wheat is worth $164.52/tonne. In this instance, we have indicated wheat at $180/tonne, and so it is uneconomic on this basis.

At the outset of these calculations, it was suggested that this analysis accounts for 85 - 90% of the economic value of an ingredient. Judgement must be made between relative calculated worth ($164.52) and market price ($180) in terms of the other nutrients or characteristics that wheat may give to the diet. For example, the pelleting characteristics of wheat would hold some value in certain markets. With other ingredients there would be a contribution of phosphorus, and this certainly would have economic value. If the current market price is less than the estimated price, then the decision is quite easy, because the ingredient is obviously economically attractive based solely on energy and protein contribution alone. It must be remembered that as the price of corn and soybean meal change, then the values for X and Y must be recalculated.

In these equations, corn and soybean meal have been used as the standards, although any pair of ingredients could be used. For example, milo and cotton seed meal, or wheat and canola meal could also be used. The values for energy and protein for these ingredients can be found in Table 2.1. These types of calculations are in fact used in the sophisticated computer programs mentioned earlier. With the help of the computer, more nutrients (other than just protein and energy) can be considered, and so the estimation of economic worth becomes more precise. Also the more sophisticated the program, the more ingredients that can be considered in the standards, rather than just corn and soybean meal.

2.7. INGREDIENT QUALITY CONTROL

Ingredients must be continuously monitored to ensure consistency of nutrient profile and presence of potential contaminants. The number and frequency of assays will depend upon the class of ingredient, historical results of analysis and to some extent the season of the year. Ingredients from new suppliers should be tested the most rigorously, and number and frequency of testing only reduced when consistency of product is established. Table 2.25 gives an overview of quality control procedures to be applied for most of the major ingredient types listed in Tables 2.1 - 2.6.

The frequency of sampling will obviously vary with the significance of a particular ingredient in the feed. For example, where fish meal is used extensively, and represents a significant proportion of dietary amino acids, then amino acid analyses may be done more frequently, and also it may be advisable to screen more frequently for gizzard erosion factors. On the other hand, where a history of consistent analyses is developed, then testing can be less frequent.

TABLE 2.25 Quality control schedule for ingredients

	Moisture	CP	Fat	Ca	P	Amino Acids	Other analyses
Corn	1	2	4	6	6	6	Molds-3;mycotoxins-4
Wheat	1	2	4	6	6	6	Pentosans-6
Barley	1	2	4	6	6	6	βglucans-4
Milo	1	2	4	6	6	6	Tannins-3
Wheat-shorts	1	2	4	6	4	6	
Rice bran	1	2	2	6	6	6	Antioxidant-6
Bakery by-product	1	1	1	1	1	4	Salt-3
Canola meal	1	2	4	6	6	6	Glucosinolates-5
Soybean meal	1	1	3	3	3	5	Urease-4
Soybeans	1	3	3	6	6	6	Urease-3
Corn gluten meal	1	2	4	6	6	5	
Cottonseed meal	1	1	4	6	6	6	Gossypol-3
Peanut meal	1	1	4	6	6	6	Aflatoxin-3
Sunflower meal	1	1	4	6	6	6	
Meat meal	1	1	1	1	1	3	Salmonella-4
Fish meal	1	1	1	1	1	3	Gizzard erosion-3
Poultry by-product	1	1	1	3	3	4	
Feather meal	1	1	3	6	6	3	
Fats and oils	1		1				Fatty acids-3;stability-4
Limestones	1			1			
Phosphates	1			1	1		Vanadium-6
Salt	1						

1 = Each delivery; 2 = weekly; 3 = monthly; 4 = 6x each year; 5 = 3x each year; 6 = 2x each year

For assay results to be meaningful, ingredients must be sampled accurately. For bagged ingredients, at least 4 bags per tonne, to a maximum of 20 samples per delivery, should be taken, and then these sub-samples pooled to give one or two samples that are sent for assay. It is always advisable to retain a portion of this mixed sample, especially when assays are conducted by outside laboratories. For bulk ingredients, there should be about 10 sub-samples taken from each truck or rail car load, and again this mixed to give a representative composite for assay.

Specialized testing of ingredients

There are a number of rapid tests available for evaluating ingredients. In some instances these tests are specific to certain ingredients and to specific nutrients and/or antinutrients within an ingredient. Alternatively some tests are more generic and can be applied across a number of ingredients. The decision to carry out any of these tests is based on significance of the ingredient in the diet, and so the relative contribution of constituents under test. Developing historical data on ingredients is also a useful way of determining the need and frequency of various testing procedures. The following tests or methodologies are assumed to be in addition to more extensive chemical testing that will routinely be used for most nutrients, such as those described in Table 2.25.

a) Bulk density

Bulk density of cereals is correlated with energy value and protein content. In N. America, the usual measurement is bushel weight, while the common metric equivalent is kg/hl. Weight of 100 kernels of cereal is also used as an indicator of bulk density. Under normal growing conditions, as bulk density declines, there is usually reduction in energy level, mainly associated with reduction in starch content of the endosperm, while protein content increases when most protein is associated with the outer bran or pericarp layers. The following Table 2.26 shows commonly used bulk density values for cereals.

Table 2.26 Whole cereal bulk density

Cereal	Density (kg/hl)
Corn	72.2
Wheat	74.0
Barley	63.0
Oats	46.0
Milo	55.0
Rice	52.0
Triticale	62.0
Rye	70.0

Bulk density of other ingredients is also useful knowledge when designing storage space and monitoring bulk density again gives an indication of variability in manufacturing procedures (Table 2.27).

Bulk density will vary with moisture content, and this should be taken into account during measurement. Density is easily measured by weighing the cereal or feed into a container of known volume. The smaller the container, the greater the care needed in standardizing the filling and especially the packing of the ingredient.

Table 2.27 Bulk density of common ground ingredients

Ingredient	Density (kg/hl)	Ingredient	Density (kg/hl)
Alfalfa	22.0	Soybean meal	63.0
Wheat Bran	20.0	Canola meal	65.0
Wheat-shorts	30.0	Groundnut meal	60.0
Corn gluten	60.0	Meat meal	65.0
Soybeans	76.0	Feathermeal	50.0
Cottonseed meal	60.0	Poultry by-product	55.0
		Fish meal	60.0

b) Near Infra Red Analysis (NIRA)

NIRA offers the potential for very rapid analyses of ingredients and feeds. The technique has the potential to assay any component that is composed of organic compounds.

The system has the potential to measure metabolizable energy as well as more simple components such as fat, moisture, protein and fiber. Analysis relies on measuring how much light energy is absorbed when the sample is bombarded with light at very specific wavelengths.

The basis of NIRA is chemometrics, which is the application of mathematics to analytical chemistry. The technique is an integration of spectroscopy, statistics, and computer sciences. Mathematical models are constructed that relate chemical composition (that is active chemical groups) to energy changes in the near infra red region of the spectrum which ranges from 700 to 2500 nm wavelength.

In this region of the spectrum we measure mainly hydrogen vibrations of chemical bonds in which hydrogen is attached to atoms such as nitrogen, oxygen, or carbon. Because most feedstuffs are opaque, NIRA uses reflectance instead of transmittance. The reflected light of a sample (diffuse reflectance) is used to indirectly quantitate the amount of energy absorbed in a sample. NIRA measures the absorption of infra red radiation by sample components, for example, peptide bonds at specific wavelengths in the near infra red spectrum. Other components of the sample absorb energy as well, however, and have the effect of interfering. This effect is eliminated by mathematical treatment of the spectral data and by multiple linear regression or other statistical procedures.

Because each molecule usually exists in its lowest energy state, absorption of energy will raise its energy state by "some" degree. Such energy absorption occurs at a wavelength that is characteristic for that particular molecule. Energy absorption in the fundamental infra red region is very strong, but also very specific for certain molecular groupings. For example, water has a characteristic absorption at the same wavelength as does starch. Strong, but specific fundamental wavelengths would be difficult to differentiate for these two components. This does not mean to say that infra red analysis does not have a place in feed analysis. For example, with pure nutrients, (amino acids, vitamins?) the use of light reflectance in the fundamental range may offer potential for very specific analysis of purity. For samples of mixed composition, whether it be ingredients or complete feeds, then a more subtle analysis must be used to differentiate all the various chemical groupings. In the weaker absorbing NIR range of wave lengths, it is "secondary" absorption wavelengths that are considered - these are most often referred to as "overtones". By considering a spectrum of wavelengths, a characteristic pattern of absorption (reflectance) energy is given for each major component of the sample. Chemometrics then involves calculation of correlation coefficients at each wavelength and simultaneously selecting both the best fit with the nutrient under study, and also the "best fit" at all other absorption frequencies so as to remove all interference problems.

The usefulness of NIRA, therefore depends entirely on the careful and conscientious calibration of the equipment. To some extent this exercise has been simplified through introduction of so-called scanning machines that cover a wide band of NIR. Prior to this technology, fixed filter (wavelength) equipment only was available, and so *a priori* knowledge of likely absorption bands or tedious testing of numerous wavelengths was essential in order to develop useful calibrations.

Developing calibrations for components such as moisture, fat, crude protein and fiber is a very straightforward procedure, and these calibrations can be combined within a single program, such that from each ingredient scan, these various analyses are conducted concurrently. For most commercial machines, ensuring consistent fineness of grind and controlling moisture content of samples resolves much of the variation associated with operating procedure.

Determination of ME with NIRA provides a considerable challenge. Firstly, there is a need for an extensive range of diets of determined analysis to be used for calibration. The conventional bioassay for ME is both time consuming and very expensive, and these facts have undoubtedly limited investigation to date in this area. Secondly, ME *per se* provides a complex problem for NIRA, because energy contribution is not confined to one nutrient but rather is represented by a range of molecular bondings and configurations. Usefulness of NIRA to predict ME therefore depends upon careful *in vivo* analysis of a range of diets preselected in terms of anticipated ME, nutrient contribution and ingredient composition. These latter parameters are of importance if "universal" calibrations are to be developed. Similarly, great care must be taken in the mathematical manipulation of spectral coefficients. Over the last few years Valdes and Leeson (1992 - 94) have developed a number of such calibrations for feeds and ingredients. Table 2.28 shows some of these results for ingredients.

TABLE 2.28 Prediction of metabolizable energy by NIRA (kcal/kg)

Ingredient	Determined	NIRA prediction
Corn	3380	3370
Barley	2720	2670
Wheat	3275	3225
Soybean meal	2340	2320
Bakery meal	2990	3005
Tallow[1]	8690	8680
Poultry fat[1]	9020	8840
Corn oil[1]	9660	9530
Palm oil[1]	7300	7700

[1]Separate calibration for fats — Adapted from Valdes and Leeson (1992,94)

TABLE 2.29. Prediction of amino acids in fish meal

	% Amino acid	
	Assay	NIRA prediction
Methionine	1.50	1.6 ± 0.06
Cystine	0.60	0.6 ± 0.07
Lysine	3.70	4.0 ± 0.30
Tryptophan	0.60	0.5 ± 0.03
Threonine	2.20	2.3 ± 0.09
Arginine	3.40	3.4 ± 0.09

Valdes and Leeson (unpublished)

There is also potential for NIRA to predict amino acids in ingredients (Table 2.29) as well as antinutrients such as glucosinolates or trypsin inhibitors. As with all NIRA analyses, the accuracy of such predictions are greatly influenced by the time and precision involved in calibration using samples of known composition.

c) *Urease testing of soybeans and soybean meal*

Levels of the enzyme urease are used as an indicator of trypsin inhibitor activity. Urease is much easier to measure than is trypsin inhibitor and both molecules show similar heat-sensitivity characteristics. A rapid qualitative screening test for urease can be carried out using conversion of urea to ammonia in the presence of an indicator.

A qualitative test for urease activity can be carried out using a simple colorimetric assay. Urea-phenol red test solution is brought to an amber color by using either 0.1NHCl or 0.1N NaOH. About 25g of soybean meal is then added to 50 ml of indicator in a petri dish. After 5 minutes the sample is viewed for presence of red particles. If the sample has no red particles showing, the mixture should stand another 30 minutes, and if again no red color is seen, it suggests overheating of the meal. If up to 25% of the surface is covered in red particles, it is an indication of acceptable urease activity, while 25 - 50% coverage suggest need for more detailed anlaysis. Over 50% incidence of red colored particles suggests an underheated meal.

d) *Protein solubility*

Plant proteins are normally soluble in weak alkali solution. However if these proteins are heat treated, as normally occurs during processing of many ingredients, the solubility of protein will decline. Dale and co-workers at Georgia have developed a fairly rapid test which seems to give a reasonable estimate of protein solubility and hence protein quality in soybean meal. The assay involves adding just 1.5g of soybean meal to 75 ml of 0.2% potassium hydroxide solution, and stirring for 20 minutes. Soluble proteins will be in the liquid phase, and so all or a portion of the centrifuged liquid is assayed for crude protein, and protein content relative to the original 1.5g sample calculated according. By knowing the crude protein content of the original sample of soybean meal, percentage solubility can easily be calculated. Typical results, as shown by Dale and Araba are given in Table 2.30. As heating time is increased, there is a decrease in protein solubility. Values of 75 - 80% solubility seem to be ideal, with higher values suggesting underheating, and lower values overheating of the protein.

TABLE 2.30 Protein solubility of samples of soybean meal heated for various times

Heating time	Urease (ph change)	Protein solubility (%)	Bird performance	
			Wt gain (g)	Feed:gain
0 (Raw)	2.40	99.2	343[d]	2.44[c]
5 min	2.04	87.7	429[c]	2.29[bc]
10 min	0.23	79.1	481[ab]	2.00[a]
15 min	0	74.9	496[a]	2.09[ab]
20 min	0	71.8	500[a]	2.03[a]

Dale and Araba (1987)

e) *Protein and amino acid dye-binding*

Proteins will bind with a number of dyes and so this provides the basis for colorimetric assays. These dye-binding techniques can be used to test protein *per se*, or used to test for protein in various extractions involved in assays of solubiiity or digestibility. Dye-binding can therefore replace the Kjeldahl analysis, depending upon sensitivity needs. The most commonly used methods are as follows:

Cresol Red - J. Amer. Assoc. Anal. Chem. 43:440
Orange G - J. Nutr 79:239
Coomassie Blue - Anal. Biochem. 72:248

Lysine also reacts with certain dyes to give a colorimetric assay. Carpenter suggested that if the e-amino group of lysine is free to react with dye, then the lysine can be considered as "available". The most commonly used dye is Fluoro-2,4 dinitrobenzene (FDNB), which gives a yellow/orange color when combined with lysine.

f) *Fish meal gizzard erosion factor*

In some countries, fish meal is an economical feed ingredient to use in poultry diets. As previously described, some samples of fish meal will cause severe gizzard erosion in young birds. Where fish meal is an integral part of a broiler diet, then it is common to carry out a chick growth test with each shipment of fish meal. About 50 chicks are fed a broiler starter diet, usually without any fish meal, for 5 - 7 days. At this time, the diet is mixed with 40 - 50% of the test fish meal, and this diet fed for another 7 - 10 days. Birds are then sacrificed and the gizzards examined for erosion, often using a subjective scale as follows:

1) very mild erosion, with good gizzard color
2) mild erosion, with evidence of destruction of the lining in some areas
3) erosion in localized areas, with cracks in the thinner lining
4) severe erosion, cracking, thinning and discoloration
5) sloughing of the gizzard lining with hemorrhage

Because 40 - 50% fish meal is used, some gizzard erosion is expected with most samples. Scores of 4 - 5 are often used to reject samples, although this decision will to some extent depend upon the level of fish meal to be used in the commercial diet.

g) Sorghum tannins

Tannins are detrimental to protein utilization, and so levels should be minimized in poultry diets. Sorghum is a potential source of tannin, and this is usually found in the outer seed coats. Unfortunately there is not a clear relationship between seed coat color and tannin content. High tannin sorghums are usually darker in color, but some dark colored sorghums are also low in tannin. The tannins are present in the testa, which is the layer immediately beneath the outer pericarp. One quick test is therefore to cut into the seed and observe for presence of a pigmented (tannin) testa. More recently, a bleach test has been developed which again shows presence or not of a pigmented testa. About 20 g (tablespoon) of sorghum is mixed with 5 g potassium hydroxide and 75 ml of household bleach. The mixture is shaken until the KOH dissolves, and then set aside for 20 minutes. Sorghum grains are then strained, rinsed with water and placed on a paper towel. The KOH will remove the outer pericarp, and expose the testa. High tannin grains will appear dark brown/black while low tannin sorghum will be bleached white/yellow.

h) Gossypol in eggs

Feeding gossypol to laying hens can result in discoloration of both the yolk (green-brown) and albumen (pink). Gossypol is usually found in cottonseed meal and as described previously for this ingredient there are ways to minimize the effects of this compound by diet modification. However egg discoloration occurs periodically, and cottonseed meal or cottonseed oil is often suspected. Placing egg yolks in a petri dish with ammonia quickly causes varying degrees of brown discoloration depending upon gossypol content.

i) Fat assays

Fat quality is best assessed by measurement of moisture, impurities and individual fatty acids. However there are a number of less extensive tests that can be used to give some idea of fat composition and quality. Fat titre is a measure of hardness, and simply relates to melting point. The break-point between tallows and greases is about 40°C. The higher the melting point, the more saturated the fat. Titre should obviously be consistent for an individual class of fat or fat blend from any one supplier. Iodine value can also be used as a measure of hardness. Each double bond (unsaturated) will take up a molecule of iodine, and so higher values mean more unsaturation, which in turn should relate to lower titre (Table 2.31).

Iodine value is greatly influenced by levels of palmitic, oleic and linoleic acid in most fats and oils. Generally, as titre increases by 10 units over the range of 50 - 100, then palmitic acid content decreases by about 2%. Also as a rule of thumb iodine value = 0.9 x % oleic acid + 1.8% linoleic acid.

TABLE 2.31 Iodine value and titre of common fats		
	Iodine Value	Titre °C
Tallow	45	45
Lard	65	40
Poultry fat	80	35
Vegetable oil	120	15

j) Hulls in rice by-products

Rice bran, sometimes called rice pollards, is used extensively in rice growing areas of the world. The major variable affecting nutritive value, is the content of hulls, which are essentially indigestible for poultry. A major component of hulls is lignin, and this reacts with the reagent phloroglucinol to produce a color reaction. The reagent is produced by combining 1g of phloroglucinol with 80 ml 2M HCl and 20 ml ethanol. The rice by-product is mixed 1:2 with reagent and held at about 25°C for 10 minutes. Development of red color will be directly proportional to hull content. Actual hull content and a color score card are necessary to "calibrate" the assay.

k) Mineral solubility

Neutralizing mineral salts with various acids can be used to give some idea of mineral availability, and when an assay is monitored over time, then information on rate of solubility is also obtained. Hopefully all mineral sources will be totally available to the bird, although at least with calcium sources there is concern about solubility occurring too quickly. With laying hens for example, a moderate rate of solubility is preferable to very rapid solubilization, because the former more closely matches the prolonged duration of need for calcium supply to the shell gland.

Limestone solubility can easily be measured by monitoring pH of the mineral in dilute acid. After recording the original pH of a 90 ml aliquot of 0.1N HCl, 10g of limestone is gradually added, and without stirring, pH measured after time intervals of say 10, 20, 30 and 60 minutes. Limestone will result in an increase in pH, as H^+ ions are liberated from solution. A pH change of +0.1 relates to a 20% solubility, while changes of 0.2, 0.3, 0.4, 0.5, 0.6, 0.7, 0.8, 0.9 and 1.0 relate to about 37, 50, 60, 70, 75, 80, 84, 87 and 90% solubility respectively. A pH change of +2.0 means 99% solubility. A high solubility after 60 minutes is expected from a quality limestone, whereas the rate of achieving 95 - 99% solubility will give an indication of the rate of calcium release in the proventriculus. Particle size and particle porosity are the factors most likely to affect rate of change of solubility. Optimum eggshell quality, and perhaps bone development in young birds, are dependent upon a consistent pattern of calcium solubility.

Neutralization of ammonium citrate has also been used to assess solubility of phosphate sources and also of manganese and zinc salts.

2.8. FEED ADDITIVES

A range of additives is used in most poultry diets and, in general, do not contribute any nutrients *per se*. Most additives are used to improve physical diet characteristics, feed acceptability or bird health. The following discussion is not intended to discuss efficacy of the various products, rather it is hoped to highlight various implications of their use in terms of diet formulation, ingredient compatibility and/or general feeding management.

a) *Pellet Binders*

When pellet quality is of concern, a pellet durability index is often ascribed to ingredients and this is considered during formulation. This index may range from 55 - 60 for corn soy diets that are notoriously difficult to pellet, to 85 - 90 for wheat-based diets with pellet binders. With corn-based diets, it is often necessary to use synthetic pellet binders in order to achieve desirable pellet quality. In most instances, the need for a good pellet is necessary to placate the purchaser of the feed, because the bird *per se* is often tolerant to a range of quality in terms of growth rate and feed efficiency. The turkey poult is perhaps the most sensitive to pellet or crumble quality, where growth rate can be markedly influenced by both pellet size and the proportion of fines. The pelleting process is discussed in more detail in the following section on Feed Mixing and Pelleting.

A number of pellet binders are available, although they are used at considerably different inclusion levels, and such levels should be clearly specified for each product. When wheat or wheat by-products are used at less than 10% of the diet, then a binder will often be necessary if high pellet durability is desired. The two major types of binders have lignosulfonate or colloidal clays as the base product, with inclusion levels of around 5 - 12 kg/tonne. There have been reports of colloidal-clay type products binding some B-vitamins in the gut, and so making them unavailable to the bird. The colloidal-clay products may also aid in reducing apparent moisture content of the bird's excreta and more recently some forms of clay have been shown to have activity in binding aflatoxin. The lignosulfonate pellet binders often contain 20 - 30% sugars, and so contribute to diet energy level. Studies show lignosulfonate binders to have ME values of 900 - 2200 kcal/kg depending upon sugar content. Because these binders are often used at 1 - 1.2% of the diet, then energy contribution is meaningful at 10 - 25 kcal/kg of diet.

b) *Anticoccidials*

Anticoccidials are used in diets for most meat birds and young breeding stock that are reared on litter floors. Over the past 20 years, the so-called ionophore anticoccidials have predominated and they have proved most efficacious in controlling clinical symptoms of coccidiosis. From a nutritional viewpoint, some care must be taken in selection of these products as they can influence metabolism of the bird under certain situations.

Monensin has been a very successful anticoccidial, and seems to work well with both broiler chickens and turkeys. Monensin like most ionophores has an affinity to bind metal ions, the most important in terms of bird metabolism being sodium and potassium. Lasalocid also binds metals, although its major affinity is for potassium and secondly, sodium. Most ionophores also increase the permeability of membranes to H^+ ions, a factor that may be of significance in acid-base balance. For this reason, there needs to be more work conducted on ionophores and heat-stressed birds. Ionophores have been shown to alter mineral availability, although this should not be of concern under commercial conditions where most minerals are present in excess of requirements. Studies show that the effect of ionophores on mineral metabolism is not always consistent for various minerals. For example, monensin may lead to increased tissue level of certain minerals, while lasalocid has the opposite effect, yet for another mineral these effects are reversed.

Ionophores, and monensin in particular, seem to have an adverse effect when used in conjunction with low protein (methionine) diets. When low protein diets, or feed restriction are employed for birds less than 21 d of age, alternatives to ionophores should be considered in an attempt to alleviate potential growth depression, loss of uniformity, and poor feathering. However, with normal diet protein levels, the ionophores do not have a measurable effect on TSAA requirement. Ionophores and monensin in particular do impart some growth depression in young birds, although this seems to be completely overcome with compensatory growth during the withdrawal or finisher period. For monensin, a 5 - 7 d withdrawal is optimum for compensatory gain, assuming that no major coccidiosis challenge will occur during this time. With minimal challenge, a non-medicated diet is recommended, while in situations of high challenge, an alternate anticoccidial may be necessary.

There has also been some controversy on the relationship between wet litter and certain ionophore products. Lasalocid, in particular, has been associated with wet litter, and as such, recommendations are often given for reducing diet sodium levels when this anticoccidial is used. Under such conditions, adjustment of chloride levels is often ignored, and as a consequence performance is sub-optimal. The whole area of ionophores and water intake has not been fully resolved other than the fact that birds fed monensin do seem to produce drier manure.

Non-ionophore anticoccidials are not used extensively in chicken broiler production, although their use is often recommended in shuttle programs. Nicarbazin is an anticoccidial that seems to work well in such shuttle programs, although again there are some potential problems with this product. Nicarbazin seems to accentuate the undesirable effects of heat stress, and if inadvertently added to layer or breeder diets at normal anticoccidial levels, can cause loss in reproductive performance. Nicarbazin fed to brown-egg birds turns their eggshells white within 48 hr although this is completely reversible when the product is withdrawn from the feed. Even low levels of nicarbazin can cause some loss in shell color, and mottling of egg yolks.

Amprolium is used extensively in diets for growing breeder pullets, because unlike the ionophores, it allows some build-up of immunity. Amprolium induces a thiamin deficiency in the developing oocyst, and as such, has been queried with respect to thiamin status of the bird. In most instances thiamin deficiency will not occur in birds, although cases have been reported of combinations of amprolium and poorly processed fish meal that is high in thiaminase enzyme, leading to thiamin deficiency in young birds.

c) *Antibiotics and Growth promoters*

There has been a gradual reduction in the use of antibiotics *per se*, although growth promoting agents are still used extensively. The mode of action of most growth promoting agents is comparable to that of antibiotics in terms of beneficial modification of gut microflora. In this context, the type of dietary ingredients used may influence the efficacy of these products because microbial activity is influenced by digesta composition. There has been insufficient work conducted in this area, *eg.* the beneficial effect seen when antibiotics are used with ingredients such as rye. It is unlikely that growth promoters result in increased digestibility of feed, rather improvements in feed efficiency are a consequence of increased growth rate and hence reduced days to market. Over the past few years, there has been criticism about the use of antibiotics in poultry feeds, especially with respect to the potential for build-up of micro-organisms resistant to a specific antibiotic, and subsequent transfer of this resistance to known pathogens. In this context, the use of antibiotics such as penicillin, that are also used in human medicine, come under very close scrutiny.

A number of ingredients are still referred to as having "unidentified growth factors". On this basis ingredients such as alfalfa meal, distiller's solubles, bakery yeasts and animal proteins are often added at 1 - 2% of the diet. Any "unexplained" response to these ingredients most often relates to their containing trace levels of vitamin or antibiotic residues.

d) Antifungal agents

In many regions of the world, molds and associated mycotoxins are major problems, affecting both growth and reproductive performance. Mycotoxins produced by both aerobic field molds and anaerobic storage molds can accumulate, often undetected, in a range of feedstuffs. A number of antifungal agents are available, most of which are based on organic acids. By altering the pH of the feed it is hoped to control mold growth, although it must be remembered that any mycotoxins already present in feed will not be destroyed by these antifungal agents. Apart from their cost, these organic acids can be problematic in accelerating the corrosion of metal feeders and feed lines.

Gentian violet is also used extensively as an antifungal agent, and in this context, its efficacy is governed by factors that determine the efficacy of organic acids *ie:* time, temperature, moisture and feed particle size. Gentian violet also has some bacteriostatic activity, and as such, is often used to maintain a beneficial gut microflora, comparable to an antibiotic. In recent years, there has been some interest in use of aluminosiliates (zeolites) as an "adsorbent" of aflatoxin. Unfortunately, relatively high levels of aluminosilicates must be used, and these provide no other nutrients, and may in fact act as chelating agents for some essential minerals. However where aflatoxin contamination is common, then adding 15 - 20 kg alumnosilicates per tonne of feed may be necessary in order to minimize the effect of this mycotoxin.

In addition to, or as an alternative to using such antifungal agents, there is a potential for minimizing mold growth through formulation, diet preparation and feeding management. There seems little doubt that the feed surface area is directly related to fungal activity - simply, the greater the surface area of feed exposed, the greater the possibility of fungal spore colonization. This fact is the most likely cause for the increase in mold growth seen with feed as it travels from the mill to the feed trough because particle size is invariably reduced. Up to a 50% increase in fines can occur with high-fat pelleted broiler diets between the time of pelleting and consumption by the bird. At the same time, there is a 100% increase in the potential (and most often the occurrence) of fungal activity. In areas of potential mycotoxin contamination, there is obviously an advantage to maintaining as large a pellet or crumble size as possible. The heat generated during pelleting has been shown to sterilize feed to some extent, because fresh pellets have low fungal counts. However, pelleting will not destroy mycotoxins already formed prior to pelleting, and warm moist pellets are an ideal medium for fungal growth. In this same context, there is an advantage to reducing the surface area of feed that is exposed to the atmosphere. Research has shown increased fungal activity in feed taken from trough vs tube feeders with the former having more feed exposed to the atmosphere.

With toxins such as aflatoxin, there is a benefit to increasing the protein content of the diet, and in particular, sulphur amino acids. It is possible that sulphates may also be beneficial in helping to spare sulphur amino acids that are catabolized during aflatoxicosis. Due to the specific enzyme system involved with aflatoxicosis, selenium at up to 0.4 ppm may be beneficial in overcoming major adverse effects of this mycotoxin. There have also been reports of niacin increasing the metabolism of aflatoxin B, and so decreasing overall toxicity.

It appears that diet modification and feed management can be manipulated to minimize chances of mycotoxicosis. However such measures will not likely be 100% effective, and it should always be remembered that most fungal growth can be reduced if moisture content of grains and feeds is kept below 14 - 15%.

e) Probiotics

Probiotics, unlike antibiotics, imply the use of live microorganisms rather than specific products of their metabolism. Not being specific molecules therefore, they are difficult to quantitate and even more difficult to describe in terms of proposed mode of action. Probiotics can be classified into two major types - viable microbial cultures and microbial fermentation products. Most research has centred on *Lactobacilli* species, *Bacillus subtilis* and some *Streptococcus* species. Similar to the situation with antibiotics, the mode of action is still unclear although the following have been suggested: a) beneficial change in gut flora with reduction in population of *E. Coli*; b) lactate production with subsequent change in intestinal pH; c) production of antibiotic-type substances; d) reduction of toxin release (suppression of *E. Coli*). With these potential routes of activity, it is perhaps not too surprising that research results are inconsistent. In most instances, the feeding of live cultures modifies the gut microflora of birds usually with increases of *lactobacilli* at the expense of coliforms. A healthy animal has a preponderance of lactic acid producing bacteria, and so it is only under situations of "stress", when coliforms often increase in numbers, that probiotics will be of measurable benefit. In this context there is interest in use of live cultures administered (orally) to day-old poultry as a means of preventing harmful bacteria such as salmonella from colonizing the gut.

With potential instability in most feeds for many *lactobacillus* species, there has been recent interest in probiotics based on *Bacillus subtilis* species, because they possess a viable spore that has greater stability than do most lactic acid producing cultures. Regardless of somewhat inconclusive results, it appears that probiotic use is increasing, and that the animal industry looks to such products as the substitutes for conventional antibiotics. These products seem ideal candidates for genetic manipulation which has been inferred by a number of researchers in this area. By using genetic engineering, some researchers suggest that bacteria can be reformed to carry more desirable gene characteristics, including the production of digestive enzymes and antimicrobial substances.

f) *Yeasts*

Yeasts, or single-celled fungi, have been used in animal feed and the human food industry for many years. Brewer's yeast was a common feed ingredient in diets for monogastric animals prior to the identification of all the B-vitamins. Today some nutritionists still incorporate these inactivated microbes as a source of so-called "unidentified growth factor". More recently there has been an interest in the use of live yeast cultures. Most often these cultures contain the yeast themselves and the medium upon which they have been grown. Most yeast cultures are derived from *Saccharomyces* species, in particular, *Saccharomyces cerevisiae*. As with probiotics, their mode of action in enhancing animal performance is not well understood. Yeasts may beneficially alter the inherent gut microflora, possibly through controlling pH. The presence of living yeast cells may also act as a reservoir for free oxygen, which could enhance growth of other anaerobes. At the present time, there does not seem to be any move to manipulate yeast for specific purposes related to animal nutrition. To some extent, this relates to scant knowledge on mode of action, and so, should more facts be uncovered in this area so-called "designer" yeast may be considered.

g) *Enzymes*

Enzymes have been added to poultry diets ever since workers at Washington State University showed improvement in digestibility of barley and rye-based diets when various enzymes were used. Since that time, corn-soybean diets predominated, and these were assumed to be highly digestible and so there was little interest in feed enzyme application. Over the past five years, this area of nutrition has gained interest and activity due to economics of small grain use and also because of a better understanding of mode of action and availability of various enzymes. Enzymes are now being manufactured specifically for feed use, and can be broadly categorized as carbohydrases, proteinases and lipases. Increasing the digestibility of various carbohydrate fractions of cereals and plant proteins has received most attention, although there is growing interest in the potential for improving digestibility of both plant and animal proteins, and of saturated fatty acids for young birds. Currently enzymes are used most commonly to aid digestion of diets containing wheat, barley and rye, and here improvements are seen in dry matter digestibility and also in consistency of the excreta. There is also current interest in enzymes designed specifically to improve soybean meal digestibility.

The term non-starch poylsaccharides (NSP) is now frequently used to describe what in the past has been referred to as fiber. Most animals have very limited ability to digest fiber, because they lack enzymes necessary to cleave these large and complex molecules. In animals such as the pig, and in ruminants, it is the resident microbial populations that synthesize cellulase etc. which are enzymes that allow for varying degrees of fiber digestion. If we can improve digestion of these complex carbohydrates, we not only increase potential energy utilization, but also remove any negative aspect that these products may have on gut lumen activity and excreta consistency.

The NSP content of cereals and other by-product feeds is usually inversely proportional to their conventional energy level. These NSP components are most often associated with the hull and underlying aleurone layers. In order for normal endogenous enzymes to contact the starchy endosperm, these outer layers must be disrupted or chemically degraded. The NSP comprises the hemicellulose, pectin and gum fractions, the main components being arabinose, xylose, mannose, galactose and glucuronic acids. Oligosaccharides also contain raffinose, stachyose and cellibiose. Because these polymers are more soluble than cellulose, they can interact freely with other nutrients in the digesta, but unfortunately tend to imbibe water producing a viscous medium (Table 2.32).

TABLE 2.32 Non-starch polysaccharides in selected ingredients (%)

Ingredient	Cellulose	Arabinoxylan	Pectin	β-Glucan
Corn	2.5	5.0	0.1	-
Wheat	2.5	6.0	0.1	1.0
Barley	4.8	7.0	0.2	4.0 - 5.0
Soybean meal	5.0[1]	0.5	12.0	-

[1]depending on hull fraction returned

Addition of feed enzymes could therefore improve NSP availability, and as important, reduce the negative impact that these undigested residues have on digesta viscosity. Normal digestion requires unimpeded movement of enzyme, substrate and digestion products throughout the digesta and especially close to the absorptive gut wall. As the viscosity of the digesta increases, the rate of diffusion decreases, and this causes reduced digestibility of all substrates. The undigested viscous digesta subsequently translates to very sticky excreta which causes problems of litter management. Reduction in digesta viscosity is therefore highly correlated with efficacy of enzymes that can digest substrates such as β-glucans. In oats and barley the bulk of the NSPs are β-glucans, whereas in wheat and rye, arabinoxylans predominate. Enzymes tailored for barley therefore contain β-glucanase enzymes, while those designed to improve wheat digestibility should contain xylanase and arabinoxylanase enzymes. For example, the commercial product Roxazyme® designed to improve utilization of wheat contains around 18,000 units β glucanase, 8,000 units cellulase and 26,000 units xylanase per gram.

While most enzyme work has involved wheat and barley, there is also potential for improving digestibility of soybean meal. After oil extraction, soybean meal contains 6 - 8% oligosacharides, the predominant components being raffinose, stachyose and cellibiose. These oligosacharides can be extracted with ethanol, and such treated meals show some 10 - 20% improvement in dry matter digestibility. This degree of improved digestibility may therefore be possible with enzymes designed specifically for degradation of these oligosaccharides.

Phytase enzyme is also available commercially, and will improve the utilization of phytate phosphorus in plant material. Normally very little phytate phosphorus is utilized, especially by the young bird, and so exogenous feed phytase is a potential means of making this expensive nutrient available to the bird, and also greatly reducing the excretion of phosphorus into the environment. The impetus for phytase use started in Europe mainly due to concerns about phosphorus excretion by poultry and swine. After 10 - 15 d of age, using phytase enzymes, birds can grow quite well on diets devoid of supplemental phosphates. These phytase enzymes also liberate calcium and other minerals from the phytase complex, and so this can be taken into account during formulation. In a conventional corn-soybean meal broiler starter diet, phytase enzyme may release as much as 0.1 - 0.2% calcium equivalent to the diet. Unless this calcium release is factored into formulation, it is possible to induce Ca:P imbalance, especially when phosphorus levels are low.

Adding enzymes to commercial poultry feed requires some planning, because many are unstable at high temperatures and especially in situations of high humidity. Pelleting, and certainly extrusion/expanding will destroy some feed enzyme supplements. Currently this problem can be overcome by adding enzymes, as a spray, after pelleting. In the future it may be possible to develop heat-resistant enzyme preparations or protect current products such that they can better withstand feed processing conditions.

h) Pigments

The yellow to orange color in avian fatty tissue is caused by various carotenoid pigments. These pigments control the color of the egg yolk, as well as the shanks and beaks of layers, and also the skin color that may be important in meat birds. The xanthophylls are the most important carotenoids in poultry nutrition, and natural ingredients rich in these compounds are alfalfa meal, corn gluten meal and marigold petal (Table 2.33).

TABLE 2.33 Xanthophyll content of selected ingredients (mg/kg)

Ingredient	Xanthophyll
Corn	20
Wheat	4
Milo	1
Alfalfa meal	175
Corn gluten meal	275
Marigold petal	7,000

Corn contains much more xanthophyll than do other cereals, although high levels of pigmentation can only be achieved with natural ingredients, by including other products such as alfalfa and corn gluten meal.

The various xanthophylls differ in their effect on skin and yolk pigmentation. β-carotene has little pigmenting value, although pigments such as zeaxanthin as found in corn is more easily deposited, while there is very high incorporation rate of synthetic products such as β-apo-8-carotenoic ethyl ester. The zeaxanthin in corn tends to impart the darker orange-red colors, whereas the luteins as found in alfalfa cause a more yellow color. Pigments are destroyed by oxidation, and so addition of antioxidants such as ethoxyquin to feed, and general feed management applied to fat protection also apply to preservation of pigments. Coccidiosis, malabsorption and certain mycotoxins will all reduce pigment absorption. Pigmentation in the young meat bird is directly proportional to pigments fed throughout growth. For the laying hen however, yolk color is a consequence of pigments consumed in the layer feed, and also the transfer of pigments accumulated in the skin and shanks when the bird was immature. This transfer of pigments to the ovary occurs regardless of diet pigments, and is responsible for the "bleaching" effect of the shanks and beak of yellow-skinned birds over time.

Because many of the naturally carotenoid-rich ingredients are low in energy, it is difficult to achieve high levels of pigmentation of meat birds without using synthetic sources. Canthaxanthin, astaxanthin and β-apo-8-carotenoic acid (where allowable in poultry diets) can be used to impart the spectrum from yellow to orange/red coloration in either skin or egg yolk.

i) Flavoring agents

The chicken is not usually considered to have much ability to select feed, based on flavor, or organoleptics *per se*. The chicken has only about 24 taste buds in comparison to 9,000 in humans and 25,000 in cattle. Relatively few studies have been conducted with flavoring agents for poultry, and for this reason, care must be taken in extrapolating data from other species. For example, sucrose octaacetate solution is reported to be readily accepted by birds, but universally rejected by humans. There seems little scope for use of flavoring agents with broiler chickens and turkeys that already seem to be eating at near physical capacity. However there may be some potential with breeders for isolation of agents that are distasteful to birds, as an aid in limiting their feed intake. Flavor agents may be beneficial in masking any unpalatable ingredients, and for maintaining a constant feed flavor during formulation changes. Flavors may also be useful tools in masking any undesirable changes in drinking water during medication. It is conceivable that use of a single flavor agent in both feed and medicated water may prevent some of the refusals seen with medicated water, especially for turkey poults.

j) Worming compounds

Most floor grown birds are exposed to infection from various species of worms. In many instances, such challenge can be prevented or minimized with the use of anthelmintic agents. Products based on piperazine and hygromycin have been used most commonly over the last 15 - 20 years. Piperazine used in diets for laying birds has been shown to result in discoloration of the yolk. When administered at 28 d intervals, one report indicated about 4% incidence of discolored yolks which appeared as irregular areas of olive to brownish discoloration. Such yolk discoloration is most pronounced in summer months especially after prolonged storage at regular egg cooler temperatures. The mottling of yolks seen with another commercial product has been compared to the mottling seen with calcium-deficient birds, suggesting a similar mode of action. However we are unaware of any published reports relating worming compounds to calcium deficiency and problems with shell quality.

k) Odor control

Various extracts of the Yucca plant are claimed to reduce ammonia levels in poultry houses. A soluble component of the Yucca plant seems able to bind ammonia, and so preventing its release from manure, which is especially important in confinement housing systems. Most poultry will react adversely to 50 ppm ammonia, and this is in contrast to the level of 20 - 30 ppm which is the usual detection range for humans. Products such as Deodorase® added to feed at 100 - 150g/tonne have been shown to reduce environmental ammonia levels by 20 - 30%, and this has been associated with improved growth rate and reduced mortality.

2.9. TOLERANCES AND WITHDRAWAL OF FEED ADDITIVES

Many of the feed additives used for stimulation of growth or the prevention and control of a specific disease are toxic if included in the feed at too high a level and the difference between the required level and the toxic level is often quite small. Toxicity can show up as slightly reduced growth or egg production and may not necessarily be accompanied by increased mortality, thus making drug toxicity difficult to detect.

Feed labelled "non-medicated" should not contain drugs because they may be quite toxic for some classes of chickens or species other than the one for which they were intended. Similarly, care must be taken to ensure that contamination does not occur when a non-medicated feed is mixed immediately after one containing drugs. Also some drugs are electrostatic and the operator must be certain that the feed processing equipment is properly grounded to avoid drug particles adhering to the side of the mixer.

Some drugs are incompatible with others and care should be taken in following the manufacturer's directions not only as far as the level of inclusion in the feed is concerned, but also instructions as to when a drug can be fed or if any interactions occur with other feed additives. There are usually specific government regulations concerning the use of drugs in feed and withdrawal periods required or recommended prior to the marketing of poultry products. These should be studied and adhered to, because drug residues are to the detriment of all concerned. Following is a list of selected additives used in Canada with corresponding withdrawal times (Table 2.34). **Note:** these additives levels and withdrawal times may vary in different countries and regulations for specific compounds change frequently over time.

2.10. FEED TOXINS AND CONTAMINANTS

Producing poultry feed that is free of toxins and contaminants is obviously the goal of all feed mills. However this ideal situation is difficult to attain, because many natural feed ingredients will contain toxins that are inherent in the feedstuff or have "naturally" contaminated the feedstuff prior to feed preparation. Mycotoxins are perhaps the best example of such "natural" toxins, and together with many plant lectins can cause poor bird growth and reduced egg production. In addition to these biological contaminants there is also concern about accidental inclusion of such products as polychlorinated biphenyls, pesticides, fungicides etc.

TABLE 2.34 Canadian guidelines for ingredient levels and withdrawal periods for selected feed additives

Product type	Common Name	Active Ingredient	Inclusion (g/t)	Withdrawal (d)
Anticoccidial	Amprol	Amprolium	60-250	0
	Amprol Plus	Amprolium/Ethopabate	125/4	1
	Amprol HiE	Amprolium/Ethopabate	125/40	3
	Avatec	Lasalocid	105	0
	Coban	Monensin	99	0
	Coyden	Clopidol	125	0
	Coxistac	Salinomycin	60	0
	Cygro	Manduramycin	5	5
	Deccox	Decoquinate	30	0
	Maxiban	Narasin/Nicarbazin	40/40	4
	Monteban	Narasin	70	0
	Nicarb	Nicarbazin	80-200	4
	Robenz	Robenidine	33	6
	Stenorol	Halofuginone	3	5
	Zoamix	Zoalene	125	0
Growth promotion	Aureomycin	Chlortetracyclin	5.5	0
	Baciferm	Bacitracin MD	4.4	0
	Flavomycin	Bambermycin	2	0
	Lincomix	Lincomycin	2.2	0
	Penicillin	Procaine penicillin	2.2	0
	Progen	Arsenilic acid	99	5
	Stafac	Virginiamycin	11	0
	Terramycin	Oxytetracycline	5.5	0
	Zinc Bacitracin	Bacitracin Zn	4.4	0
	3-Nitro	3-nitro-4-hydroxy phenylarsonic acid	50	5

a) Mycotoxins

Mycotoxins are now virtually ubiquitous in poultry diets, and with ever decreasing levels of testing sensitivity, they are isolated as common contaminants of most grains and some vegetable protein ingredients. We still do not know the causes of high levels of mold growth occurring in pre-harvest grains. Certainly such aerobic molds are more prevalent in hot-humid conditions, and insect damage to the standing crop seems to provide a route of entry for the mold. Unfortunately visual inspection of harvested grains can be misleading in regard to mycotoxin content. Likewise, merely because grains appear moldy, does not mean to say that they are contaminated with harmful toxins. In storage, the major factors affecting mold growth are again temperature and humidity. The higher the temperature, the greater the chance for mold growth. However such mold growth rarely occurs in grains containing less than 14 - 15% moisture. Unfortunately many grain silos are not waterproof, or grains are not aerated, such that pockets of moisture cause microclimates ideal for mold growth. Following is a review of the major mycotoxins affecting meat birds and egg layers.

Aflatoxin

Produced by *Aspergillus flavus* mold, aflatoxin is one of the most potent carcinogens known. Usually present in cereals in ppb levels, acute toxicity will occur at 1.2 ppm. Aflatoxin B_1 is the most common form of the toxin, the B designation relating to the fact that the toxin fluoresces a blue color when exposed to ultraviolet light, and so this can be used in the screening of ingredients. Blue fluorescence occurs with other components, and so this simple test screens out negative samples, but needs subsequent chemical analysis for confirmation. Aflatoxin is found in most cereals, although corn and milo are the most common hosts. As with any mold, *Aspergillus* growth is greatly reduced when corn or milo moisture levels are less than 15%.

Aflatoxin is a potent hepatotoxin, and so varying degrees of liver breakdown occur. As toxicity develops, normal liver function declines, and reduced growth rate is quickly followed by death. Toxicity is enhanced by the presence of other toxins such as ochratoxin and T2. The effects of aflatoxin are also much worse if birds are infected with aspergillosis. There also seems to be a nutrient interaction, because toxicity is more severe when diets are low in either crude protein or methionine or when the diet contains marginal levels of riboflavin, folic acid or vitamin D_3. There is no treatment for acute aflatoxicosis, although because of the liver disruption, giving higher levels of antioxidants and/or selenium seems to slow the onset of symptoms and speed up recovery if aflatoxin is removed from the diet.

There are a number of effective preventative measures, although all of these are not economical. Treating infected grains with ammonia, hexane or hydrogen peroxide have all been shown to reduce aflatoxin levels. Under commercial conditions adding binding agents to the feed seems to reduce the adverse effects of aflatoxin. To date, aluminosilicates and certain bentonite clays have proven effective. For example adding 10 - 15 kg/tonne of hydrated sodium-calcium aluminosilicate has been shown to drastically reduce mortality in broilers and turkeys fed diets containing 0.5 - 1.0 ppm aflatoxin. Such aluminosilicates do not seem to bind other mycotoxins however.

Tricothecenes

Three mycotoxins, namely T_2, DAS (Diacetoxyscirpenol) and DON (Deoxynivalenol or vomitoxin) are included in this group. All of these mycotoxins are produced by Fusarium species molds such as *Fusarium graminearum* and *Fusarium roseum*. The tricothecenes affect protein metabolism and have the characteristic feature of causing mouth lesions in most animals. However DON does not seem to be particularly harmful to poultry, unlike the situation in pigs and other mammals, and birds can tolerate up to 20 ppm of this mycotoxin. T_2 and DAS however are more toxic, causing problems at 2 - 4 ppm. The adverse effect of tricothecenes is made even worse by the presence of aflatoxin or ochratoxin, and seems to be worse in young broilers fed ionophore vs non-ionophore anticoccidials. There are no really effective treatments, and while the addition of relatively high levels of antioxidants may slow the disruption of protein synthesis, they are not effective long-term. Adsorbents and binding agents are generally ineffective.

Ochratoxin

As with other mycotoxins, there are a number of forms of ochratoxin, although ochratoxin A (OA) is by far the most significant for poultry. OA is produced by a number of molds, with *aspergillus* and *penicillium* species being most commonly involved. OA is toxic at 2 ppm and as with tricothecenes, it has an adverse effect on protein synthesis. However OA also affects kidney function, and so the classical signs are swollen kidneys and associated increased water intake with wet excreta. Secondary visceral gout, which appears as urate deposits over the viscera, is common with OA toxicity, due essentially to failure of uric acid clearance by the kidney tubules. OA toxicity is compounded by the presence of aflatoxin, DON and T_2 toxicosis, and also made worse by feeding diets high in vanadium (usually as a contaminant of phosphates or limestone). There are no effective preventative measures, although birds sometimes respond to diet manipulation in the form of increasing crude protein levels. There are also reports of beneficial response to increasing diet vitamin C levels, especially in egg layers.

Other mycotoxins

There are a diverse group of other mycotoxins that periodically cause problems for poultry. Their occurrence is less frequent than the major mycotoxins already discussed, and in some instances exact toxicity levels have not been clearly established. Table 2.35 summarizes these mycotoxins in terms of effect on poultry and their probable threshold for toxicity.

Table 2.35 Effect of minor mycotoxins on poultry

Mycotoxin	Effect	Toxicity	Comments
Fumonisin	Degeneration of nerve cell lipids	>80 ppm	Diet thiamine levels important
Cyclopiazonic acid	Mucosal inflammation	50 - 100 ppm	Often present along with aflatoxin
Oosporin	Kidney damage, gout	>200 ppm	Most commonly found in corn
Citrinin	Kidney damage	>150 ppm	Commonly associated with ochratoxin
Ergot	Tissue necrosis	>0.5%	Wheat and rye
Fusarochromanone	Tibial dyschondroplasia	>50 ppm	Fusarium species
Moniliformin	Acute death	>20 ppm	Mechanism unknown
Zearalenone	Reproduction, vitamin D_3 metabolism	>200 ppm	Can affect shell quality

b) Plant toxins

A number of cereals and vegetable protein crops contain natural toxins that can affect bird performance.

Cyanides

While there are a number of potential feed ingredients that contain natural cyanides, cassava, or manioc, is probably the most common and contain relatively high levels of this toxin. Cassava meal is derived from the ground tuberous root of the cassava plant. Ingestion of this material by animals can result in enlarged thyroids, due to the presence in the meal of cyanogenic glucosides, the main one being linamarin. These glucosides are concentrated in the peel of the root. On hydrolysis by the enzyme linamarase, the glucosides produce hydrocyanic acid (HCN), which is highly toxic. In addition to the enzyme in the root, glucosidic intestinal enzymes and HCl can also hydrolyse the glucosides.

Hydrocyanic acid inhibits animal tissue respiration by blocking the enzyme cytochrome-oxidase. HCN is converted to the detoxification product thiocyanate in the liver which is then excreted via the urine. This detoxification system utilizes sulphur from methionine in the conversion of cyanate to thiocyanate, thus increasing the animal's requirement for this amino acid. Thiocyanate is responsible for the goitrogenic effect of cassava, due to its effect on iodine uptake and metabolism in the thyroid, resulting in reduced output of thyroxine, the hormone regulating tissue oxidative functions. Cyanate is known to alleviate the toxicity of an excess of dietary selenium by complexing with selenium, thus making it less available to the bird. Linseed meal, which has been known for some time to alleviate selenium toxicity in animals, has been shown to contain two cyanogenic glucosides, namely linustatin and neolinustatin. These compounds are closely related in structure to linamarin and thus on hydrolysis yield HCN.

The cyanide content of cassava varies with variety and can range from 75 to 1000 mg/kg of root. Crushing the root releases the enzyme linamarase which acts on the glucosides to produce HCN, which being volatile are readily eliminated during drying. Rate of drying in commercial forced air driers is important as it has been reported that at 80 to 100°C, only 10 to 15% of cyanide is removed compared to 80 to 100% detoxification occurring at 47 to 60°C but with a longer time. Steam pelleting can also assist in the volatization of free HCN.

While there are differing reports as to how much cassava meal can be incorporated into poultry diets without reducing performance, this will obviously depend on the concentration of cyanide in the meal. However, cassava meals containing up to 50 mg total cyanide/kg have been fed successfully up to 50% inclusion in broiler diets.

Glucosinolates

Glucosinolates belong to a group of anti-nutritive compounds of which over 100 different types are known to occur in members of the Cruciferae family. The genus *Brassica* is a member of this family which includes many important feeds and foods such as, rapeseed, mustard, kale, radish, cabbage, cauliflower, brussels sprouts, etc. In individual species, usually around 12 to 20 glucosinolates are found although most of these are present in small amounts.

Hydrolysis of these glucosinolates is brought about by the enzyme myrosinase, which is usually present in most glucogenic plants. In the intact plant the enzyme and its substrate are separated, but with cellular rupture (grinding, insect damage, etc.) these components are combined and hydrolysis can occur.

For many plants including rapeseed, glucosinolates can be readily divided into three main groups, based on physiological effects and hydrolysis products. By far the largest of these groups are glucosinolates that yield organic isothiocyanates on hydrolysis. These compounds are volatile and possess a range of antimicrobial, antifungal and antibacterial properties, and have a very pungent taste (mustard, horseradish, etc.). A second, but much smaller group, form potent anti-thyroid compounds on hydrolysis with 5-vinyloxazolidine-2-thione being the most common. If present in large amounts these compounds can impart an intense bitterness. Glucosinolates of the third group all contain an indole side chain, and on hydrolysis yield thiocyanate ions which are anti-thyroid or goitrogenic.

The glucosinolate contents of the various rapeseed types ranges from a high of 100 to 200 μM/g to less than 30 μM/g, while new varieties are claimed to be glucosinolate-free.

A significant research program was initiated in Canada in the early 60's to develop rapeseed varieties low in glucosinolates and erucic acid, a fatty acid known to result in detrimental metabolic problems with certain animals. In 1968 the first low erucic acid variety was licensed and shortly thereafter a low glucosinolate variety appeared. In 1974 the first double low variety, very low in erucic acid and glucosinolates, was licensed. A number of improved varieties were developed and in 1979 the name canola was adopted in Canada to apply to all double low rapeseed cultivars. For reasons not yet completely understood, reduction of total glucosinolate had little effect on the content of the indole group. Thus when expressed as a percent of total glucosinolates, this group increases from around 5 to 40% in the low glucosinolate varieties.

While the feeding value of canola meal has been markedly increased for poultry, as compared to the older rapeseed varieties, there are still some problems encountered. The occurrence of liver hemorrhages with the feeding of rapeseed meal is well documented. Unlike the fatty liver hemorrhagic syndrome, these hemorrhages are not associated with increased liver or abdominal fat contents. Certain strains of laying hens were more susceptible than others, however, with most strains it was not uncommon to see signs of the condition. While liver hemorrhages have been significantly reduced in laying hens with introduction of canola meal, problems are sometimes seen when feeding 10% or more of this product. Research to date suggests that this is the result of intact glucosinolates, rather than any of their products of hydrolysis. However, it is still unclear how glucosinolates function in the occurrence of hemorrhagic liver.

An increase in weight of the thyroid still occurs following feeding of canola meal, although severity is much reduced from that seen with the older rapeseed varieties. Thiocyanate is responsible for the goitrogenic effect noted with these products due to their effect on iodine uptake and metabolism, and so increase in thyroid size is seen. Because thiocyanate is the end product of indole glucosinolate hydrolysis, and levels of this compound are still high in canola, then this product probably accounts for the enlarged thyroids still seen with low glucosinolate meals.

A number of reports suggest that canola meal is unpalatable for poultry, and at times results in reduced feed intake leading to lower egg weight and reduced weight for age for broilers. On reviewing these reports it is evident that in many cases the poorer performance is due to a reduction in feed intake, rather than a reduced nutrient availability, which itself usually results in increased feed intake. This effect on feed intake may relate to sulphur metabolism. Canola meal contains 1.4% sulphur compared to around 0.4% in soybean meal. High dietary levels of sulphur will reduce performance by altering dietary anion-cation balance, with a resultant decrease in feed intake (See Table 2.14).

Another major problem with the feeding of rapeseed or canola meal is egg taint which is experienced in certain flocks of layers, and especially brown-egg layers containing RIR ancestory. This is the result of a single major autosomal semi-dominant gene being present which is responsible for the bird lacking the ability to oxidize trimethylamine (TMA) to TMA oxide, the odorless excretory product of TMA. While the double low varieties of canola contain very low levels of glucosinolates, there is still sufficient present, along with the soluble tannins, to impair TMA oxidation and thus tainted eggs can result. Because brown-egg layers, are quite common in many parts of the world, canola meal, in such regions, is used sparingly for layers.

Tannins

Tannins are water soluble polyphenolic plant metabolites that are known to reduce the performance of poultry when fed at moderate levels in a diet. Grain sorghum is probably the most common feedstuff which contains relatively high levels of tannin. However, faba beans, rapeseed and canola meal all contain sufficient tannins to affect poultry performance.

The growth depressing effect of tannins is undoubtedly due, in large part, to their ability to bind proteins. Tannic acid is hydrolized by the chick to gallic acid, its major hydrolytic product, and to a lesser extent to the somewhat toxic compounds, pyrocatecol and pyrogallol. A large portion of the gallic acid is methylated and excreted in the urine as 4 - 0 methyl gallic acid. This pathway offers a possible explanation as to why additions of methionine, choline and other methyl donors, have been reported to be beneficial when included in diets containing tannic acid.

Much of the work on toxicity of tannins has involved purified tannic acid. Legume and cereal tannins are of a condensed type while tannic acid is of a hydrolyzable type. Since there are conflicting reports on the degree of growth depression and the role of methionine in alleviating tannin toxicity, it follows that the predominant detoxification process may differ between these two compounds. More recent work suggests that while gallic acid is the breakdown product of both condensed tannins and tannic acid, and can be detoxified by methyl groups, the stability of condensed tannins is such that this route of detoxification may be of little importance.

:Sorghum tannins

The nutritive value of sorghum is usually considered to be 90 to 95% that of corn, due in large part, to its tannin content. There are a number of varieties of sorghum on the market which are usually classified as bird resistant or non-bird resistant varieties. These have either a low (less than .5%) or high (1.5% or higher) level of tannins. A number of toxic effects have been reported with the feeding of high tannin sorghum. These include depressed growth and feed utilization, reduced protein digestibility, lower egg production, and leg abnormalities with broilers.

A number of procedures have been utilized in an attempt to reduce the toxicity of the tannins in sorghum. These include soaking in alkali solution, and reconstitution (water soaking) which are reported to deactivate tannins and thus improve the nutritive value of the cereal. Besides the action of the methyl donors, which have been reported to improve the feeding value of high tannin sorghum, such products as polyvinylpyrrolidone, and calcium hydroxide, or a slurry of sodium carbonate, have also been reported to give positive responses. However, several crude enzyme preparations that have been tried were not effective in enhancing the feeding value of high tannin sorghum.

Tannins have also been implicated in egg yolk mottling. Yolk mottling is a condition which periodically appears in a flock, and without a direct involvement of nicarbazin, gossypol or certain worming compounds, there is no ready explanation for its appearance. While several reports have suggested tannic acid and its derivatives as possible causes, other than the addition of commercial tannic acid at levels above 1%, there appears to be no mottling seen with diets containing up to 2.5% tannins.

There are reports suggesting that tannins are bound tightly to a fraction of the nitrogen in sorghum and that this reduces protein digestibility. However, because the tannins are relatively insoluble they appear to have little influence in tying up crude protein of the other dietary ingredients. In a recent study with turkeys, a high tannin sorghum variety when used at 40% in the diet, resulted in depressed performance to 8 weeks of age. However, the feeding of a similar level to turkeys beyond 8 weeks of age had no detrimental effects. The authors suggest that a more fully developed digestive system of the older birds may be able to overcome the anti-nutritional effects of the tannins. There has been little work reported on the feeding of sorghum to ducks. A recent report showed that the feeding of a high tannin sorghum at levels in excess of 75% of the diet, resulted in depressed growth for rats and chicks but had no effect on the performance of the ducks. The authors suggest, as a possible explanation, that this might be due to the habit of the ducks to drink water immediately after eating and also the consumption of significant quantities of feed that was deposited at the bottom of the water troughs. As mentioned previously there is data to suggest that reconstitution of sorghum will reduce tannin toxicity by reducing protein binding.

While dark colored varieties of sorghum seed usually contain higher levels of tannin than lighter colored varieties, seed color, in general, is a poor indicator of the tannin content of sorghum.

:Faba Bean tannins

Raw faba beans are known to result in depressed performance of poultry while autoclaving results in a significant improvement in bird performance. Dehulling also results in improved energy value with this effect being greater than can be accounted for by reduction in fiber content. The growth depressing properties of faba beans are due to two water-acetone soluble fractions, one containing low weight polyphenolic compounds, the other containing condensed tannins, the later being the major growth inhibiting substance. These condensed tannins are similar to those found in sorghum and are concentrated in the hull fraction.

While proper heat treatment of faba beans can markedly increase their nutritive value, there appears to be some detrimental effect on intestinal villi structure regardless of the degree of heat treatment or the fraction of seed consumed. This has led to reports that factors, other than those usually considered, such as protease inhibitors, phytates, tannins and lectins, may be contributing to the low nutritional value of faba beans. Tannin-free varieties of faba beans are available that contain less than .1% condensed tannins in their hulls compared to over 4% in the high tannin varieties. These lighter colored seeds are of improved nutritive value. Regardless of tannin content, appropriate heat treatment improves the nutritive value of faba beans.

:Rapeseed and Canola tannins

Rapeseed and canola meal have been reported to contain 2 to 3 % tannins, which is concentrated in the hull. These tannins have been shown to contribute to the egg taint problem of these meals, when fed to brown-egg layers, due to their inhibitory effect on trimethylamine (TMA) oxidase. This enzyme is responsible for the conversion of TMA to its water soluble and odorless oxide, the form in which it is usually excreted.

The original method for assaying tannin also included sinapine. Because the sinapine content of canola is around 1.5%, a value of 1.5% for total tannins is more realistic than earlier reported values of around 3%. With tannins concentrated in the hull of both rapeseed and canola, the amount of extractable tannins has been investigated and appears to range from .02 to .2%. The ability of these tannins to inhibit amylase *in-vitro* was not detected. Hence, it has been assumed that the tannins in rapeseed and canola are bound in such a manner that their influence on digestibility of other ingredients is negligible.

Lathyrism

As with many species of animals, poultry are susceptible to lathyrism, a metabolic condition caused by the consumption of legume seeds of the genus Lathyrus, of which sweet peas are a member. The seeds are rich in protein (25 to 27%) and their availability and relatively low cost, in many Asian and mid Eastern countries, often results in their use in poultry feeds. The causative agents for lathyrism are the lathyrogens, of which lathyrogen beta-aminopropionitrile (BAPN) is the principle toxin found. However, there are some synthetic lathyrogens available that have been used in studying the condition.

Lathyrism manifests itself in two distinctive forms. Firstly there is a disorder of the nervous system, leading to a crippling condition and referred to as neurolathyrism, and secondly a disorder of the collagen and elastin component of connective tissue resulting in a skeletal and/or vascular disease and referred to as osteolathyrism. Typical symptoms seen with poultry consuming significant quantities of toxins are, depressed performance, ruffled feathers, enlarged hocks, curled toes, ataxia, leg paralysis and eventually mortality.

Most of the poultry research involves specific synthesized lathrogens rather than natural seeds. BAPN has been shown to inhibit cross-linking compounds in elastin and collagen by inhibiting the enzyme lysyl oxidase, an important component in the synthesis of these compounds. It has also been reported to reduce growth rate for chicks, poults and ducklings and to reduce egg production of adults of these species. BAPN can result in defective shell membranes, so ultimately affecting shell calcification, leading to malformed and soft-shelled eggs. This effect is similar to that seen with a copper deficiency since the enzyme lysyl oxidase is a metalloenzyme that requires copper. Consequently, there are reports of copper alleviating the toxicity symptoms of BAPN.

While recommended maximum levels of inclusion of the various lathyrus seeds, to avoid metabolic problems, varies with the type of seed and the lathyrogen content of the seed, a general recommendation would be to keep the dietary level of BAPN below 50 mg/kg of diet. While the addition of lathyrogens to a laying diet results in a decrease in production after 4 to 5 days, hens seem to return to production in 10 to 14 days after their removal from the diet. For laying stock therefore lathyrogens, fed for a short period of time, do not appear to have any long term detrimental affects. However, this would not be the case for young growing birds. Interestingly there has been some research interest on the ability of BAPN to tenderize meat from spent hens. This is obviously related to its effect in altering collagen cross-linking by inhibiting the enzyme lysyl oxidase.

Gossypol

The use of cottonseed products in diets for laying hens has long been a problem for nutritionists as well as producers. As early as 1891 there were reports of mottled egg yolks resulting from the feeding of cottonseed meal to layers. In the early 1930s gossypol was identified as the compound involved in discoloration of egg yolks when hens were fed cotton seed meal. It soon became evident that there were two problems that could occur with the feeding of cottonseed meal to layers. 1. The albumen of stored eggs developed a pink color and thus the disorder became known as pink egg white: 2. Another problem, which can be present at the same time, was brown or olive pigment in the yolks. This later defect was the result of gossypol from the cottonseed pigment glands interacting with iron in the egg yolk.

Although pink albumen discoloration is known to occur spontaneously, it is usually seen with ingestion of products from plants of the botanical order, Malvales. Two naturally occurring cyclic fatty acids have been isolated from plants known to cause the above disorders. These compounds were called malvalic and sterculic acids. A color test developed many years ago by Halpen, can be used to identify cottonseed oil in vegetable oil mixtures. The test has been shown to be very specific to cyclopropenoid compounds, especially malvalic and sterculic fatty acids. The pink white albumen condition noted in stored eggs, which is common with the ingestion of either malvalic or sterculic fatty acids, results from a combination of conalbumen and egg white protein mixing with iron that diffuses from the yolk. This is due, in part, to changes in membrane permeability and an increase in yolk pH. The amount of these compounds fed, storage conditions and breed of hen, have all been shown to influence the degree and incidence of the condition.

Yolk discoloration is also caused by the ingestion of gossypol and/or malvalic or sterculic acid. However, there is a difference in incidence and degree of discoloration and mottling depending on whether intact gossypol or the free fatty acids are involved. Changes in membrane permeability and a shift in yolk and albumen pH result in water and albumen protein migrating to the yolks. The severity of the condition will depend on the amount of gossypol ingested and can lead to pasty custard-like or viscous yolks being observed. These can be seen at ovulation but the condition can be accentuated with storage. There are reports of increased embryo mortality during the first week of incubation when breeders are fed high levels of malvalic or sterculic acid, however, the levels fed must be much higher than those normally present in laying hen diets.

Although varieties of cottonseed have been developed that are gossypol free, their low yield has meant that they are not widely used in commercial production. Thus much of the cottonseed grown world-wide, still contains appreciable quantities of gossypol. Processing method can markedly reduce the gossypol content of the meal to levels less than .04% free gossypol. In addition, soluble iron salts can be added to diets containing cottonseed meal. The iron will complex with gossypol, thus reducing its toxic effects. In a recent report, broilers fed a diet with up to 30% cottonseed meal, with soluble iron added (to provide a 2:1 ratio of iron to free gossypol) resulted in no detrimental effect on weight gain or liveability.

Alkaloids

Alkaloids are found in a number of feedstuffs but by far the most important are the lupine legumes. Seeds of the plant *Crotalaria retusa* L., contain up to 4.5% of the pyrolizidine alkaloid monocrotaline and these can be a problem in cereal contamination in some areas of Asia and Australia. The older varieties of lupines were often referred to as bitter lupines, due to the presence of significant quantities of quinolizidine alkaloids, mainly lupanine. These alkaloids affect the central nervous system causing depressed laboured breathing, convulsions and death from respiratory failure. Newer varieties of lupines now being grown are very low in alkaloids (less than .02%) and have been shown to be well tolerated by poultry.

One of the most common sources of alkaloids finding their way into animal feeds is grain contaminated with ergot. Samples of ergot can run as high as .4% total alkaloids. Chickens receiving 1 to 2% of ergot in their diet can show symptoms ranging from depressed growth to necrosis of the extremities, staggers, ataxia, tremors and convulsions.

c) Autointoxication

Autointoxication could be defined as self poisoning as it is endogenous in origin and results from the absorption of waste products of metabolism or from products of decomposition in the intestine. High fiber diets fed to young chicks can cause obstruction of the digestive tract with subsequent absorption of products of decomposition or metabolic wastes. Litter consumed by chicks or overconsumption of green grass or plants can also lead to gut impaction problems.

The chilling or overheating of chicks can lead to vent pasting and occlusion resulting in stasis of the intestine contents with autointoxication being the end result. Birds suffering from autointoxication are anorexic, show increased water consumption, followed by weakness and prostration. A generalized toxaemia may result leading to nervous symptoms prior to death.

d) Bacterial toxins

Although losses in birds due to bacterial toxins are not of great economic importance, they do occasionally result in heavy losses in a particular flock. The main organism affecting poultry is *Clostridium botulinum.* No significant lesions are found in botulism poisoning and a positive diagnosis is usually based on identification of the organism and its toxin.

Botulism is caused by the toxin produced from the *C. botulinum* organism under anaerobic conditions. *C. botulinum* is a saprophyte found in soil and dirt and can also be found in intestinal contents and feces. The mere presence of the organism is sufficient to cause disease or to be of diagnostic significance. Growth of the organism, in anaerobic conditions, results in the production of toxins. Botulism can result from birds eating carcasses of birds which have died from the disease and also fly larvae from such carcasses. The toxins present in the meat are ingested by larva rendering them extremely poisonous. Symptoms may appear within a few hours to a day or two after contaminated food is eaten. The common symptom noted is paralysis, with the leg and wing muscles first affected. If the neck muscles are affected the head hangs limp, hence the name "limberneck" which has been used to refer to the disease. In mild cases leg weakness, ruffled feathers and soft pasty feces may be noted. The severity of the disease depends on the amount of toxin consumed. However, death usually occurs as this toxin is very potent. Losses in birds are most commonly due to type A and C toxins. Type A, is common in the mountainous regions of North and South America, while type C is world wide in distribution.

For many years a disease of wild ducks and other aquatic birds was common in the western part of North America. It is now known that this is due to botulism poisoning. Insect larvae in an aquatic environment may die as the result of anaerobic conditions caused by decaying vegetation. When these larvae are eaten by birds, botulism organisms invade tissues and produce toxins. Prevention relates to proper management procedures that eliminate dead and decomposed carcasses around a poultry house. A good rodent and fly control program is also essential as is screening of the building to eliminate entry of wild birds.

e) Chemotherapeutic drugs

While the use of various pharmaceutical compounds has contributed significantly to the development of the modern poultry industry, their misuse can result in toxicity problems. Some of the more common drugs that can result in problems if used at toxic levels are:

Sulfonamides

Toxicity is manifested by signs of ruffled feathers, paleness, poor growth and increased blood clotting time. Hemorrhages in skin and muscle may be noted and necrosis of the liver, spleen, lungs and kidney are often seen.

Nitrofurans

Toxicity results in depressed growth and hyperexcitability, resulting in chicks cheeping and dashing about. Enteritis and congestion of the kidneys and lung, along with body edema and cardiac degeneration may be noted.

Nicarbazin

Toxicity in chicks results in birds being listless and showing signs of ataxia, with incoordination and a stilted gait. Fatty degeneration of the liver may be noted. The most common problem with nicarbazin is its effect on laying hens. Brown eggs will be depigmented and yolk mottling may be noted with white and brown eggs.

Monensin

Recommended levels result in little effect on performance but higher levels of intake can result in significant mortality, especially with turkeys.

f) Toxic seeds

Phytotoxins can be considered as any toxic substance derived from plants including roots, stems, leaves, flower and seeds. Some plants are toxic throughout the whole growing season while others are only toxic during certain stages of development. The majority of toxic plants are relatively unpalatable and are usually avoided by birds. However, with the absence of succulent feed, range birds will consume sufficient foliage or seeds to result in poisoning. Some of the more common poisonous plants are as follows:

Black locust (*Robinia pseudoacacia*)
The toxin is the glycoside robitin-alectin (hemagglutinin). It has been reported that the leaves of black locust are toxic during early July and August in the N. Hemisphere and cause mortality with chickens if consumed at this period. Symptoms noted are listlessness, diarrhoea, anorexia and paralysis with death occurring within several days. Hemorrhagic enteritis may also be seen.

Castor bean (*Ricinus communis*)
Many legume seeds contain a protein fraction which is capable of agglutinating red blood cells. These compounds are referred to as lectins and they vary widely in their degree of specificity to types of red blood cells and also their degree of toxicity. Such legumes must be degraded by heat treatment in order to detoxify them and so enhance their nutritive value. Castor bean was one of the first such legumes to be investigated and a lectin called ricin was isolated which is extremely poisonous. However, the steaming of castor meal for 1 hour at 15 lbs of pressure will reduce the toxicity of the meal to 1/2000 of its original level. A progressive paralysis starting with the legs and progressing to complete prostration may be noted. With the exception of blood-stained mucus in the dropping, clinical signs are indistinguishable from those of botulism. A pale swollen mottled liver is often seen with petechial hemorrhages present on the heart and visceral fat.

Coffee bean seed (*Cassia occidentalis; C. obtusifolia*)
Mechanical harvesting methods have increased the danger of contamination of corn and soybeans with coffee bean plants which are frequently found in relatively large numbers in the southern USA. At all levels of incorporation of the anthraquinone lectins in coffee seeds, egg production and weight gain are reduced. Platinum colored yolks and profuse diarrhoea are also noted with layers. Birds fed 2 to 4% of the coffee seeds become ataxic or partially paralysed before death. Muscle lesions are similar to those seen with vitamin E deficiency. Death often occurs due to an hyperkalemic heart failure. Production will return to normal with the removal of the contaminated feed.

Corn cockle (*Argostemma githago*)

Corn cockle is often harvested with wheat and so can become incorporated into poultry feeds. The diet must contain 5% or more of corn cockle to show toxic symptoms, which are caused by githagenin, a plant saponin. General weakness, with a decreased respiration and heart rate may be noted often associated with diarrhea. Hydropericardium and edema of the intestine can be seen along with petechial hemorrhages in the myocardium and congestion and degeneration of the liver.

Coyotillo (*Karwinskia humboldtiana*)

This plant is indigenous to south west Texas and Mexico. The fruit and seed are toxic to poultry and 3 to 4 days after ingestion generalized toxaemia signs can be noted, followed by paralysis and death.

Cacao (*Theobroma cacao*)

High levels of cacao bean wastes (in excess of 7% of the diet) are required to show toxic symptoms caused by the toxin theoromine. Such symptoms include nervous and excitable birds. Birds die in convulsions and usually are on their back with legs drawn tightly against their body. The comb is often cyanotic.

Crotalaria seed

A few species are toxic to poultry the most problematic being *C. spectabilis* and *C. giant* (*striata*). The toxin is a pyrolizidine alkaloid, designated, monocrotaline Crotalaria is a small black or brownish seed and is a contaminant in corn and soybeans in the southeast USA. One percent in a chick diet can result in death by 4 weeks of age. Birds become huddled having a pale comb and diarrhea and may exhibit a duck-like walk. With young birds abdominal fluid and edema, similar to that seen with ascites may be noted. With mature birds there is a drop in egg production and massive liver hemorrhages may be noted. The lesions are similar to those reported for toxic fat and salt poisoning.

Daubentonia seed (*Daubentonia longifolia*)

This seed can be a problem in the southern USA. As little as 9 seeds can cause death in 24 - 72 hours. The comb can be cyanotic, with the head hanging to one side. Emaciation and diarrhea may also be noted. Severe gastroenteritis, ulceration of the proventriculus and degeneration of the liver are not uncommon.

Glottidium seed (*G. vesicarium*)

This seed is often found in the south-eastern USA. Clinical symptoms are a cyanotic comb and wattles, ruffled feathers, emaciation and yellow diarrhea. Necrotic enteritis as well as liver and kidney degeneration are also common observations.

Death camas (*Zygadenus*)

This is a green range plant with an alkaloid toxin called nuttallii. Consumption of 5 to 10 g by a chicken can result in clinical symptoms in 12 hours, that include incoordination, diarrhea and prostration followed by death.

Vetch (*Vicia sativa*)

Vetch belongs to the Leguminosae family which is related to the legumes Lathyrus, Pisum and Ervum. It is common in the north-west USA and produces a cyanogenic glucoside called viciana, which is converted by the enzyme vicianase into hydrocyanic acid. Problems comparable to lathyrism are observed. These are excitability, incoordination, respiratory problems and convulsions.

Milkweed

Two common species are, *Asclepias tuberosa* and *A. incarnata*. They contain the bitter glucoside, ascclepdin, which is toxic to birds. Symptoms vary widely depending on the quantity of material consumed. The first sign is usually lameness, developing quickly into complete loss of muscle control. The neck becomes twisted with the head drawn back. In some cases symptoms gradually subside. In fatal cases symptoms become more progressive and prostration, coma and death result. No characteristic lesions are seen on necropsy.

Algae

Certain types of algae, including *Microcystis aeruginosa*, which readily grows in many lakes, can become concentrated by wind and deposited on shore or in shallow water. Degradation of this material produces toxins which have been responsible for losses in wild and domestic birds. The condition is usually noted in July or August in the N. Hemisphere. Toxicity is proportional to the amount of toxin consumed. Death can result in 10 to 45 minutes for mature ducks and chickens. Clinical symptoms include restlessness, twitching, muscle spasms, convulsions and death. These symptoms are similar to those seen with strychnine poisoning.

2.11 FEED MANUFACTURE

Until quite recently, the average poultry feed contained relatively few synthetic ingredients and the smallest amount of any addition amounted to 0.5% or more. Some natural ingredients, however, are being gradually replaced and supplemented by extremely small quantities of synthetic and purified ingredients, especially the vitamins, trace minerals, pigments and various pharmacological compounds. Consequently, the proper mixing of feed requires more and more technical know-how each year. Improper mixing can result in variation in the quality of feed and vitamin or mineral deficiencies, thus resulting in lack of protection against disease or chemical or drug toxicity.

Vitamin-Mineral Premixes

Micro-ingredients should be properly premixed before being added to a feed. It is desirable to have similar physical characteristics among ingredients to be premixed. The diluent suggested for use in the vitamin-mineral premixes is ground yellow corn or wheat middlings (Table 2.36). The corn should be of medium grind for best results. If the carrier is too coarse, it is not possible to obtain good distribution of the supplements, while too fine a carrier leads to dustiness and caking. For mineral mixes, limestone or kaolin (china clay) make satisfactory carriers. Where premixes are being stored for relatively short periods of time, the vitamin and mineral premix can be combined. However, where mixes are to be stored for more than 2 weeks, it is advisable to make separate vitamin and mineral mixes. Also, if premixes are to be shipped long distances and thus subjected to a great deal of handling, and perhaps high temperature, it is advisable to make separate vitamin and mineral mixes. This helps to reduce the physical separation of nutrients and leads to less vitamin deterioration.

When vitamin-mineral premixes are prepared in quantity ahead of time, they should be clearly labelled and stored in a cool dry place for future use. With the addition of an antioxidant and the margins of safety provided in most premixes, they can be held for two to three months under ideal conditions. Rather than suggesting the use of products with specific potencies to supply the vitamins and other nutrients (Table 2.36) the units or weights of the compounds have been indicated and the decision as to product use is left to the individual. Some feed manufacturers are capable of making premixes from more highly potent vitamin and mineral preparations than others; this usually results in a cost saving compared with the use of more dilute preparations. The choice of potency of products for use in the premixes should be governed, to a large extent, by the personnel and facilities available. Because vitamin and mineral supplements represent a relatively small part of the total cost of a diet, margins of safety are being added in most cases (Table 2.36). Lower levels can be used with satisfactory results under ideal conditions.

The direct addition of vitamin, antibiotic or other supplements to the feed, each of which must often be included at less than 1 kg/tonne, is not usually recommended. These micro-ingredients should be suitably premixed first, so that at least 2 kg/tonne is added. It is generally recommended that vitamin-mineral premixes be added to the mixer after about one-half of the other ingredients have been included. The time required for a satisfactory mix is very important and varies considerably, depending upon the equipment used. Usually 2 - 3 minutes is the optimum for horizontal mixers and up to 5 minutes for vertical machines although mixing times are being continually reduced with newer equipment. This can vary with the type of mixer and manufacturer's specifications should always be followed.

The segregation of ingredients in a mixed feed can occur due to handling after mixing. This can be a problem when mash feeds containing no added fat are blown into bulk bins. However, care in unloading and a cyclone on top of the bulk tank, will help overcome the problem. This is usually not as great a problem when the feed is pelleted or crumbled.

Vitamin stability

Naturally occurring vitamin E is quite unstable, particularly in the presence of fat and trace minerals, however, vitamin E added as a supplement usually is in a highly stable form (eg. gelatin coated beadlet containing an antioxidant).

Vitamin A in fish oil and pro-vitamin A compounds in yellow corn are easily destroyed in the typical mixed ration. Most dehydrated green feeds are now treated with an antioxidant that helps prevent the destruction of the pro-vitamin A compounds during storage. Today, most poultry feeds contain supplementary gelatin-coated synthetic vitamin A which is quite stable. The inclusion of antioxidants in the feed helps to retain the potency of vitamins A and E in mixed feed.

Vitamin D_3 is much more available to poultry than is vitamin D_2, hence the latter should never be used in poultry feeds. Vitamin D_3 supplements are available in a dry, stabilized form. These products are reported to be stable when mixed with minerals.

TABLE 2.36 Vitamin-Mineral Premixes

- all premixes should be made up to 5 - 10 kg by the addition of a carrier such as wheat middlings
- the amounts shown below are the levels of nutrients to be added per tonne of finished feed

	Chicken				Turkey			Waterfowl		
VITAMINS	**Starter**	**Grower**	**Laying**	**Breeder**	**Starter**	**Grower**	**Breeder**	**Starter**	**Grower**	**Breeder**
Vitamin A (M.IU)	10.0	8.0	7.5	13.0	10.0	8.0	13.0	10.0	8.0	10.0
Vitamin D_3 (M.IU)	3.3	3.3	3.3	3.3	3.3	3.3	3.3	1.5	1.3	1.4
Vitamin E (T.IU)	30.0	20.0	25.0	30.0	40.0	25.0	40.0	20.0	15.0	15.0
Riboflavin (g)	6.0	5.0	5.0	8.0	6.0	5.0	8.0	5.0	4.0	5.5
Thiamine (g)	4.0	4.0	4.0	4.0	4.0	4.0	4.0	4.0	4.0	4.0
Pyridoxine (g)	3.3	3.3	3.3	5.0	3.3	3.3	5.0	3.3	3.3	3.3
Pantothenic acid	15.0	10.0	10.0	15.0	15.0	12.0	15.0	12.0	10.0	10.0
Vitamin B_{12} (g)	.015	.012	.015	.015	.015	.012	.015	.015	.010	.015
Niacin (g)	50.0	30.0	40.0	50.0	50.0	40.0	50.0	50.0	40.0	50.0
Vitamin K (g)	2.0	2.0	2.0	3.0	2.0	2.0	3.0	1.5	1.5	1.5
Folic acid (g)	1.0	1.0	1.0	1.0	1.0	0.5	1.0	0.5	0.25	0.5
Biotin (g)	0.2	0.2	0.2	0.2	0.3	0.2	0.3	0.2	0.1	0.2
MINERALS										
Manganese (g)	70.0	70.0	70.0	70.0	70.0	70.0	70.0	70.0	70.0	70.0
Zinc (g)	60.0	60.0	60.0	60.0	60.0	60.0	60.0	60.0	60.0	60.0
Copper (g)	8.0	8.0	8.0	8.0	8.0	8.0	8.0	8.0	8.0	8.0
Selenium (g)	0.3	0.3	0.3	0.3	0.3	0.3	0.3	0.3	0.2	0.2
Iron (g)	70	70	70	70	70	70	70	70	70	70

***Note the values shown are amounts of pure ingredients. All vitamin premixes should contain 125g Ethoxyquin/kg**

Calcium pantothenate may be destroyed in the presence of supplements containing acid ingredients such as niacin, arsenilic acid and 3-nitro. The calcium chloride complex of calcium pantothenate is more stable than ordinary calcium pantothenate under acid conditions. Recent work has shown that thiamin, folic acid, pyridoxine and some vitamin K supplements can be relatively unstable in the presence of trace mineral supplements. This is especially true where the minerals are supplied as sulphate salts, hence, special consideration must be given to the above-mentioned vitamins when premixes contain both vitamins and minerals.

Most of the other vitamins are fairly stable. However, care should be taken in storing vitamins to ensure their potency. Always store in a cool, dry, lightproof space or container. While vitamin supplements are an extremely important part of a well balanced diet, animals usually have sufficient body stores to meet their requirements for several days. Modern poultry farms receive feed deliveries on a weekly or even more frequent basis. Failure to incorporate the vitamin premix in a delivery of feed will likely have little or no effect on the performance of most classes of poultry, assuming the "next delivery" contains the vitamin supplement. For breeding birds this may not be true, especially for the vitamin riboflavin, which could well affect hatchability if hens are fed a deficient diet for 7 to 10 days.

Pelleting

The usual pelleting process involves treating ground feed with steam and then passing the hot, moist mash through a die under pressure. The pellets are then cooled quickly and dried by means of forced air. Sufficient water should be applied so that all feed is moistened. Pelleting at too low a temperature, or with too little steam, results in a "shiny pellet", due to increased friction on the pellet going through the die. Often such pellets are only the original mash enclosed in a hard capsule and have not benefited from the "cooking" process brought about by moisture and heat.

Optimum moisture content of a feed required for good pelleting will vary with the composition of the feed, however, a range of 15 to 18% moisture is usually desirable. Feeds containing liberal quantities of high fiber ingredients will require a higher level of moisture while feeds low in fiber will require less moisture. A good pellet when hot can be reduced to two-thirds of its length without crumbling. Such feed has been "steam-cooked" and holds together well. Rations can be pelleted at any temperature up to 88°C that will allow for maximum production per hour without any major fear of vitamin destruction or decreased bird performance.

Some manufacturers have experienced difficulty in obtaining good pellets when pelleting a corn-soybean diet containing added fat. Products such as lignosol or bentonite are reasonably effective as binding agents, however, they have little nutritive value, and so, one should consider whether the advantage of introducing such material into pelleted or crumbled diets warrants the cost. The inclusion of 10 to 15% of wheat, wheat middlings or to a lesser extent barley will give a pellet of satisfactory hardness. When these ingredients are too expensive, the addition of about 2% of extra water to the mash will aid in producing a better pellet. If this procedure is followed, however, extra drying of the pellets is required so that mold growth does not occur during storage. Work in our laboratory has indicated that molasses may be used as a pellet binder. Besides aiding in pelleting, molasses unlike other binders also contributes energy to the diet and so inclusion levels of 1 to 2% in certain diets may be beneficial.

In addition to the advantages of less feed wastage and ease of handling, pelleted diets are more efficiently utilized by poultry. While some of this improvement is due to chemical changes brought about by heat, moisture and pressure, a significant part of the enhanced efficiency is due to birds spending less time in eating pellets and so a reduction in energy requirements by the bird. This situation is demonstrated in Table 2.37.

TABLE 2.37 Time spent eating mash and pelleted diets

	Av. time spent eating (min/12 hr day)		Av. feed consumed (g/bird/12 hr)	
	Mash	Pellets	Mash	Pellets
Turkeys (38-45 days of age)	136	16	62	57
Chickens (21-28 days of age)	103	34	38	37

Jensen et al (1962)

The need for good quality pellets is often questioned by feed manufacturers, since regrinding of pellets or crumbles, and feeding these to birds, has little apparent effect on performance (Table 2.38).

TABLE 2.38 Performance of birds fed a corn-soybean meal diet as mash, pellets or reground pellets

Feed type	Body wt (g)	Feed:Gain
Mash	407	2.06
Pellets	471	1.90
Reground pellets	461	1.96

There seems little doubt that good quality crumbles and pellets can be advantageous for improving the growth rate of turkeys. However, pellet quality seems of less importance with broiler chickens, especially where high energy diets are considered. More important in the pelleting process, is the treatment of feed with steam and pressure, although it is realized that in certain markets it is difficult to sell feed that is not of "ideal" pellet quality.

Expanding, Extrusion and Thermal Cooking

Extrusion has been used for a number of years to produce dry cereal snack foods and more recently, various pet foods. Extrusion usually employs higher temperatures and pressure than does normal steam pelleting, and so there is greater potential for starch gelatinization and theoretically higher digestibility. Extrusion is however much slower than conventional pelleting, and initial capital cost is very high.

Expanders and compacters, bridge the gap between conventional pelleting and extrusion, in terms of processing conditions, capital cost and speed of throughput. Feed exiting an expander is not forced through a die as such, and so the final product is more of a "cake". Where expanders are used in poultry operations, the expanded feed is often pelleted and so expansion is an additional process in feed manufacture. However there is the possibility of feeding non-pelleted, expanded feed, simply by breaking up this "cake" material.

Thermal cooking offers the most extreme processing conditions, where high temperatures can be maintained for very long periods of time, relative to pelleting, extrusion or expansion. Thermal cooking will result in the best possible starch gelatinization etc., and will also give the best control over microbial content.

While all heat processing conditions are going to reduce microbial counts in feed there will be a concomitant loss of heat-sensitive nutrients such as some vitamins and amino acids. In this context synthetic amino acids may be more susceptible to heat processing than those naturally present in other ingredients. One recent study suggested some 6% loss of total methionine in an extruded broiler starter that contained 0.18% supplemental methionine. For most vitamins, other than vitamin C and MSBC, normal pelleting conditions are expected to result in 8 - 10% loss of potency. Extrusion however, which usually employs much higher temperatures, can lead to 10 - 15% loss of most vitamins. Under any heat-treatment conditions there will always be significant loss ($\simeq$ 50%) of vitamin C, and up to 30 - 50% loss of MSBC (Table 2.39).

TABLE 2.39 Effect of steam pelleting, extrusion and expansion on loss of vitamin potency

	Percentage loss of vitamin potency		
Vitamin	**Pelleting (82°C, 30 sec)**	**Expander (117°C, 20 sec)**	**Extrusion (120°C, 60 sec)**
Vitamin A(beadlet)	7	4	12
Vitamin D_3(beadlet)	5	2	8
Vitamin E	5	3	9
MSBC	18	30	50
Thiamin	11	9	21
Folic acid	7	6	14
Vitamin C	45	40	63
Choline chloride	2	1	3

Adapted from Coehlo, (1994)

2.12 WATER AS A NUTRIENT

Water, is the most critical nutrient that we consciously supply to birds, yet in most instances, it is taken completely for granted. Water often receives attention only when mechanical problems occur. Water is by far the largest single constituent of the body, and represents about 70% of total body weight. Of this body water, about 70% is inside the cells of the body and 30% is in the fluid surrounding the cells and in the blood. The water content of the body is associated with the protein content. This means that as a bird ages, and its body fat content increases, then its body water content as a percent of body weight will decrease. The bird obtains its water by drinking, by eating and by catabolism of body tissues which is a normal part of growth and development.

Water intake

a) *Drinking water*

Water intake of a bird increases with age, although it decreases per unit of body weight. Drinking behaviour is closely associated with feed intake, and so any factors affecting feed intake will indirectly influence water intake. At moderate temperatures birds will consume twice as much water by weight, as they eat as feed. Any nutrients that increase mineral excretion by the kidney will influence water intake. For example salt or an ingredient high in sodium, will increase water intake. Similarly, feeding an ingredient high in potassium such as molasses or soybean meal, or calcium/phosphorus sources contaminated with magnesium, will result in increased water intake. Such increases in water intake are of no major concern to the bird itself, but obviously result in increased water excretion and so wetter manure. Environmental temperature is perhaps the major factor influencing fluctuation in water intake. Layers at 30°C will consume twice as much water as birds housed at 15°C. Because feed intake decreases at higher temperatures, then the water:feed ratio increases dramatically under these conditions leading again to wetter manure. Table 2.40 indicates average water consumption of various poultry species maintained at 20 or 32°C. These figures indicate approximate water usage values and will vary with the stage of production, health and feed composition.

In calculating the water needs of egg producing stock, it should be realized that water intake is not constant throughout the day, rather it varies depending upon the stage of egg formation (Fig 2.1). This data clearly shows a peak in water consumption immediately following egg laying, and again, at the time just prior to the end of a normal light cycle. This means that water needs must be accommodated during these peak times (around 10 - 11 am and 6 - 8 pm) within a 6 am - 8 pm light cycle, because most birds will be in the same stage of egg formation as dictated by the light program.

TABLE 2.40 Daily ad-lib water consumption of poultry (Litres per 1000 birds)

		20°C	32°C
Leghorn pullet	4 wk 12 wk 18 wk	50 115 140	75 180 200
Laying hen	50% prod 90% prod	150 180	250 300
Non-laying hen		120	200
Broiler breeder pullet	4 wk 12 wk 18 wk	75 140 180	120 220 300
Broiler breeder hen	50% prod 80% prod	180 210	300 360
Broiler chicken	1 wk 3 wk 6 wk 9 wk	24 100 240 300	40 190 500 600
Turkey	1 wk 4 wk 12 wk 18 wk	24 110 320 450	50 200 600 850
Turkey breeder hen		500	900
Turkey breeder tom		500	1100
Duck	1 wk 4 wk 8 wk	28 120 300	50 230 600
Duck breeder		240	500
Goose	1 wk 4 wk 12 wk	28 250 350	50 450 600
Goose breeder		350	600

These figures indicate approximate water usage values and will vary with the stage of production, health and feed consumption

Fig. 2.1 Water consumption of laying hens in relation to time of oviposition. (from Mongin and Sauveur, 1974)

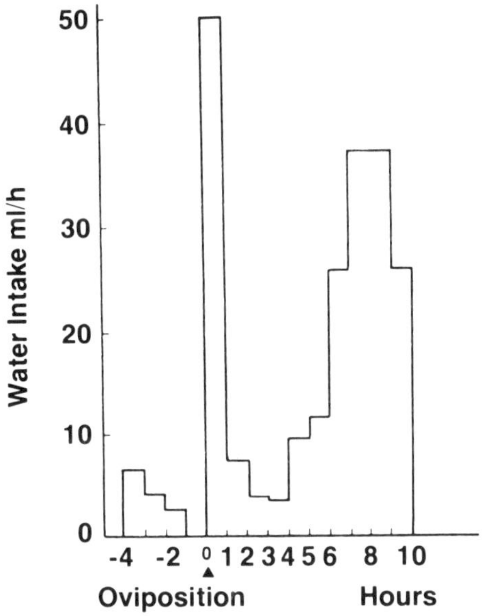

b) Dietary water

The contribution of feed is not usually considered in calculating water balance, yet most feeds will contain around 10% of free water. Other so-called bound water may become available during digestion and metabolism, such that 7 - 8% of total requirements can originate from the feed.

c) Metabolic water

Water is created in the body as a by-product of general metabolism. If fats are metabolized, then about 1.2 g of water are produced from each gram of fat. Likewise protein and carbohydrate will yield about 0.6 and 0.5 g per gram respectively. Total metabolic water can be more easily estimated from the bird's energy intake because on average 0.14g of water is produced for each kcal of energy metabolized. This means that for a laying hen consuming 280 kcal ME/day, about 39 g of metabolic water will be produced. Feed and metabolic water together therefore account for about 20% of total water needs, and so are very important in the calculation of water balance.

Water output

a) Excreta

The quantities of water excreted in the feces and urine, vary with differences in water intake. Broiler chickens produce excreta containing about 60 - 70% moisture, while that produced by the laying hen contains about 80% moisture. For the laying hen at least, the quantity of water excreted in the feces is about four times that excreted as urine. Undoubtedly this loss is subject to considerable variation with the amount and nature of undigested feed.

b) Evaporative loss

Evaporation is one of four physical routes by which poultry can control their body temperature. Due to its molecular structure and bonding, water has an unusually high latent heat of vaporization. Some 0.5 kcals of heat are required to vaporize one gram of water. Evaporative heat loss takes place through the body surface and the respiratory tract. The fowl has no sweat glands, consequently evaporation via the skin is restricted. Loss in this manner overwhelmingly occurs via the moist surface layer of the respiratory tract to the inspired air which is "saturated" with water vapor at body temperature. Evaporation rate is therefore proportional to respiratory rate. Heat loss through evaporation represents only about 12% of total heat loss in the broiler chicken housed at 10°C, but this increases dramatically through 26 - 35°C where it may contribute as much as 50% of total heat loss from the body. High environmental humidity will reduce efficiency of evaporative heat loss through the skin more so than via the lungs, if only because the latter can be heated to 41°C compared to a skin temperature of 27°C. At high temperatures, evaporative water loss will approximate water intake and so this obviously imposes major demands on the ventilation systems.

Water balance and dehydration

Under normal physiological conditions for adult birds, water intake and output are controlled to maintain a constant level of water in the body. A positive water balance is found in the growing bird to accommodate growth. With drinking water being supplied *ad libitum* under most commercial conditions, dehydration due to lack of drinking water should not occur. The adverse effects of short term reduced water intake are often a result of a concomitant reduction in feed intake.

The turkey poult is most susceptible to dehydration resulting from drinking water deprivation, and mortality occurs when drinking water is re-introduced to the poults. Poults 11 days of age, subjected to a 48 hour period of water deprivation, showed 83% mortality following reintroduction of *ad libitum* cold water; in most cases death occurred within 30 minutes. Poults 18 days of age showed less mortality which was somewhat delayed (2 - 34 hours) while turkeys at 8, 12 and 21 weeks of age subjected to the same conditions showed no mortality. The exact reason for these problems is not fully understood. Poults deprived of water show reduced body temperature, and when water is introduced, body temperature continues to decrease for 30 minutes or so. Poults often drink large amounts of water following dehydration, and it has been suggested that the problem relates to simple water intoxication and associated dilution of electrolytes in the body. If young poults are dehydrated for whatever reason, then administration of electrolytes in the water may be beneficial. This problem does not seem to occur with other species.

Drinking water temperature

Water offered to birds is usually at ambient temperature. This means that for laying birds housed under controlled environmental conditions, the temperature of drinking water is held fairly constant, while for broiler chickens, water temperature decreases with age corresponding to a reduction in brooding temperature. It is only for the first few days of a chick's life that drinking water temperature is specified, where traditional management recommendations suggest the use of "warm" water. However, there is little documented evidence supporting this recommendation. Birds drink more water at higher environmental temperatures, yet the cooling of water may result in even higher intakes. Table 2.41 outlines the results of a small scale study conducted with layers housed at 33°C. When birds received cool water for a 4 week period, they were able to maintain peak egg production, possibly due to a less severe reduction in feed intake. Under commercial conditions, with long runs of water pipe, it is obviously very difficult to duplicate these conditions. However, it does show the importance of trying to keep the water as cool as possible, and in this regard, the usual practice of placing water tank reservoirs on high towers in direct sunlight should be seriously questioned.

TABLE 2.41 Layer performance at 33°C with hot vs cold drinking water

	Water temperature	
	33°C	2°C
Feed/bird/day (g)	63.8	75.8
Egg production (%)	81.0	93.0
Egg weight (g)	49.0	48.5

Water restriction

Most birds should have continuous access to clean water. Some breeders recommend water restriction of laying hens as a means of preventing wet manure problems, especially in hot climates, although serious consideration should be given to other preventative measures prior to this last resort. Production may drop as much as 30% when hens are deprived of water for 24 hours, and it may take as long as 25 to 30 days before production returns to normal. Similar results have been reported for broilers where decreases in water supply have resulted in marked depressions in weight gain. Table 2.42 shows the results of a controlled test where water restriction was imposed on broilers. It can be noted that a marked drop in feed intake was observed, with the greatest reduction occurring with the first 10% reduction in water intake.

TABLE 2.42 Effect of water restriction on weekly feed consumption of broilers (g)

	Degree of water restriction (%)					
Age (weeks)	0	10	20	30	40	50
2	200	168	168	150	168	141
4	363	358	372	327	308	290
6	603	531	494	472	440	431
8	776	667	644	612	572	522
Total	3516	3171	3052	2836	2740	2581

***All birds received water ad libitum for first week. (Data from Kellerup et al. 1971)**

The effect of an accidental 48 hour cut in water supply of layers is shown in Fig 2.2. Production dropped off very quickly to virtually 0%, although interestingly a few birds maintained normal production. Most birds that resumed production within 28 d achieved normal output for their age, although there was an indication of improved shell quality.

Fig. 2.2 Effect of a 48-hour period of water deprivation on egg numbers.

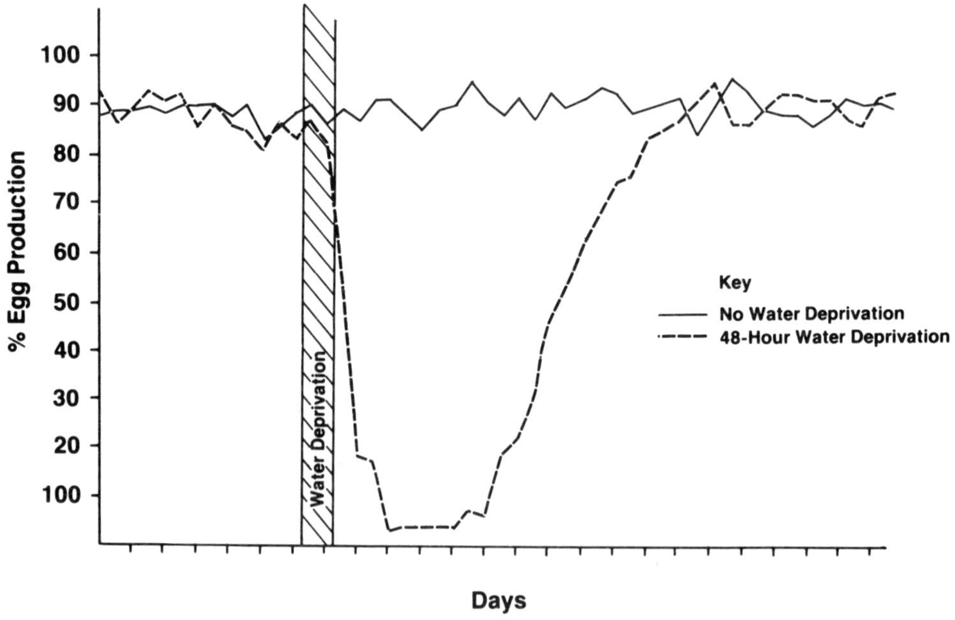

For certain classes of stock, intentional water restriction is used as a management tool. To date, this is most common with broiler breeders fed on a skip-a-day program. Water restriction may occur on both feed-days and off-feed days. Restriction on off-feed days is done because it is assumed that birds will over-consume water on these days due to hunger or boredom. However it seems as though breeders do not drink that much water on an off-feed day (Table 2.43).

TABLE 2.43 Water intake of 13 week-old broiler breeders

	Water restricted each day	Water restricted only on feed days	Ad-lib water
Water consumed on a feed day	175 ml	182 ml	270 ml
Water consumed on off-feed day	108 ml	109 ml	36 ml
Average	141 ml	145 ml	153 ml

All birds drank the same average amount of water over a 2 day feeding schedule regardless of water treatment. When birds are given free-choice water, they obviously overconsume on a feed-day, but drink little on an off-feed day. These data suggest the need for water restriction of skip-a-day fed birds, although special attention on feed-days, rather than off-feed days, will be advantageous in preventing wet litter.

Water quality

The quality of water is something that should be given careful consideration when evaluating the performance of a flock. Chemical contaminants are the most serious problem affecting water quality. However, poultry usually adjust to high levels of certain minerals after a period of time, and so only in a relatively small number of cases, does the mineral content of water significantly affect the performance of a flock. There are certain areas where the water salinity is high enough to adversely affect flock performance. In such cases, it may be necessary to remove some of the added salt from the diet. However, this should be done only after careful consideration to ensure that there will be a sufficient salt intake because performance can be severely reduced if dietary salt levels are too low.

Any bacterial contamination of water is an indication that surface water is entering the water supply and steps should be taken to correct the situation. Alternatively the water may be chlorinated to eliminate contamination. Another problem that can exist with water is a build-up of nitrates or nitrites. Such contamination is usually an indication of run-off from animal wastes or fertilizers leaking into the water system. Although the standard for human water supply is 10 to 20 ppm of nitrate nitrogen, higher levels can usually be tolerated by animals. Levels beyond 50 ppm need to be present before water is suspected as a factor in the poor performance of poultry. As nitrites are 10 times more toxic than nitrates, and because bacteria in the intestinal tract and in the water supply can convert nitrates to nitrites, levels of these two contaminants in the water supply should be kept to a minimum. Superchlorination of the water will quickly oxidize nitrites to nitrates thereby reducing their toxicity. Before initiating a superchlorination program, check with a local pathologist to ensure a proper level of chlorination in order not to interfere with the performance or efficiency of vaccines or other drugs.

Table 2.44 outlines standards for drinking water in terms of mineral levels. Toxicity and loss of performance will vary dependent upon bird age and class of stock, but in general these values can be used as guidelines to indicate the possibility of toxicity with birds consuming such water for prolonged periods.

TABLE 2.44 Concentration of water minerals above which problems may occur with poultry (ppm)

Total soluble salts (hardness)	1500
Chloride	500
Sulphate (SO_4)	1000
Iron	500
Magnesium	200
Potassium	500
Sodium	500
Nitrate (NO_3)	50
Arsenic	0.01
pH	6.0 - 8.5

In the last few years, there has been an interest in the treatment of water for poultry. In large part, this is carried out in an attempt to prevent problems of mineral deposits occurring in pipelines, boilers and automatic waterers, rather than preventing toxicity problems *per se*. Such treatments involve orthophosphates which sequester calcium and magnesium, thereby preventing precipitation in the water supply. In most situations, these systems will not unduly alter the water composition as far as the bird is concerned. As a last resort, some producers use water softeners, and in these situations, there is some cause for concern regarding to the bird's health. These softeners contain an active column of resin, that has the ability to exchange one ion (mineral) for another. Over time, the resin column becomes saturated with the absorbed minerals (usually calcium and magnesium salts) that are extracted from the water, and so it must be flushed and re-charged with the donor mineral. In most softeners, this recharging process involves sodium from common salt. This means that sodium is replacing other minerals in the water, because sodium salts readily dissolve, and will not leave mineral scale in the equipment. The amount of sodium that is pumped into the water supply is therefore in direct proportion to the hard minerals extracted from the water. In areas of very hard water, one can expect higher levels of sodium in water reaching the birds, and vice-versa in areas of lower water hardness. The question that is often asked is, do these altered levels of sodium affect the bird?

Sodium is an essential nutrient for the bird, and in most instances, the bird's requirement is met with total diet equivalents of around 0.18% (or 0.4% salt). Levels slightly in excess of this will often lead to mild diarrhea and wet droppings, while a 100% increase will definitely lead to wet droppings and may cause mortality in some young birds, especially turkeys. Egg production and growth will also be influenced at these levels, while higher levels may lead to mortality (often due to dehydration) even in older birds. Under normal conditions, salt levels in water are negligible, and provide less than 1% of the bird's requirement. Some natural sources of water contain high levels of sodium, and in these instances, it is often necessary to adjust diet levels accordingly. As previously mentioned, the amount of sodium pumped into the water by softeners, depends on natural water hardness, and so each farm must be considered individually. Potential sodium toxicity depends on the frequency of salt addition in relation to water usage. Because most classes of birds drink about the same amount of water in relation to their feed intake, (and their salt % in the diet is about the same) we can make some general conclusions regarding potential toxicity, and subsequently the need to adjust diet salt levels to correct or prevent problems. Table 2.45 was constructed from information provided in the NAS publication "Nutrients and Toxic Substances in Water for Livestock and Poultry" and from a consideration of the bird's normal daily expected intake of salt from the diet. This data applies to birds maintained under moderate environmental temperatures.

Table 2.45 Bird response to salt usage in water softeners (20°C barn temperatures)

Water Softener Salt Usage	Water Usage (gallons)						
	500	1000	2500	5000	10,000	50,000	100,000
10 kg	×	●	○	■	■	■	■
20 kg	×××	×	●	○	■	■	■
30 kg	□	××	●	●	■	■	■
40 kg	□	×××	×	●	○	■	■
80 kg	□	□	××	×	●	■	■
160 kg	□	□	□	××	×	○	■

■ No problems

○ May wish to adjust diet sodium level in order to prevent wet droppings

● Wet droppings likely when birds first introduced, adjust diet sodium levels

× Wet droppings likely, adjust diet sodium levels

×× Wet droppings, some production loss likely; adjusting diet sodium levels will only be partially successful

××× Wet droppings, production loss, mortality with younger stock Water intake of sodium overwhelms diet considerations

□ Mortality likely with all classes of stock

Table 2.45 therefore suggests that there should be no problems in using 30 kg of water softener salt per 10,000 gallons of water. At 40 kg per 10,000 gallons, there is some cause for concern, and the producer may wish to consult a nutritionist to investigate the potential of adjusting dietary levels of salt. At 80 kg in 5,000 gallons for example, problems are likely, and salt levels in the diet need adjustment. If 80 kg is used per 1000 gallons, high mortality is expected, and dietary salt adjustment will not correct the problem. In this situation, some form of water treatment other than a water softener (that uses salt for recharging) must be considered. Laying birds are likely to be the most susceptible, because they drink correspondingly more water in relation to feed intake. For birds at peak production therefore, problems may occur at slightly lower levels. Care must also be taken in areas where sodium levels in the water are naturally very high. In these situations, water softeners may be inadvisable.

General management considerations with water

Where continuous flow water troughs are used for caged birds, one must be sure that the birds at the end of the trough obtain sufficient water. A rise in house temperature will result in increased water consumption, and unless the water supply can be adjusted accordingly, shortages of water may result for the birds at the far end of the line. It has also been demonstrated that poorly beak-trimmed birds may not be able to drink sufficient water to sustain maximum production. Where the lower beak of the bird is too long, up to 20% loss in egg production can occur, compared with properly beak-trimmed birds. When disease or stress occur, a decrease in water consumption is usually noted a day or two before a decrease in feed consumption. For this reason, managers should consider installing water meters on all water lines to each pen or cage row and have the attendant keep a daily record of water consumption. Such records can give early warning of potential problems with the flock.

A note of caution should be given when moving pullets to laying cages. It has been observed on a number of occasions that when pullets are moved near to point of lay, they seem to have more difficulty adjusting to new watering equipment. This is especially true where nipple waterers are used. If possible, pullets should be moved to laying cages 2 to 3 weeks before production commences because the immature bird has reduced water turn-over. With ever decreasing age at maturity, this management technique is now more difficult to implement.

CHAPTER 2

Selected References

Coelho, M.B. 1994. *Vitamin stability in premixes and feeds: a practical approach. BASF Technical Symposium.* Indianapolis. May 25. p 99-126.

Lane, R.J. and T.L. Cross 1985. *Spreadsheet applications for animal nutrition and feeding.* Reston Publ., Reston, Virginia.

Leeson, S., G. Diaz and J.D. Summers 1995. *Poultry Metabolic Disorders and Mycotoxins. Publ. University Books.* Guelph, Ontario, Canada.

McDowell, L.R. 1989. *Vitamins in Animal Nutrition.* Academic Press, N.Y.

National Academy of Sciences 1973. *Effect of processing on the nutritional value of feeds.* NAS. Washington, D.C.

National Academy of Sciences 1974. *Nutrients and Toxic Substances in Water for Livestock and Poultry.* NAS Washington, D.C.

National Academy of Sciences 1980. *Mineral tolerances of domestic animals.* NAS. Washington, D.C.

National Academy of Sciences 1987. *Vitamin tolerance of animals.* NAS. Washington, D.C.

National Academy of Sciences 1994. *Nutrient requirements of poultry.* 9th Rev. Ed. NAS. Washington, D.C.

Novus 1994. *Raw Material Compendium.* 2nd Edition. Publ. Novus Int., Brussels.

Pesti, G.M. and B.R. Mitter 1993. *Animal Feed Formulation.* Publ. Van Nostraud Reinhold, N.Y.

Sibbald, I.R. 1983. *The TME system of feed evaluation.* Agriculture Canada 1983-20E. Animal Research Centre, Ottawa, Canada.

Sibbald, I.R. 1987. *Examination of bioavailable amino acids in feedstuffs for poultry and pigs: a review with emphasis on balance experiments.* Can. J. Anim. Sci. 67:221-301.

Valdes, E.V. and S. Leeson 1992. *Near Infrared Reflectance analysis as a method to measure metabolizable energy in complete poultry feeds.* Poult. Sci. 71:1179-1187.

Wiseman, J., F. Salvador and J. Craigon 1991. *Prediction of the apparent metabolizable energy content of fats fed to broiler chickens.* Poult. Sci. 70:1527-1533.

CHAPTER 3. FEEDING PROGRAMS FOR GROWING EGG-STRAIN PULLETS

3.1. DIET SPECIFICATIONS AND FORMULATION

Diets are usually formulated to meet the needs of a given class of poultry under specific environmental situations. This is not to imply that the ultimate has been achieved in ration formulation to meet exacting nutrient requirements. Indeed, we are a long way from achieving that goal. However, we have progressed to the point where one given set of requirement values is no longer acceptable for all strains within a given class of poultry housed under a range of environmental conditions. A single requirement value under such conditions is not only inefficient as far as nutrient utilization is concerned, but in many cases does not allow the true genetic potential of the bird to be expressed.

The following sections outline diet nutrient recommendations and examples of diet formulations for growing pullets. It is not suggested that these are the only diet situations that should be considered, since it is realized that specific environmental and management practices dictate the need for flexibility. It is hoped that the reader will look through the discussion on factors influencing feeding programs that follows this initial reference material prior to making final decisions on diet specifications. The following section outlines examples of diets that may be used for various ages of pullets from day of age through to the prelay period. They are not intended to be all encompassing or to suit conditions in all environments and geographical locations. Rather, they are intended to show relative balance and contribution of selected ingredients, and how these can be readily manipulated. In these diets, we have dictated the need for a 10 kg vitamin-mineral premix. This inclusion level can obviously be changed depending upon concentration of the premix that is locally available.

TABLE 3.1 Diet specifications for growing Leghorn pullets

	Chick starter			Chick grower			Pre-lay
Approximate protein level (%)	16.0	18.0	20.0	13.0	15.0	17.0	17.0
Amino acids (% of diet)							
Arginine	0.86	0.94	1.03	0.66	0.78	0.92	0.80
Lysine	0.80	0.90	1.00	0.62	0.72	0.85	0.70
Methionine	0.36	0.41	0.45	0.29	0.34	0.39	0.35
Methionine + cystine	0.58	0.66	0.72	0.47	0.55	0.65	0.60
Tryptophan	0.17	0.18	0.19	0.15	0.16	0.18	0.17
Histidine	0.30	0.33	0.36	0.24	0.28	0.32	0.30
Leucine	1.04	1.16	1.28	0.81	0.95	1.10	1.00
Isoleucine	0.56	0.62	0.68	0.44	0.51	0.61	0.55
Phenylalanine	0.52	0.58	0.64	0.42	0.48	0.57	0.51
Phenylalanine + tyrosine	1.02	1.13	1.24	0.81	0.93	1.11	1.00
Threonine	0.50	0.56	0.62	0.40	0.47	0.55	0.50
Valine	0.62	0.69	0.76	0.49	0.67	0.68	0.67
Metabolizable energy (kcal/kg)	2850	2850	2900	2750	2850	2950	2850
Calcium (%)	1.0	1.0	1.0	0.80	0.85	0.90	2.0
Available phosphorus (%)	0.40	0.40	0.42	0.35	0.37	0.39	0.43
Sodium (%)	0.18	0.18	0.18	0.18	0.18	0.18	0.18
Vitamins (per kg of diet)							
Vitamin A (I.U.)	9000			7000			
Vitamin D_3 (I.U.)	2500			2000			
Choline equivalents (mg)	1200			1000			
Riboflavin (mg)	5.0			4.0			
Pantothenic acid (mg)	13.0			11.0			
Vitamin B_{12} (mg)	0.012			0.010			
Folic acid (mg)	0.75			0.60			
Biotin (mg)	0.20			0.15			
Niacin (mg)	50.0			40.0			
Vitamin K (mg)	2.0			2.0			
Vitamin E (I.U.)	30.0			20.0			
Thiamin (mg)	2.2			2.0			
Pyridoxine (mg)	4.0			3.5			
Trace minerals (per kg of diet)							
Manganese (mg)	70			70			
Iron (mg)	80			80			
Copper (mg)	8.0			8.0			
Zinc (mg)	60.0			60.0			
Selenium (mg)	0.3			0.2			
Iodine (mg)	0.4			0.4			

TABLE 3.2. Examples of Chick Starter Diets (kg)

Ingredients	1	2	3	4	5	6
Ground corn	656.5	---	350.0	709.5	---	374.0
Ground wheat	---	736.5	346.5	---	796.5	372.5
Soybean meal (48%)	290.0	210.0	250.0	238.0	150.0	200.0
Fat (stabilized)*	10.0	10.0	10.0	10.0	10.0	10.0
Ground limestone	15.0	15.0	15.0	15.0	15.0	15.0
Calcium phosphate (20% P)	15.0	15.0	15.0	15.0	15.0	15.0
Salt	2.5	2.5	2.5	2.5	2.5	2.5
Vitamin:mineral premix[1]	10.0	10.0	10.0	10.0	10.0	10.0
Methionine	1.0	1.25	1.10	0.88	1.25	1.10
Calculated analyses						
Crude protein (%)	19.6	19.7	19.6	17.6	17.6	17.7
Digestible protein (%)	18.0	17.9	17.8	16.0	16.0	16.1
Crude fat (%)	3.6	2.2	3.0	3.8	2.3	3.1
Crude fiber (%)	2.5	2.8	2.6	2.5	2.8	2.6
Metabolizable energy (kcal/kg)	3005	2940	2975	3045	2970	3010
Calcium (%)	0.95	0.96	0.96	0.94	0.95	0.95
Available phosphorus (%)	0.43	0.42	0.42	0.41	0.43	0.42
Sodium (%)	0.17	0.19	0.18	0.17	0.19	0.18
Methionine (%)	0.43	0.42	0.42	0.41	0.40	0.40
Methionine + cystine (%)	0.73	0.74	0.72	0.66	0.68	0.67
Lysine (%)	1.06	1.04	1.05	0.90	0.88	0.90

[1]Use additional choline if premix does not contain this vitamin

*Throughout this book, fat is added to diets at 1% to improve feed texture, reduce dust, and improve palatability. An equal amount of molasses would do a similar job. Where crumbled or pelleted feeds are used, fat also helps as a lubricant and significantly increases the life of pelleting equipment.

TABLE 3.3 Examples of chick grower diets (kg)

Ingredients	1	2	3	4	5	6
Ground yellow corn	495.0	---	288.0	737.0	---	392.0
Ground wheat	---	612.0	287.0	---	822.0	392.0
Ground barley	197.5	147.5	147.5	---	---	---
Wheat shorts	---	150.0	---	---	---	---
Wheat bran	150.0	---	150.0	---	---	---
Fat (stabilized)	10.0	10.0	10.0	10.0	10.0	10.0
Soybean meal (48%)	105.0	38.0	75.0	205.0	120.0	158.0
Ground limestone	15.0	15.0	15.0	15.0	15.0	15.0
Calcium phosphate (20% P)	15.0	15.0	15.0	20.0	20.0	20.0
Salt	2.5	2.5	2.5	2.5	2.5	2.5
Vitamin:mineral premix[1]	10.0	10.0	10.0	10.0	10.0	10.0
Methionine	0.5	0.7	0.7	0.8	1.0	0.8
Calculated analyses						
Crude protein (%)	13.9	14.1	13.9	16.2	16.6	16.1
Digestible protein (%)	12.1	12.3	12.0	14.8	14.9	14.7
Crude fat (%)	4.1	3.0	3.6	3.9	2.3	3.2
Crude fiber (%)	4.7	4.1	4.5	2.5	2.8	2.6
Metabolizable energy (kcal/kg)	2774	2790	2779	3059	2800	3030
Calcium (%)	0.95	0.95	0.95	1.04	1.06	1.05
Available phosphorus (%)	0.44	0.44	0.45	0.50	0.53	0.51
Sodium (%)	0.16	0.19	0.18	0.16	0.19	0.18
Methionine (%)	0.28	0.28	0.28	0.37	0.35	0.34
Methionine + cystine (%)	0.48	0.49	0.49	0.61	0.61	0.58
Lysine (%)	0.60	0.58	0.59	0.80	0.80	0.78

[1]Use additional choline if premix does not contain this vitamin

TABLE 3.4. Examples of Pre-laying diets (kg)

Ingredients	1	2	3	4
Ground yellow corn	645.0	---	620.0	---
Ground wheat	---	770.0	---	745.0
Wheat shorts	100.0	50.0	100.0	50.0
Soybean meal (48%)	175.0	100.0	200.0	125.0
Fat (stabilized)	10.0	10.0	10.0	10.0
Ground limestone	45.0	45.0	45.0	45.0
Calcium phosphate (20%P)	12.0	12.0	12.0	12.0
Salt	2.5	2.5	2.5	2.5
Vitamin:mineral premix[1]	10.0	10.0	10.0	10.0
Methionine	0.6	0.8	0.5	0.9
Calculated analyses				
Crude protein (%)	16.0	16.0	17.0	17.0
Digestible protein (%)	14.5	14.5	15.4	15.3
Crude fat (%)	4.2	2.5	4.1	2.5
Crude fiber (%)	2.9	3.0	2.9	3.0
Metabolizable energy (kcal/kg)	2870	2860	2850	2850
Calcium (%)	2.0	2.0	2.0	2.0
Available phosphorus (%)	0.43	0.46	0.44	0.46
Sodium (%)	0.17	0.19	0.17	0.19
Methionine (%)	0.34	0.32	0.34	0.34
Methionine + cystine (%)	0.57	0.55	0.59	0.61
Lysine (%)	0.76	0.75	0.83	0.82

[1]**Use additional choline if premix does not contain this vitamin**

3.2 FEEDING MANAGEMENT OF GROWING PULLETS

a) *General considerations*

It is generally agreed that most Leghorn and brown egg strains have changed over the last 5 - 10 years, and because of this, nutritional management is becoming more critical. In essence, these changes relate to age at maturity although it is questionable that this has changed suddenly in just a few years. In fact, what has been happening is that age of maturity has slowly been decreasing by about 1 d per year. Unfortunately, many producers are just now becoming aware of earlier maturity because their conventional programs are no longer working, and this is especially true for many strains of brown egg pullets. Moving birds to laying cages at 21 - 22 weeks is no longer feasible, and this now invariably results in management problems. Similarly, first egg appearing at 16-18 weeks means that we must critically review our rearing programs. The key to successful nutritional management today is through maximizing body weight of the pullet. Pullets that are on target or slightly above target weight at maturity will inevitably be the best producing birds for the shell-egg market.

The traditional concern with early maturity has been that it results in small egg size. Results from our early studies indicate the somewhat classical effect of early maturity in Leghorns without regard to body weight (Table 3.5).

TABLE 3.5 Pullet maturity and egg characteristics				
Age at housing	Egg Production (%)		Egg size (% large)	
	18- 20 wk	Mean to 35 wk	30 weeks	63 weeks
15 wk	32.0	92	17	44
18 wk	12.0	92	21	65
21 wk	0	91	37	69

There seems little doubt that body weight and/or body composition are the major factors influencing egg size both at maturity and throughout the remainder of the laying period. Summers and Leeson (1983) concluded that body weight is the main factor controlling early egg size (Table 3.6).

TABLE 3.6 Effect of body weight on egg size	
18 week weight (g)	Early egg weight (g)
1100	46.9[a]
1200	48.4[b]
1280	48.8[bc]
1380	49.7[c]

[a-c]means followed by different letters are significantly different

We concluded that although there is evidence to indicate that nutrients such as protein, methionine and linoleic acid can influence egg size throughout the laying cycle, these nutrients have little effect on early egg size. This is probably related to the pullet producing at maximum capacity at least up to the time of peak egg mass.

Although it is fairly well established that body weight is an important criterion for adequate early production, there is still insufficient evidence regarding optimum body structure and composition. Frame size is being discussed, and is now most frequently included in breeder management guides as a form of monitoring. It is known that most (90%) of the frame size is developed early, and so by 12 - 16 weeks of age, the so-called "size" of the pullet is fixed. While this parameter is useful as another monitoring tool and its measurement should be encouraged, we have had little success in affecting frame size without also affecting body weight. It therefore seems very difficult to produce, by nutritional modification, pullets that are below target weight yet above average frame size and vice versa. The relationship between body weight and shank size is further complicated by the fact that environmental temperature also affects bone length independent of nutrition. Table 3.7 shows the shank length of pullets grown at 18 or 30°C and fed high or low energy diets. Birds grown at 30°C had consistently longer shanks than did birds grown at 18°C, even though birds grown in the cool environment were heavier. The same effect was seen using vastly different diet energy levels of 2500 vs 3000 kcal ME/kg.

TABLE 3.7 Body weight, nutrient intake, and shank length of Leghorn pullets grown at 30 or 18°C

	Body weight (g)			Nutrient intake to 98 days		Shank length 98 days	
	28d	56d	98d	Energy (Mcal ME)	Protein (kg)	cm	cm/kg B.Wt.
Dietary ME 2500 kcal/kg							
18°C	240	601	1145[a]	13.6[a]	0.89[a]	9.96[b]	8.70
30°C	234	591	1049[b]	11.1[b]	0.73[b]	10.08[a]	9.61
	NS	NS	**	**	**	**	
Dietary ME 3000 kcal/kg							
18°C	260[a]	650[a]	1173[a]	14.7[a]	0.80[a]	9.94[b]	8.47
30°C	247[b]	559[b]	1002[b]	11.2[b]	0.61[b]	10.07[a]	10.05
	**	**	**	**	**	**	

[a,b]**Means followed by different superscripts are significantly different (P<.01).**

Pullets grown in the warmer environment ate about 25% less feed, as expected, yet exhibited longer shank bones. This reduction in nutrient intake had the expected effect of reducing body weight but frame size was not affected in absolute terms. Therefore, bone growth characteristics were maintained with some 25% reduction in intake of minerals and vitamins normally considered necessary for bone development.

It is not clear why birds held at higher temperatures have longer shank bones, although there is a possibility of altered hormone balance. For example, thyroid hormones are known to influence bone development through mediation of somatomedins and it has been shown that even though birds held at 30 vs. 22°C have reduced thyroid size, their circulating T4 levels are increased by 100%. Another factor that may be of importance is blood flow to the feet and legs during heat stress. It is well known that birds divert more blood to the legs during heat stress as a means of countercurrent cooling between the arterial and venous supply. In some birds, heat loss from the legs can be the largest contributor to overall heat loss, and it is interesting that this has been recorded to occur at 30°C since at higher temperatures evaporative losses become more important. Since the hind limbs are apparently more heavily supplied with blood at 30 vs. 18°C, and even though nutrient intake is reduced at higher temperatures, it is conceivable that the active growth plate receives a greater supply of nutrients related simply to increased blood flow.

An alternative argument is that even with 25% reduction in nutrient intake there is adequate supply of those nutrients necessary for normal bone development, although supplies of protein and/or energy may be inadequate for other tissue synthesis. It would be interesting to see if environmental temperature influences development of other parts of the skeleton and especially the keel. Different shank length standards may be necessary for summer vs winter grown pullets, especially when shank length/unit body weight is a management criterion.

It would appear that early maturing chickens reach sexual maturity at significantly younger ages, but at similar body weights compared to later maturing birds. It seems as though early maturing birds achieve a threshold level of body mass and commence production when the minimum physiological age is reached, while late maturing birds at the same age do not have the body mass required for production. Recent reports have indicated the requirement of a certain lean body mass prior to onset of maturity. With most mammals, attainment of minimum fat reserves are essential for puberty, and so it seems likely that body composition is as important as total body mass in influencing the onset of egg production. In studies involving a relatively small number of birds, we have seen no correlation between age at first egg and either percentage or absolute levels of body fat. While no clear picture has yet emerged with respect to body composition and maturity, it seems likely that birds having some energy reserve as they approach peak egg production are less prone to subsequent problems. Too frequently, a production curve as shown in Fig 3.1. is observed with commercial flocks.

Fig. 3.1 Reduction in egg production after peak, associated with small appetite and body weight.

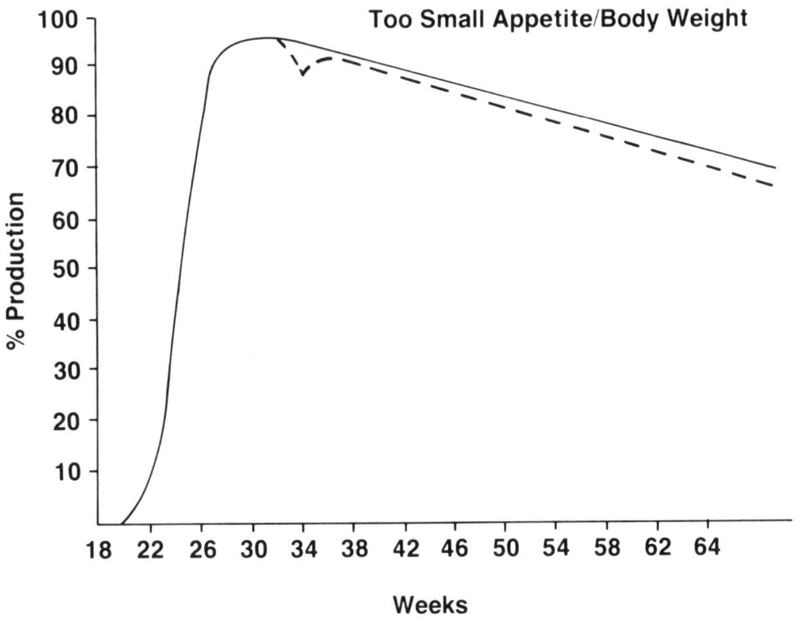

Our experience suggests that if this type of production loss is not due to an identifiable disease and/or management problem, then it most likely relates to birds being deficient in energy. It is perhaps not too surprising that birds are in such a precarious situation with respect to energy balance. Most mammals such as cows and sows must lose body weight during peak lactation in order to meet energy requirements. Perhaps the most classical case of energy deficiency at this time is seen with the turkey breeder. Due to a decline in feed intake from time of first lighting through to peak egg production, the turkey breeder necessarily loses considerable body mass in an attempt to maintain energy balance. It is likely that the same situation applies to both the Leghorn and modern brown egg type pullet. Obviously, the effect is most pronounced for underweight flocks with small appetites where energy intake is minimal. In fact, with many flocks exhibiting production characteristics as shown in Fig 3.1, it is body weight at housing that deserves immediate investigation rather than factors occurring later at the time of the production loss.

The key to solving many of our present industry problems would therefore seem to be attainment of "heavy" pullets at desired age of maturity. In this instance "heavy" refers to the weight and condition which will allow the bird to progress through maturity with optimum energy balance. It is likely that such conditions will be a factor of the flock in question, being influenced by stocking density, environmental temperature, feather cover, etc. Unfortunately attainment of desired weight for age has not always proven easy, especially where earlier maturity is desired or when adverse environmental conditions prevail. Leeson and Summers (1981) suggested that energy intake of the pullet is the limiting factor to growth rate, since regardless of diet specifications, pullets seem to consume similar quantities of energy (Table 3.8).

TABLE 3.8 Energy intake of growing pullets (8 - 15 weeks)

Diet energy-protein	15 wk Body wt (g)	Energy intake (Mcal)	Protein intake (g)
1. 2950 kcal - 14% CP	1272	9.77	464^c
2. 3100 kcal - 24% CP	1267	9.17	718^a
3. 3200 kcal - 20% CP	1291	9.51	597^b

$^{a-c}$**means followed by different letters are significantly different**

All of these birds were of comparable weight even though diet specifications were dramatically variable. As seen in Table 3.8, birds consumed similar quantities of energy, even though protein intake varied by 85%. These data suggest that if adequate protein intake is achieved, additional diet protein does little to stimulate growth rate.

In more recent studies, we have reared Leghorn pullets on diets varying in protein or energy, and again, energy intake seems to be the major factor influencing body weight (Tables 3.9 and 3.10).

TABLE 3.9 Effect of diet protein level (0-20 weeks) on pullet growth and nutrient intake

Diet protein (%)	Body wt 20 wks (g)	Energy intake 0-20 wks (Mcal)	Protein intake 0-20 wks (kg)
15	1445	24.3	1.28[d]
16	1459	22.9	1.28[d]
17	1423	22.9	1.37[cd]
18	1427	22.0	1.39[c]
19	1444	22.9	1.53[b]
20	1480	23.0	1.62[a]

(All diets 2850 kcal ME/kg)

[a-d]means followed by different letters are significantly different

TABLE 3.10 Effect of diet energy level (0-20 weeks) on pullet growth and nutrient intake

Diet energy (kcal ME/kg)	Body wt 20 wks (g)	Energy intake 0-20 wks (Mcal)	Protein intake 0-20 wks (kg)
2650	1320[c]	20.6[c]	1.40[a]
2750	1378[bc]	21.0[bc]	1.37[a]
2850	1422[ab]	21.8[ab]	1.37[a]
2950	1489[a]	22.1[ab]	1.35[ab]
3050	1468[a]	21.4[abc]	1.26[c]
3150	1468[a]	22.5[a]	1.29[bc]

(All diets 18% CP).

[a-c]means followed by different letters are significantly different

These studies indicate that growth rate is more highly correlated with energy intake than with protein intake. This does not mean to say that protein (amino acid) intake is not important to the growing pullet. Protein intake is very important, but there does not seem to be any measurable return from feeding more than 800g of protein to the pullet through 18 weeks of age. On the other hand, it seems as though the more energy consumed by the pullet, the larger the body weight at maturity. Obviously, there must be a fine line between maximizing energy intake and creating an obese pullet.

b) Maximizing nutrient intake

If one calculates expected energy output in terms of egg mass and increase in body weight, and relates this to feed intake, then it becomes readily apparent that the Leghorn must consume at least 90g/bird/day and the brown egg bird close to 100g/bird/day at peak production. With egg-type stock, feeding is to appetite and so our management programs must be geared to stimulating appetite. The practical long term solution is to rear birds with optimum body weight and body reserves as they begin production. This situation has been aggravated in recent years, with the industry trend of attempting to rear pullets on minimal quantities of feed. Unfortunately, this move has coincided with genetically smaller body weights and hence smaller appetites, together with earlier sexual maturity.

In order to maximize nutrient intake, one must consider relatively high nutrient dense diets, although these alone do not always ensure optimum growth. Relatively high protein (16-18% CP) with adequate methionine (2% CP) and lysine (5% CP) levels together with high energy levels (2800-3000 kcal/kg) are usually given to Leghorn pullets, especially in hot weather situations. However there is some evidence to suggest that high energy diets are not always helpful under such warm conditions. This situation may relate to stimulation of nutrient intake when lower energy diets are fed at high temperatures (Table 3.11).

TABLE 3.11 Influence of diet energy on growth and nutrient intake of Leghorn pullets maintained at 30 or 18°C to 18 weeks of age

	Body wt 126d (g)	Total feed intake (kg)	ME intake (Mcal)	Protein intake (g)
Temperature 18°C				
1. 2500 kcal ME/kg	1398	7.99	20.04	1330
2. 3000 kcal ME/kg	1434	6.98	21.07	1160
Temperature 30°C				
1. 2500 kcal ME/kg	1266	6.05	15.17	1010
2. 3000 kcal ME/kg	1218	5.19	15.69	870

In this recent study, Leghorn pullets were heavier at 126 d when fed the high energy diet in the cool environment, but diet had no effect at 30°C. As expected, pullets ate less of the high energy diet, and because all other nutrient levels were fixed, this results in reduced intake of all nutrients except energy. Pullets therefore ate less protein and amino acids when fed 3000 vs 2500 kcal ME/kg, and this can be critical where intake *per se* is less at 30°C. The pullets fed 3000 kcal/kg are borderline in intake of balanced protein at 870 g vs our requirement for 800 g to this age. High energy diets may therefore not always be beneficial under heat stress conditions, and intake of other nutrients such as protein and amino acids must be given priority during formulation. The Leghorn pullet eats for energy requirement, albeit with some imprecision, and so energy:protein balance is critical. All too often we see inadequate amino acid intake when high energy corn-based diets are used, the result of which is pullets that are both small and fat at maturity.

One of the most important concepts today in pullet feeding, is to offer diets according to body weight and condition of the flock, rather than according to age. For example, traditional systems involve feeding starter diets for about 6 weeks followed by grower and then perhaps developer diets. This approach does not take into account individual flock variation, and today this can be most damaging to underweight flocks. It is becoming more difficult to attain early weight for age. This means that flocks are often underweight at 4 - 6 weeks of age. This can be for a variety of reasons such as sub-optimal nutrition, heat stress, disease, etc. The worse thing that can happen to these flocks is an arbitrary introduction of a grower diet, merely because the flock has reached some set age. Today we must feed the higher nutrient dense starter until the target weight is reached. For example, Fig 3.2 shows an underweight flock at 6 weeks.

Fig. 3.2 Pullet growth in relation to feeding program.

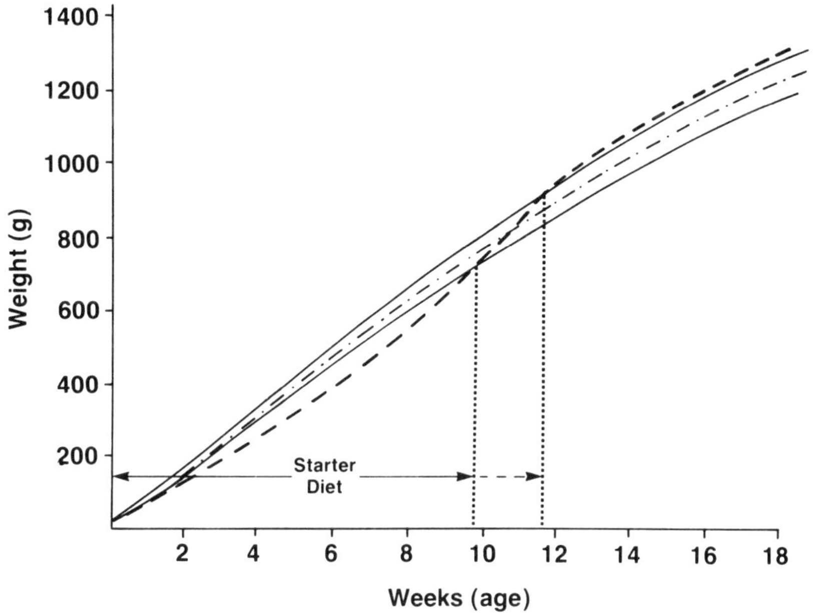

To change this flock to a grower at 6 weeks of age will cause problems because the flock will likely stay small until maturity, then be late maturing and produce a sub-optimal number of eggs, that will also be small. This type of flock can most effectively be "corrected" by prolonged feeding of the starter diet. In this situation, the birds reach the low end of the guide weight at almost 10 weeks of age (Fig 3.2). At this time, a grower diet could be introduced. Since the flock is showing a growth spurt, then feeding to almost 12 weeks could be economical - we now have a flock that is "heavy". We have therefore converted the flock from being underweight and a potential problem, to one that is slightly over weight and so ideally suited to realizing maximum genetic potential during peak production. Some producers, and especially contract pullet growers, are sometimes reluctant to accept this type of program, since they correctly argue that feeding a high protein diet for 10 - 12 weeks will be more expensive. Depending upon local economic conditions, feeding an 18% protein starter diet for 12 vs 6 weeks of age, will cost the equivalent of 2 eggs. A bird in ideal condition at maturity will produce far in excess of these 2 eggs relative to a small underweight bird at maturity.

c) *Suggested feeding program*

The following schedule is recommended for growing pullets to maturity:

Starter 18-19%CP;2750-2900 kcal ME/kg
Day old _______________ Target body weight

Grower 15-16%CP;2750-2900 kcal ME/kg
Target wt ______________ Mature body size

Prelay or layer 16-18%CP;2750-2900 kcal ME/kg
Mature body size _______ 1st egg

As previously indicated, we are not making recommendations regarding age or even dictating the body weight at which diet changes should occur. Rather the recommendations dictate the need for flexibility, and the treatment of each flock as an individual case. For example, the starter diet is to be used until target weight for age is achieved. Hopefully this will be at around 450 g when the Leghorn bird is 6-8 weeks of age. However, each flock will be subjected to varying environmental conditions, and so this may vary. The time to change to a lower-nutrient dense diet is when a desired weight-for-age is achieved, which we suggest is a weight that will be towards the top side of the breeder's growth curve. Changing at a specific weight, or a specific age in isolation can lead to disastrously underweight flocks.

In our recommendations, we suggest the lower-nutrient dense grower diets to be fed from this target weight-for-age up until the desired mature body size is achieved. Again, we are not dictating a specific mature body weight, since this may be varied at the desire of the pullet grower (see following section). Prelay diets should only be used in an attempt at conditioning the calcium metabolism of the bird (see following section) and not as a means of initiating catch-up growth. Such growth spurts rarely occur at this age, and as such, pre-lay diets are being used as a "crutch" for poor rearing management.

An argument that is often heard about the role of body weight at maturity, is that it is not, in fact, too important, because the pullet will show catch-up growth prior to first egg ie: if the pullet is small, it will take a few days longer to mature, and start production at the "same weight". This does not seem to happen, as small birds at 18 weeks are smaller at first egg (Table 3.12).

TABLE 3.12 Effect of immature body weight on development to sexual maturity

Body Weight (g)			Age at first egg (d)	Weight of first egg (g)
18 wks	1st egg	Change		
1100	1360	+260	153	40.7
1200	1440	+240	150	42.0
1280	1500	+220	149	43.7
1380	1590	+210	148	42.5

These data suggest that the smaller pullet does show some compensatory growth to the time of the first egg, although this is insufficient to allow for total "catch-up" growth. It is also interesting to note the relationship between body weight and age at first egg, and also between body weight and size of first egg. In other studies we have followed up on the growth of the pullet through a production cycle in relation to 18 week (immature) body weight. Again, there is a remarkably similar pattern of growth for all weight groups indicating that immature weight seems to "set" the weight of the bird throughout lay (Fig 3.3).

Fig. 3.3 Effect of immature body weight on subsequent body weight during lay.

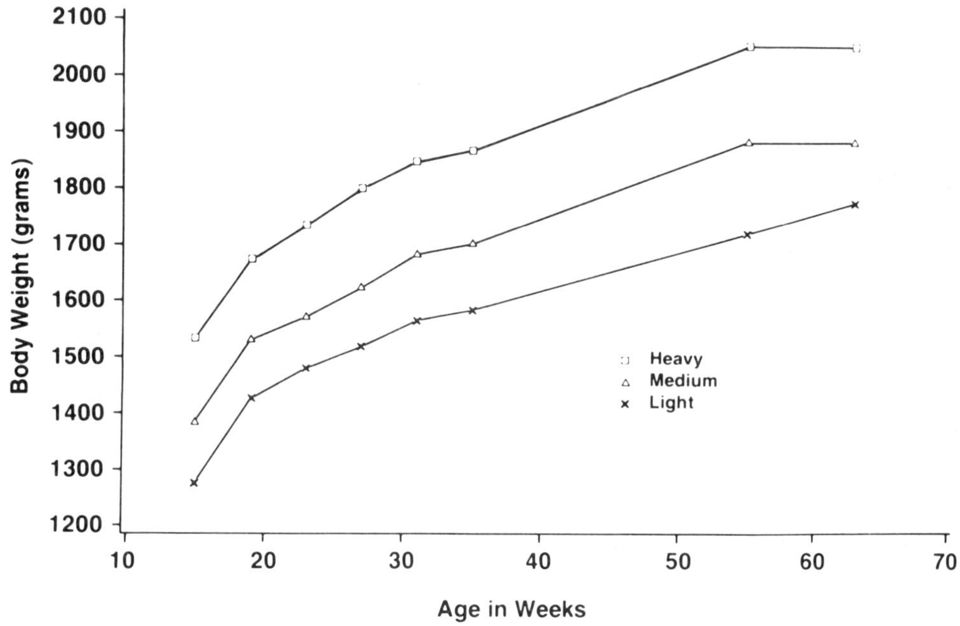

Most importantly from a production viewpoint, is the performance of birds shown in Fig 3.3 When the lightest weight birds were fed diets of very high nutrient density (20% CP, 3000 kcal ME/kg) they failed to match egg production and egg size of the largest weight pullets that were fed very low nutrient dense diets (14% CP, 2600 kcal ME/kg). These results emphasize the importance of body weight in attaining maximum egg mass output.

The actual body weights to be achieved during rearing will obviously vary with breed and strain. Most Leghorn strains should weigh around 400g, 900g and 1300g at 6, 12 and 18 weeks respectively. Similarly, the brown egg birds should weigh around 500g, 1000g and 1500 g at these ages. As a rule of thumb, these weights for age can be used as guidelines for anticipated diet change.

Discussion to date has focused on the role of body weight and appetite of the growing pullet. While rearing programs such as reverse-protein (Leeson and Summers, 1979) have application where delay in maturity is required (usually due to inadequate light control) nutritional management programs today must allow for maximum early growth so as to attain breeder's recommended weight goals as soon as possible. This type of nutritional management obviously entails accurate monitoring of body weight, a task that has too often been neglected with Leghorn pullets.

d) Manipulation of mature body size

In the preceding discussion, we have outlined the importance of maximizing body weight at sexual maturity, and the reasons for this have been explained. Under certain conditions, it is realized that some tempering of mature body size may be economically advantageous. Because body size has a dramatic effect on egg size, large birds at maturity can be expected to produce large eggs throughout their laying cycle. Depending upon the pricing of various egg grades, a very large egg may be uneconomical to produce, and in most instances tempering of egg size of birds at 40-65 weeks of age is often difficult to do without some accompanying loss in egg numbers. Because body weight controls feed intake and egg size, an easier way of manipulating life-cycle egg size is through the manipulation of mature body size. If the maximum possible egg size is desired, then efforts must be made to realize the largest possible mature weight. However, where a smaller overall egg size is economical, then a smaller pullet is desirable. Such light weight pullets can be achieved by growing birds slower through the growth cycle, or more economically by light-stimulating pullets at an earlier age.

Fig 3.4 gives a schematic representation of the above concept. In this scenario, birds are on the heavy side of the breeder's weight guide, and so if moved at 18 weeks, would be heavier than the ideal weight and be expected to produce very large eggs. If this is not economical in the laying house, then these birds should be moved at the "ideal weight" which in this scenario means moving at 17 rather than 18 weeks of age. Moving the bird, and light stimulating at 17 vs 18 weeks will have no adverse effect on performance, as light stimulation is still at the desired weight goal (that has been achieved one week earlier than anticipated).

Fig. 3.4 Early move and light stimulation.

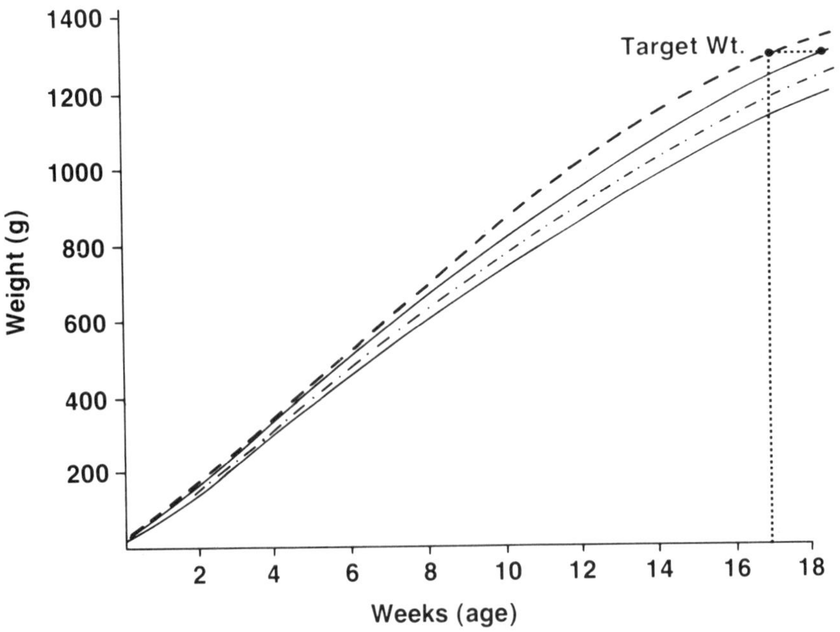

Remember that "early maturity" is not a problem for birds that have ideal body weight and condition. Early maturity and light stimulation will only result in subsequent small egg size and increased incidence of prolapse, if the bird is small at this age. This concept is preferred over attempts at trying to slow the bird down during growth, in order to temper mature body size (Fig 3.5).

Fig. 3.5 Potentially harmful adjustment.

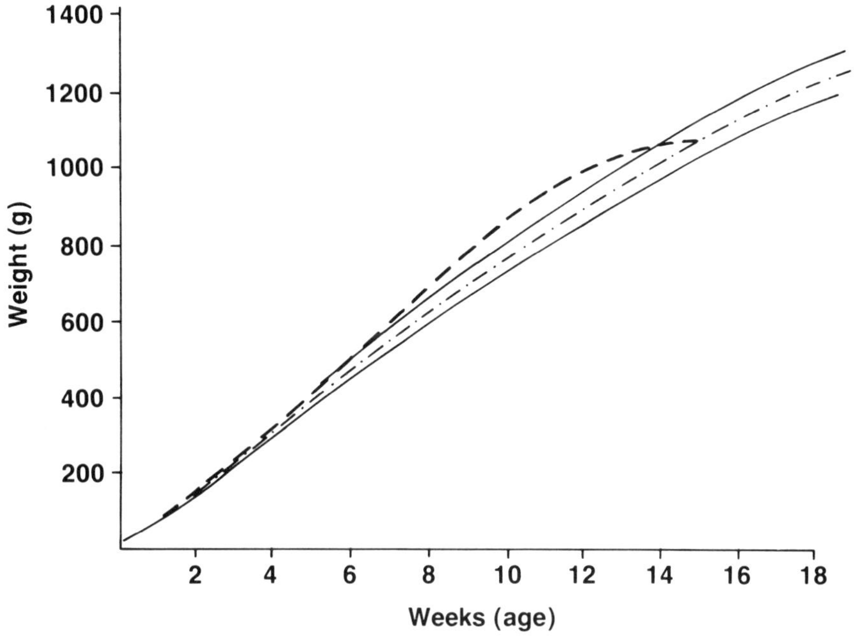

Such adjustments are invariably brought about by use of very low nutrient dense diets and/or use of restricted feeding. Both these practices have the desired "effect" of slowing down mean growth, but at the great cost of loss of pullet uniformity. Loss of uniformity means loss in control of feeding for egg production (see Chapter 4).

e) *Pre-lay nutrition*

Prelay diets are often used to try and manipulate body size or to bring about a transitional change in the birds calcium metabolism prior to maturity.

(i) Prelay calcium

There is still considerable confusion and variation practiced in the levels of calcium given to birds prior to egg production. During the laying cycle, the bird utilizes its medullary bone reserves, in the long bones of the leg, to augment its diet supply when a shell is being formed.

Because egg production is an "all or none" event, the production of the first egg obviously places a major strain on the bird's metabolism, when it has to contend with a sudden 2 g loss of calcium from the body. Some of this calcium will come from the medullary bone, and so the concept has arisen of building up this bone reserve prior to first egg. This obviously means higher levels of calcium in pre-lay diets. There are basically three options for calcium feeding around the time of maturity:

- Use of 1% calcium grower diets until around 5% egg production

As previously mentioned, the largest weight pullets in a flock will likely mature earlier, and so it is these birds that may be disadvantaged with inadequate levels of calcium at this time. If such birds receive a 1% calcium grower diet at the time they are producing their first few eggs, they will only have a sufficient calcium reserve to produce 2 - 3 eggs. At this time, they will likely stop laying, or less frequently continue to lay and exhibit cage layer fatigue. If these earlier maturing birds stop laying, they do so for 4 - 5 days, and then try to start the process again. The bird goes through very short clutches, when at this time she is capable of a very prolonged 30 - 40 egg first clutch. Advocates of prolonged feeding of grower diets suggest that it makes the bird more efficient in the utilization or absorption of calcium, such that when she is eventually changed to a layer diet, improved efficiency continues for some time, and so the bird has large quantities of calcium available for shell synthesis. Fig 3.6 indicates that percentage calcium absorption from the diet does decline with an increased level of calcium in the diet.

Fig. 3.6 Relationship between calcium intake and calcium retention.

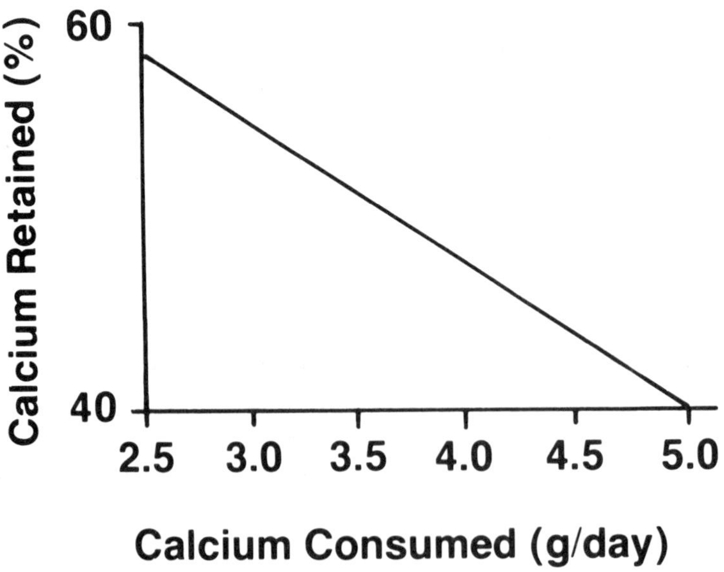

However, there is no scientific evidence to suggest that efficiency of utilization is affected, and in fact calculations from Fig 3.6 indicate that as the calcium level in the diet is increased, calcium retention increases even though percentage retention has declined. If 1% calcium grower diets are used as pullets mature, these diets should not be used after appearance of first egg, and to 0.5% production at the very latest. It must be remembered that under commercial conditions, it is very difficult to precisely schedule diet changes, and so decisions for diet change need to precede actual time of diet change, such that production does not reach 5 - 10% before birds physically receive the calcium enriched diets.

- Use of 2% calcium pre-lay diets

Specialized pre-lay diets are a compromise, in that they provide more calcium than do most grower diets, but still not enough for sustained production. The concept of using so-called pre-lay diets is to build up the medullary reserves without adversely influencing general mineral metabolism. However, as previously discussed with grower diets, 2% calcium pre-lay diets are inadequate for sustained egg production, and should not be fed past 1% egg production. The main disadvantage of prelay diets is that they are used for a short period of time, and many producers do not want the bother of handling an extra diet at the layer farm. There is also a reluctance by some producers with multi-age flocks at one site to use pre-lay diets, where delivery of diets with 2% calcium to the wrong flock on site can have disastrous effects on production.

- Early introduction of 3.5 - 4.0% calcium layer diets

In terms of calcium metabolism, the most effective program is early introduction of the layer diet. Such high calcium diets allow sustained production of even the earliest maturing birds. As previously mentioned, higher calcium diets fed to immature birds, lead to reduced percentage retention, although absolute retention is slightly increased (Table 3.13).

TABLE 3.13 Effect of % diet calcium fed to birds immediately prior to lay, on calcium retention

Diet Ca (%)	Daily Ca retention (g)	Excreta Ca (% dry matter)
0.9	0.35	1.4
1.5	0.41	3.0
2.0	0.32	5.7
2.5	0.43	5.9
3.0	0.41	7.5
3.5	0.51	7.7

Feeding layer diets containing 3.5% calcium, prior to first egg, therefore results in a slight increase in calcium retention of about 0.16 g/d relative to birds fed 0.9% calcium grower diets at this time. Over a 10 d period, however, this increased accumulation is equivalent to the output in 1 egg.

Early introduction of layer diets is therefore beneficial in terms of optimizing the calcium balance of the bird. However there has been some criticism levelled at this practice. There is the argument that feeding excess calcium prior to lay imposes undue stress on the bird's kidneys, since this calcium is in excess of the immediate requirement and must be excreted. In the study detailed in Table 3.12, we do show increased excreta calcium. However kidney histology from these birds throughout early lay revealed no changes due to pre-lay calcium feeding. Recent evidence suggests that pullets must be fed a layer diet from as early as 6 - 8 weeks of age before any adverse effect on kidney structure is seen (see following section on urolithiasis). It seems likely that the high levels of excreta calcium shown in Table 3.12 reflect fecal calcium, suggesting that all excess calcium may not even be absorbed into the body, merely passing through the bird with the undigested feed. This is perhaps too simplistic a view, since there is other evidence to suggest that excess calcium may be absorbed by the immature bird at this time. Such evidence is seen in the increased water intake and excreta water content of birds fed layer diets prior to maturity (Figs 3.7 and 3.8).

Fig. 3.7 Effect of introducing a 4% calcium layer diet at 112 days (——) and at 144 days (- - - -) on daily water intake.

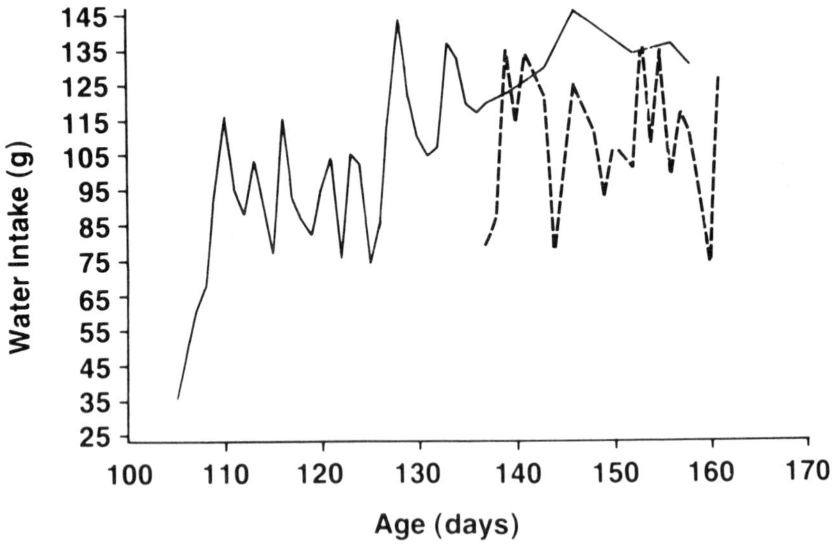

Fig 3.8 Effect of introducing a 4% calcium layer diet at 112 days (–) and at 144 days (- - - -) on daily excreta moisture.

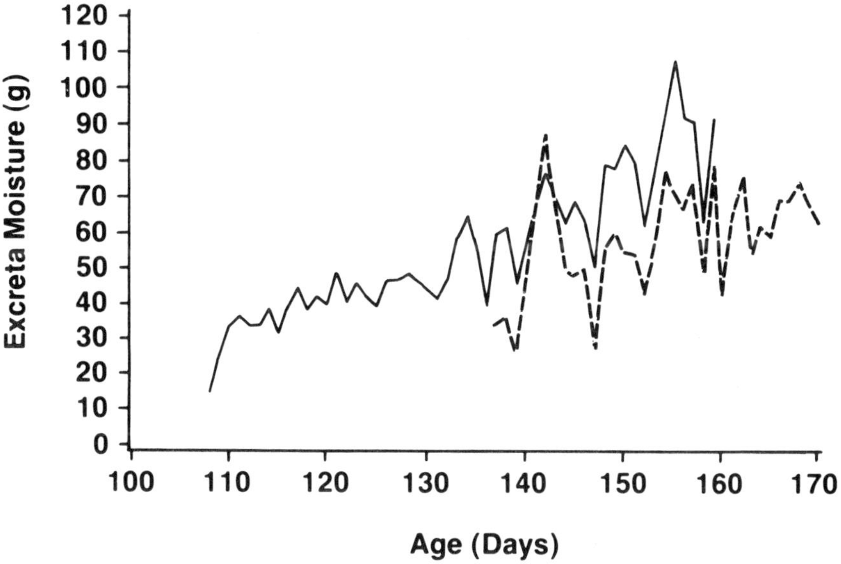

Early introduction of a layer diet seems to result in increased water intake, and a resultant increase in excreta moisture. Unfortunately this increased water intake and wetter manure seems to persist throughout the laying cycle of the bird, (Table 3.14).

TABLE 3.14 Effect of prelay calcium level on excreta moisture (%)

Prelay diet Ca	Bird age (days)			
	147	175	196	245
1%	71.4	78.7	75.3	65.5
2%	71.6	77.2	73.9	63.9
3%	72.1	77.7	74.1	63.9
4%	77.0	80.0	76.0	69.4

These data suggest that birds fed high calcium layer diets during the pre-lay period will produce manure that contains 4 - 5% more moisture than birds fed 1% calcium grower or 2% calcium prelay diets. There are reports of this problem being most pronounced under heat stress conditions. A 4 - 5% increase in manure moisture may not be problematic under some conditions, although for those farms with a chronic history of wet layer manure, this effect may be enough to tip the balance and produce a problem.

In summary, the calcium metabolism of the earliest maturing birds in a flock should be the criterion for selection of calcium levels during the pre-lay period. Prolonged feeding of low-calcium diets is not recommended. Early introduction of layer diets is ideal, although where wet manure may be a problem, a 2% calcium pre-lay diet is recommended. There seems to be no problem with the use of 2% calcium pre-lay diets, as long as birds are consuming a high calcium layer diet not later than 1% egg production.

ii) Pre-lay Body Weight and composition

Prelay diets are often formulated, and used, on the assumption that they will improve body weight and/or composition, and so correct problems arising with the previous growing program. Body weight and body condition should not really be considered in isolation, although at this time we do not have a good method of readily assessing body condition in the live pullet. For this reason our main emphasis at this time is directed towards body weight.

The most important criterion used during rearing is pullet body weight as described previously. Each strain of bird has a characteristic mature body weight that must be reached or surpassed for adequate egg production and egg mass output. In general, prelay diets should not be used in an attempt to manipulate mature body size. The reason for this is that for most flocks it is too late at this stage of rearing to meaningfully influence body weight - all too often pre-lay diets are used as a crutch for poor rearing management.

However, if underweight birds are necessarily moved to a layer house, then there is perhaps a need to manipulate body weight prior to maturity. With black-out housing, this can sometimes be achieved by delaying photostimulation - this option is becoming less useful in that Leghorns and brown egg strains seem now to be maturing early without any light stimulation. If pre-lay diets are then necessarily used in an attempt to correct rearing mismanagement, then it seems as though the bird is most responsive to energy. This fact likely fits in with the effect of estrogen on fat metabolism, and the significance of fat used for liver and ovary development at this time. While such high nutrient density pre-lay diets may be useful in manipulating body weight, it must be remembered that this late growth spurt (if it occurs) will not be accompanied by any meaningful change in skeletal growth. This means that in extreme cases, where birds are very small in weight and stature at approximately 16 - 18 weeks of age, then the end result of using high-nutrient dense prelay diets may well be pullets of correct body weight, but of small stature. These short shank length pullets seem more prone to prolapse/pick-out, and so this is another example of the limitations in use of classical pre-lay diets.

While body composition at maturity may well be as important as body weight at this age, it is obviously a parameter that is difficult to quantitate. There is no doubt that energy is likely the limiting nutrient for egg production for all strains of bird, and around peak production, feed may not be the sole source of energy. Labile fat reserves at this time, are therefore essential to augment feed sources that are inherently limited by low feed intake. These labile fat reserves become critical during situations of heat stress or general hot weather conditions. Once the bird starts to produce eggs, then its ability to deposit fat reserves is greatly limited. Obviously if labile fat reserves are to be of significance, then they must be deposited prior to maturity. As with most classes of bird, the fat content of the pullet can best be manipulated through changing the energy:protein balance of the diet. If labile fat reserves are thought necessary, then high energy, high fat pre-lay diets should be considered. As previously stated, this scenario could well be beneficial if peak production is to coincide with periods of high environmental temperature.

The requirements for a specific body composition at the onset of maturity have not been adequately established. With mammals, onset and function of normal estrus activity is dependent on attainment of a certain body fat content. In humans for example, onset of puberty will not occur if body fat content is less than around 14%. No such clear cut relationship has emerged with egg layers. Work conducted with broiler breeders, in fact indicate a more definite relationship between lean body mass and maturity, rather than fat content and maturity.

iii) Subsequent egg weight and egg composition

It seems as though egg size is ultimately controlled by the size of the yolk that enters the oviduct. In large part this is influenced by body weight of the bird, and so factors described previously for body weight can also be applied to concerns with egg size. There is a general need for as large an early egg size as is possible, especially with breeder hens. Most attempts at manipulating early egg size have met with limited success. Increased levels of linoleic acid in prelay diets may be of some use, although levels in excess of the usual 1% found in most diets produce only marginal effects on early egg size. From a nutritional standpoint, egg size can best be manipulated with diet protein, and especially methionine concentration. It is logical, therefore to consider increasing the methionine levels in prelay diets.

For breeder hens, one also has to consider egg composition in relation to successful embryo development. It is well known that hatchability of eggs from young breeders is lower. The reasoning for this suboptimal hatch seems to relate to "maturity" of embryonic membranes, and movement of nutrients from the yolk and albumen to the embryo. However, part of this problem may also relate to inadequate passage of certain nutrients from the breeder hen into the egg. For example, it is known that young breeders do not deposit normal quantities of biotin into the egg - normal biotin concentration in the egg is apparently not achieved until production of the 8 - 10th egg. If this is a general effect with a number of key nutrients, then it would seem worthwhile to study the effect of prelay nutrient intake on egg composition in relation to embryonic needs.

iv) Pre-pause

In recent years there has been interest in some countries of so called pre-pause feeding programs. The idea behind these programs is to withdraw feed, or feed a very low nutrient dense diet at time of sexual maturity. This somewhat unorthodox program is designed to "pause" the normal maturation procedure, and at the same time to stimulate greater egg size when production resumes after about 10 - 14 days. This type of prelay program is therefore most beneficial where early small egg size is undesirable.

Pre-pause can be induced by simply withdrawing feed, usually at around 1% egg production. Under these conditions pullets immediately lose weight, and fail to realize normal weight-for-age when refed. Egg production and feed intake normalize after about 22 weeks, although there is 1 - 1.5g increase in egg size. Fig. 3.9 shows the production response of Leghorn pullets fed only wheat bran from 18 weeks (1% egg production) through to 20 weeks of age. This data is presented on an equalised physiological basis, rather than equal age basis.

Fig. 3.9 Early egg production of pullets fed wheat bran at 1% egg production or at 18 weeks of age.

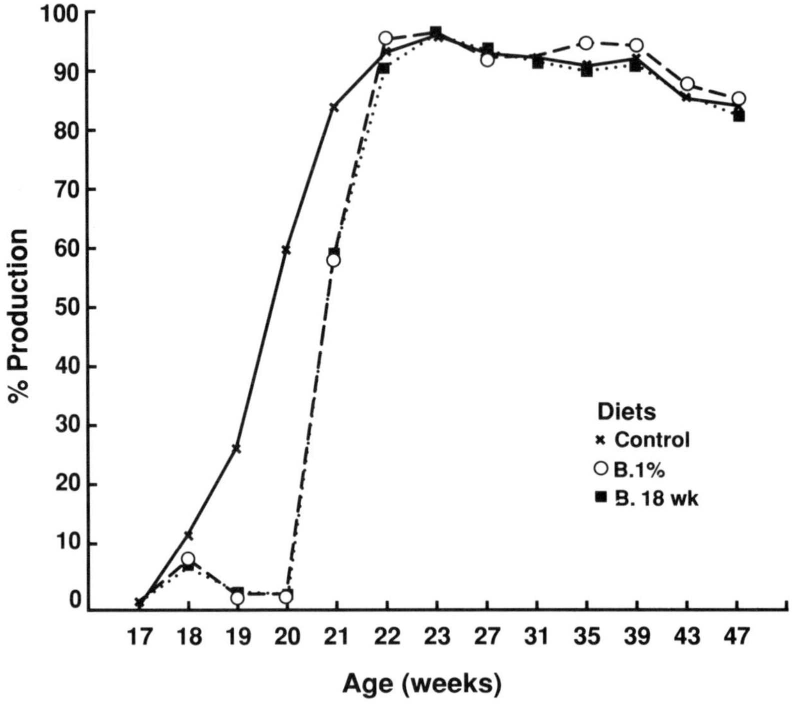

The most noticeable effects of a pre-pause diet such as wheat-bran, are very rapid attainment of peak egg production and an increase in egg size once refeeding commences. These effects (Fig. 3.9) are undoubtedly due to increased feed intake. This management system could therefore be used to better synchronize onset of production (due to variance in body weight), to improve early egg size or to delay production for various management related decisions. The use of such pre-pause management will undoubtedly be affected by local economic considerations.

v) Urolithiasis

Kidney dysfunction often leads to problems such as urolithiasis, and this most commonly occurs during the late growing or early egg production phase of the pullet. While infectious bronchitis can be a confounding factor, urolithiasis is most often induced by diet mineral imbalance in the late growing period. At post-mortem, often one kidney is found to be enlarged and contain mineral deposits known as uroliths. Some outbreaks are correlated with a large increase in diet calcium and protein in layer vs grower diets, coupled with the stress of physically moving pullets at this time, and being subjected to a change in the watering system (usually onto nipples in the laying cages). The uroliths are most often composed of calcium-sodium-urate.

The occurrence is always more severe when growing pullets are fed high calcium diets for an extended period prior to maturity. For example, urolithiasis causing 0.5% weekly mortality, often occurs under experimental conditions when pullets are fed layer diets after 10 - 12 weeks of age (relative to maturity at 18 - 19 weeks). However there is no indication that early introduction of a layer diet for just 2 - 3 weeks prior to maturity is a causative factor.

Because diet electrolytes can influence water balance and renal function, it is often assumed that an electrolyte excess or deficiency may be predisposing factors in urolithiasis or gout. Because salts of uric acid are very insoluble, then the excretion of precipitated urate salts could serve as a water conversation mechanism, especially when cations are excreted during salt loading or when water is in short supply. When roosters are given saline water (1% NaCl) and fed high-protein diets, uric acid excretion rates are doubled relative to birds offered the high protein diet along with non-saline drinking water. Because uric acid colloids are negatively charged, they attract cations such as Na, and so when these are in excess, there is an increased excretion via urates, presumably at the expense of conventional NH_4 compounds. There is some evidence of an imbalance of Na+K-Cl levels influencing kidney function. When excess Na+K relative to Cl is fed, a small percentage of the birds develop urolithiasis. It is likely that such birds are excreting a more alkaline urine, a condition which encourages mineral precipitation and urate formation.

Urolithiasis therefore seems to be most problematic in laying hens fed high levels of calcium well in advance of sexual maturity. Although the situation is often confounded with IBV infection, it seems obvious that no more than 1% calcium should be fed to Leghorn birds prior to maturity. Feeding pre-lay (2% Ca) or layer diets containing 3% calcium for 2 - 3 wk prior to first egg is not problematic, and surprisingly, uroliths rarely form in adult male breeders fed high calcium diets. High levels of crude protein will increase plasma uric acid levels, and potentially provide conditions conducive to urate formation. Certainly numerous mycotoxins influence kidney function, and so general mill management regarding quality control and/or use of feed additives to suppress their harmful effects would likely be beneficial.

In humans at least, urolith formation can be controlled by adding urine acidifiers to the diet. Studies with pullets show similar advantages. Adding 1% NH_4Cl to the diet results in a more acidified urine, and uroliths rarely form under these conditions. Unfortunately this treatment results in increased water intake, and associated wet manure. One of the potential problems in using NH_4Cl in laying hens, is that it induces a metabolic acidosis and this is detrimental to eggshell quality especially under conditions of heat stress. Such treatment also assumes the kidney can clear the increased load of H^+, and for a damaged kidney, this may not always occur. As a potential urine acidifier without such undesirable side effects, several researchers have studied the role of Alimet® a methionine analogue. From 5 - 17 wk of age, pullets were fed diets containing 1 or 3% calcium in combination with 0, 0.3 or 0.6% Alimet®. Birds fed the untreated high calcium diet excreted alkaline urine containing elevated calcium concentrations together with urolith formation and some kidney damage. Feeding 0.6% Alimet® acidified the urine, but did not cause a general metabolic acidosis. Alimet® therefore reduced kidney damage and urolith formation without causing acidosis or increased water consumption.

It is concluded that urine acidification can be used as a prevention or treatment of urolithiasis, and that this can be accommodated without necessarily inducing a generalized metabolic acidosis. From a nutritional viewpoint, kidney dysfunction can be minimized by not oversupplying nutrients such as calcium, crude protein and electrolytes for too long a period prior to maturity.

f) Lighting programs

Light can have a dramatic influence on the growth and body composition of the growing pullet, and so light programs must be taken into account in the selection of feeding programs. In terms of pullet management, daylength has two major effects, namely the development of reproductive organs and secondly a change in feed intake. It is well known that birds reared on a step-up or naturally increasing daylength will mature earlier than those reared on a constant daylength. Similarly if birds are subjected to a step-down daylength much after 12 weeks of age, they will likely exhibit delayed sexual maturity. The longer the photoperiod, the longer the time that birds have to eat feed, and so usually this results in heavier birds. Figs 3.10 and 3.11 show the growth rate and feed intake of birds reared on constant 8 vs 14 h of light to 19 weeks of age.

Pullets reared on a constant 14h daylength were about 60g heavier at 19 weeks, essentially due to increased feed intake. Longer photoperiods may therefore be beneficial in hot weather situations where a depressed feed intake in pullets is often problematic. As the daylength for the growing pullet is increased, we can expect a reduction in age at maturity. Research data suggests earlier maturity with constant rearing daylengths up to 16 - 18 hours per day, although longer daylengths such as 20 - 22 hours per day seem to delay maturity. Another potential problem with longer daylength during rearing, is that it allows less potential for light stimulation when birds are moved to laying cages. However in parts of the world around equatorial regions where maximum daylength fluctuates between 11 - 13 hours, many birds are managed without any light stimulation. In fact, under such hot weather high light intensity conditions, excessive stimulation often results in prolapse and blowouts. In these situations if light stimulation is given, it should follow rather than lead the onset of egg production. It seems that for modern strains of birds, light stimulation at "maturity" is not always necessary for adequate layer performance. In a recent trial, we have shown some advantages to constant 14h photoperiods "day-old to death" vs an 8 h rearing photoperiod followed by 14 h layer photoperiod (Table 3.15).

TABLE 3.15 Effect of rearing daylength on subsequent layer performance

Photoperiod				
Rearing	Laying	336d egg prod.	Egg weight (g)	Shell deformation (μm)
8h	14h	271[a]	58.4[b]	26.5[a]
14h	14h	256[b]	60.3[a]	25.4[b]

Pullets that were grown on constant 14h light and not given any extra daylength at maturity produced fewer eggs, with this difference due mainly to reduced peak production. However, this flatter peak was associated with a significant increase in egg size, and a dramatic improvement in shell quality (lower eggshell deformation). The reason for improved shell quality is not clear, although, we have seen this with other flocks that fail to show adequate sustained peaks - maybe giving up a few eggs at peak is a means of improving shell quality. The increased egg size for birds on the constant 14h photoperiod is undoubtedly due to birds being heavier at maturity, thus eating more feed throughout the laying period.

Fig. 3.10 Effect of photoperiod duration on body weight of pullets.
--- 14-hr photoperiod, —— 8-hr photoperiod.

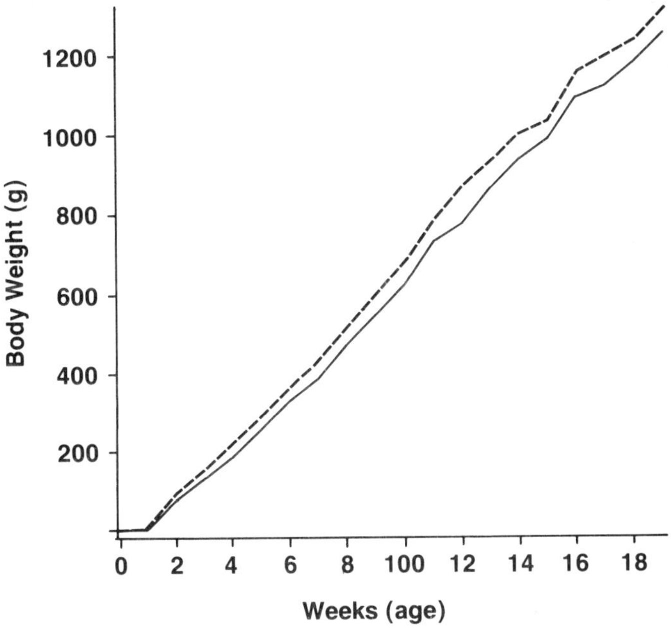

Fig. 3.11 Effect of photoperiod duration on feed intake of pullets.
--- 14-hr photoperiod, —— 8-hr photoperiod.

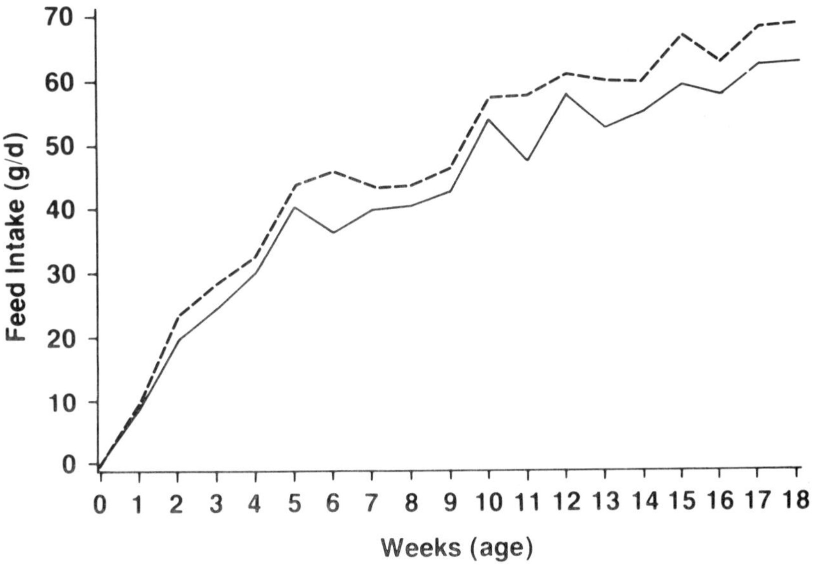

If birds are light stimulated prior to first egg, then age at light stimulation will have an effect on age at first egg. Our data suggest that between 98-140 d of age, for each 1 d decrease in age at light stimulation, first egg will occur about 0.5 d earlier (Fig 3.12). This means that light stimulating a pullet at 120 d rather than 140 d, will likely result in earlier maturity by about 10 days. At this time, it is important to re-emphasize the previous discussion concerning adequacy of body weight and body condition before considering earlier light stimulation. Another program that can be used to stimulate growth is "step-down" lighting (Fig 3.13).

Fig. 3.12 Age at light stimulation (8-14 hr) and sexual maturity.

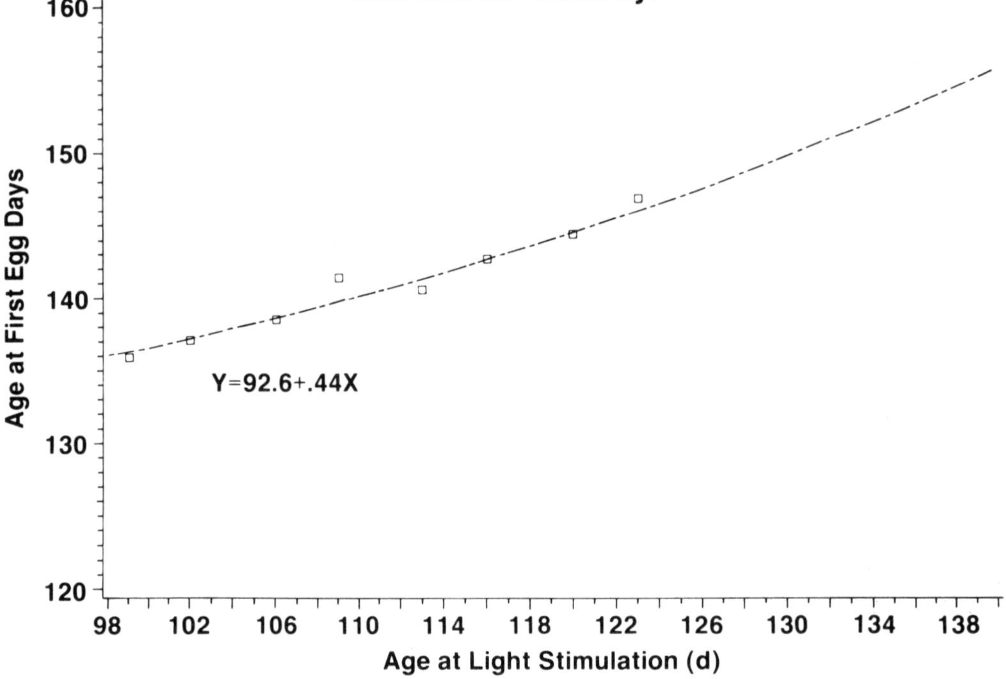

In Fig 3.13, birds are exposed to a 23 hr day for the first week of growth, then daylength is reduced by about 1 hour each week until 10 hr per day is achieved, at which time it is held constant. When birds are in open sided houses, then obviously the 10 hr constant photoperiod from 10 - 18 weeks as shown in Fig 3.13 can be modified to suit the maximum natural daylength expected over this period. Birds can then be photostimulated at 18 weeks (or desired mature body weight), or perhaps maintained on constant 13 - 14 hr as discussed previously. The step-down program has the advantage of allowing the young pullets to eat feed for considerably longer times each day during their early development. In hot weather conditions, this long daylength means that birds are able to eat more feed during cooler parts of the day. The system should not be confused with older step-down lighting programs that continued step-down until 18 - 20 weeks; these older programs were designed to delay maturity. With this newer program, maturity will not be affected as long as the step-down regime is stopped by 10 - 12 weeks of age, *ie:* before the pullet becomes very sensitive to changes in daylength. The step-down lighting program is one of the simplest ways of increasing growth rate in pullets, and is practical with both controlled environment and open sided buildings.

Fig. 3.13 Step-down lighting.

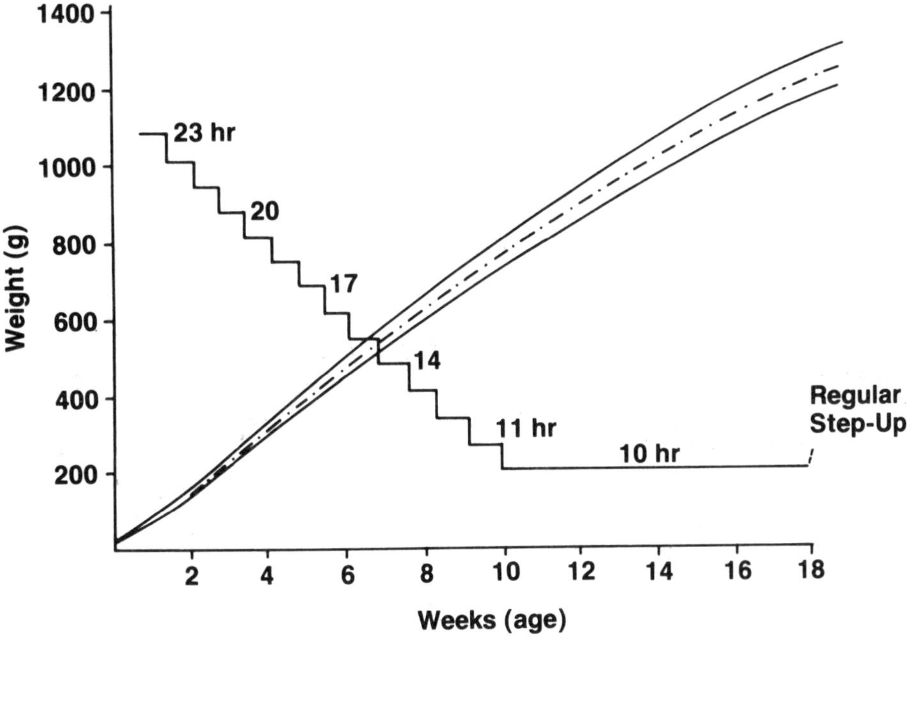

g) Brown-egg pullets

There is very little information available on specific nutrient needs of brown egg pullets, and whether or not they need diets any different to those used for white egg birds. It is generally assumed that white and brown egg pullets are similar in their nutrient needs relative to body weight. Brown egg pullets are usually heavier than white egg birds, although this difference seems to be decreasing over the last few years. For example, in the past it has been fairly common practice to start physical feed restriction after 10 - 12 weeks of age, in order to control growth rate. Today, with many strains of pullet, this feed restriction is unnecessary, and in fact may be detrimental in hot weather conditions. The principle of feeding management of brown egg pullets is essentially the same as described for the slightly smaller white egg bird as outlined in the previous sections. Achieving target weight-for-age must be the major criterion of the growing program, because this ensures the best chance of realizing the bird's genetic potential as a layer.

If physical feed restriction is necessary, due to birds becoming overweight, then bird uniformity becomes a major concern. With a mild restriction program, birds can be allowed to "run-out" of feed one day per week and, usually this will do little harm to uniformity. If it is necessary to impose a greater degree of feed restriction, on a daily basis, then it is important to ensure rapid and even feed distribution, much as subsequently discussed for broiler breeders (Chapter 6). Feed restriction should be relaxed if birds are subjected to any stresses such as beaktrimming, vaccination, general disease challenge or substantial reduction in environmental temperature. An alternative management procedure for overweight birds, is to schedule an earlier light stimulation and move to layer cages (see Fig. 3.4). There is an indication that young brown egg pullets may not adjust feed intake too precisely in response to adjusting diet energy level (Table 3.16).

TABLE 3.16 Response of brown and white egg pullets to diet energy level

	Body weight (g)			Feed intake (g)		
Brown egg	**42d**	**84d**	**126d**	**0-42d**	**42-84d**	**84-126d**
2750 kcal/kg	410	1090	1590	1010	2700	3240
3030 kcal/kg	450	1160	1660	1020	2700	3070
White egg						
2750 kcal/kg	380	953	1362	940	2490	3100
3030 kcal/kg	360	940	1375	910	2360	2790

As energy level is increased, at a fixed protein level (Table 3.16) a reduction in growth rate is sometimes seen because protein and amino acid intake are limited. Brown egg pullets seem to change their feed intake very little under these conditions, and consequently there is improvement in growth rate. In another study, pullets were fed diets at 2750 or 3000 kcal ME/kg. Over the 126 d growing period, brown egg pullets consumed 6% more energy when fed the high energy diet (20.6 vs 19.4 Mcal). Contrary to this increased energy intake, white-egg pullets consumed about 18 Mcal ME regardless of energy level in the diet. An alternative scenario in explaining these results, is that the heavier brown-egg pullet has reduced amino acid needs, and so when fed high energy diets there is less effect on amino acid intake relative to needs. In a series of studies we have shown the brown egg pullet to grow quite well on very low levels of lysine relative to that recommended by most breeders (Table 3.17).

TABLE 3.17 Response of brown-egg pullets to diet lysine			
0-42d		84-126d	
Diet lysine (%)	Body wt (g)	Diet lysine (%)	Body wt (g)
0.58	393	0.38	1467
0.68	434	0.46	1512
0.78	437	0.54	1464
0.88	434	0.62	1488
0.98	436	0.70	1515
1.08	414	0.78	1470
1.18	407	0.86	1521
(diets at 2930 kcal ME/kg)			

Up to 42 d of age, the lysine requirement of the pullet seems to be 0.58 - 0.68% of the diet, which is substantially less than values of 0.9 - 1.0% as recommended by most breeders. From 84 - 126 d during the later phases of growth, there was no response to growth rate with more than 0.46% diet lysine. These experimental results suggest that under moderate environmental temperatures, it may be inadvisable to use high energy diets for growing brown egg pullets. On the other hand, assuming their response to diet energy is independent of temperature, then it may be easier to stimulate growth of these pullets under heat stress conditions, simply by increasing the nutrient density of the diet.

CHAPTER 3

Selected References

Coelho, M.B. 1994. *Vitamin stability in premixes and feeds: a practical approach. BASF Technical Symposium.* Indianapolis. May 25. p 99-126.

Leeson, S. 1986. *Nutritional considerations of poultry during heat stress.* World's Poult. Sci. J. 42:619-681.

Leeson, S. 1991. *Growth and development of Leghorn pullets subjected to abrupt changes in environmental temperature and dietary energy level.* Poultry Sci. 70:1732-1738.

Leeson, S. and L.J. Caston 1993. *Does environmental temperature influence body weight;shank length in Leghorn pullets?* J. Appl. Poult. Res. 2:253-258.

Leeson, S. and J.D. Summers 1979. *Step-up protein diets for growing pullets.* Poult. Sci. 58:681-686.

Leeson, S. and J.D. Summers 1981. *Effect of rearing diet on performance of early maturing pullets.* Can. J. Anim. Sci. 61:743-749.

Leeson, S. and J.D. Summers 1984. *Effects of cage density and diet energy concentration on the performance of growing Leghorn pullets subjected to early induced maturity.* Poult. Sci. 63:875-882.

Leeson, S. and J.D. Summers 1984. *Influence of nutrient density on growth and carcass composition of weight-segregated Leghorn pullets.* Poult. Sci. 63:1764-1772.

Leeson, S. and J.D. Summers 1985. *Early reproductive characteristics of Leghorn pullets reared on least-cost diets formulated to protein and/or amino acid specifications.* Can. J. Anim. Sci. 65:205-210.

Leeson, S. and J.D. Summers 1985. *Response of growing Leghorn pullets to long or increasing photoperiods.* Poult. Sci. 64:1617-1622.

Leeson, S. and J.D. Summers 1989. *Performance of Leghorn pullets and laying hens in relation to hatching egg size.* Can. J. Anim. Sci. 69:449-458.

Leeson, S. and J.D. Summers 1989. *Response of Leghorn pullets to protein and energy in the diet when reared in regular or hot-cyclic environments.* Poult. Sci. 68:546-557.

Leeson, S., J.D. Summers and L.J. Caston 1993. *Growth response of immature brown-egg strain pullets to varying nutrient density and lysine.* Poultry Sci. 72:1349-1358.

Lewis, P.D. and G.C. Perry 1995. *Effect of age at sexual maturity on body weight gain.* Br. Poultry Sci. 36:854-856.

Martin, P.A., G.D. Bradford and R.M. Gous 1994. *A formula method of determining the dietary amino acid requirements of laying type pullets during their growing period.* Br. Poultry Sci. 35:709-724.

McNaughton, J.L., L.F. Kubena, J.W. Deaton and F.N. Reece 1977. *Influence of dietary protein and energy on the performance of commercial egg-type pullets reared under summer conditions.* Poult. Sci. 56:1391-1398.

Stockland, W.L. and L.G. Blaylock 1974. *The influence of temperature on the protein requirement of cage-reared replacement pullets.* Poult. Sci. 53:1174-1187.

Summers, J.D. 1993. *Influence of prelay treatment and dietary protein level on the reproductive performance of WL hens.* Poultry Sci. 72:1705-1713.

Summers, J.D. and S. Leeson 1983. *Factors influencing early egg size.* Poult. Sci. 62:1155-1159.

Summers, J.D. and S. Leeson 1994. *Laying hen performance as influenced by protein intake to sixteen weeks of age and body weight at point of lay.* Poultry Sci. 73:495-501.

CHAPTER 4. FEEDING PROGRAMS FOR LAYING HENS

4.1 DIET SPECIFICATIONS AND FORMULATIONS

Diet specifications for laying hens and light breeder hens are shown in Table 4.1. There is a considerable range of daily feed intake patterns shown by laying hens, and so it is important to select and formulate diets based on expected feed intake, such that the daily intake of specified nutrients are achieved. The large range of daily feed intakes encountered with laying hens is caused by variation in age at sexual maturity, inherent body weight and environmental effects such as temperature and bird density. Most Leghorn strains of bird will now start to mature on intakes of 80 - 85 g/day, and it is quite difficult to formulate diets for these birds that will ensure adequate intakes of all nutrients. Meeting the bird's energy needs is perhaps most critical at this time. Through the period of peak egg numbers, it is important that the bird not be deficient in energy, and so high, rather than low energy diets are usually preferred. However energy level can be altered, within reasonable limits, and the bird will adjust its feed intake accordingly. Maintaining the balance of other nutrients to energy is therefore an important concept in layer nutrition.

TABLE 4.1 Diet specifications for layers and light breeders

	Laying hens Feed intake/day (g)						Light breeder hen Feed intake/day (g)			
	120	**110**	**100**	**90**	**80**	**70**	**110**	**100**	**90**	**80**
Approximate protein level (%)	14.0	15.5	17.0	19.0	20.5	22.1	16.0	18.0	20.0	21.5
Amino acids (% of diet)										
Arginine	0.60	0.68	0.75	0.82	0.90	0.98	0.73	0.80	0.89	0.98
Lysine	0.56	0.63	0.70	0.77	0.84	0.91	0.67	0.74	0.81	0.89
Methionine	0.31	0.34	0.37	0.41	0.47	0.56	0.35	0.39	0.43	0.49
Methionine + cystine	0.53	0.58	0.64	0.71	0.80	0.91	0.61	0.68	0.75	0.84
Tryptophan	0.12	0.14	0.15	0.17	0.18	0.20	0.15	0.16	0.18	0.20
Histidine	0.14	0.15	0.17	0.19	0.25	0.25	0.16	0.18	0.20	0.22
Leucine	0.73	0.82	0.91	1.00	1.09	1.18	0.86	0.96	1.06	1.17
Isoleucine	0.50	0.57	0.63	0.69	0.73	0.82	0.60	0.67	0.74	0.81
Phenylalanine	0.38	0.42	0.47	0.52	0.57	0.61	0.45	0.50	0.55	0.61
Phenylalanine + tryosine	0.65	0.75	0.83	0.91	0.99	1.08	0.79	0.88	0.97	1.07
Threonine	0.50	0.57	0.63	0.69	0.73	0.82	0.60	0.67	0.75	0.81
Valine	0.56	0.63	0.70	0.77	0.82	0.91	0.67	0.74	0.81	0.89
Metabolizable energy (kcal/kg)	2700	2700	2800	2850	2850	2900	2800	2800	2850	2900
Calcium (%)	3.00	3.25	3.50	3.60	3.80	4.00	3.3	3.5	3.7	4.0
Available phosphorus (%)	0.35	0.40	0.40	0.42	0.45	0.47	0.40	0.40	0.44	0.48
Sodium	0.17	0.18	0.18	0.19	0.20	0.22	0.18	0.18	0.19	0.20
Vitamin (per kg of diet)										
Vitamin A (I.U.)	7500						8000			
Vitamin D_3 (I.U.)	2500						2500			
Choline equivalents (mg)	1200						1400			
Riboflavin (mg)	4.5						8.0			
Pantothenic acid (mg)	10.0						15.0			
Vitamin B_{12} (mg)	0.010						0.014			
Folic acid (mg)	0.75						1.0			
Biotin (mg)	0.15						0.2			
Niacin (mg)	40.0						60.0			
Vitamin K (mg)	2.0						2.0			
Vitamin E (I.U.)	25.0						30.0			
Thiamin (mg)	2.0						2.2			
Pyridoxine (mg)	3.0						4.0			
Trace Minerals (per kg of diet)										
Manganese (mg)	70						70			
Iron (mg)	80						80			
Copper (mg)	8.0						8.0			
Zinc (mg)	60						60			
Selenium (mg)	0.3						0.3			
Iodine (mg)	0.4						0.4			

TABLE 4.2 Examples of layer diets (kg)

Ingredients	1	2	3
Ground corn	596.0	653.0	657.0
Wheat shorts	---	---	120.0
Soybean meal (48%)	280.0	234.0	110.0
Fat (stabilized)	20.0	10.0	10.0
Limestone	78.0	78.0	78.0
Calcium phosphate (20% P)	11.5	11.5	11.5
Salt (iodized)	3.5	3.5	3.5
Vitamin:mineral premix[1]	10.0	10.0	10.0
Methionine	1.0	0.5	0.5
Calculated analyses			
Crude protein (%)	18.6	16.9	13.0
Digestible protein (%)	17.0	15.4	11.7
Crude fat (%)	4.4	3.6	4.2
Crude fiber (%)	2.3	2.3	2.8
Metabolizable energy (kcal/kg)	2860	2850	2770
Calcium (%)	3.30	3.26	3.25
Available phosphorus (%)	0.41	0.40	0.40
Sodium (%)	0.19	0.19	0.18
Methionine (%)	0.42	0.34	0.28
Methionine + cystine (%)	0.70	0.59	0.46
Lysine (%)	1.02	0.88	0.56

[1]Use additional choline if premix does not contain this vitamin

TABLE 4.3 Examples of layer diets (kg)

Ingredients	4	5	6	7
Ground yellow corn	---	---	300.0	333.0
Ground wheat	667.0	689.0	352.0	329.0
Ground barley	---	100.0	---	100.0
Alfalfa meal	30.0	30.0	15.0	15.0
Soybean meal (48%)	180.0	68.0	215.0	110.0
Fat (stabilized)	20.0	10.0	15.0	10.0
Limestone	78.0	78.0	78.0	78.0
Calcium phosphate (20% P)	11.5	11.5	11.5	11.5
Salt (iodized)	2.5	2.5	2.5	2.5
Vitamin:mineral premix[1]	10.0	10.0	10.0	10.0
Methionine	1.0	0.75	0.75	0.5
Calculated analyses				
Crude protein (%)	17.9	13.9	17.8	13.8
Digestible protein (%)	16.0	11.9	16.1	12.3
Crude fat (%)	3.2	2.3	3.3	3.2
Crude fiber (%)	3.3	3.8	2.8	3.3
Metabolizable energy (kcal/kg)	2768	2754	2800	2800
Calcium	3.3	3.3	3.3	3.3
Available phosphorus (%)	0.43	0.42	0.41	0.40
Sodium (%)	0.19	0.19	0.18	0.18
Methionine (%)	0.37	0.29	0.36	0.28
Methionine + cystine (%)	0.66	0.51	0.54	0.49
Lysine (%)	0.96	0.66	0.95	0.65

[1]Use additional choline if premix does not contain this vitamin

TABLE 4.4 Examples of diets for light breeders (kg)

Ingredients	1	2	3	4
Ground yellow corn	624.0	417.0	434.0	---
Ground wheat	---	---	---	690.0
Ground barley	---	230.0	258.0	---
Dehydrated alfalfa meal	---	---	15.0	30.0
Soybean meal (48%)	260.0	240.0	185.0	155.0
Fat (stabilized)	15.0	15.0	10.0	25.0
Ground limestone	76.0	73.0	73.0	76.0
Calcium phosphate (20%P)	11.5	11.5	11.5	11.5
Salt	3.5	3.5	3.5	3.5
Vitamin:mineral premix[1]	10.0	10.0	10.0	10.0
Methionine	0.5	0.5	0.5	1.0
Calculated Analyses				
Crude protein (%)	17.9	17.9	15.9	17.6
Digestible protein (%)	16.4	16.1	14.1	15.2
Crude fat (%)	4.0	3.7	3.3	3.7
Crude fiber (%)	2.4	3.8	4.1	3.3
Metabolizable energy (kcal/kg)	2830	2788	2740	2814
Calcium (%)	3.20	3.13	3.10	3.24
Available phosphorus (%)	0.41	0.42	0.41	0.43
Sodium (%)	0.16	0.16	0.16	0.19
Methionine (%)	0.36	0.35	0.32	0.36
Methionine + cystine (%)	0.63	0.63	0.57	0.63
Lysine (%)	0.96	0.95	0.81	0.90

[1]Use additional choline if premix does not contain this vitamin

4.2 GENERAL CONSIDERATIONS

The development of feeding programs for layers cannot be implemented without consideration of the rearing program as discussed in the previous chapter. Unfortunately, many egg producers purchase point-of-lay pullets from independent pullet growers, and here the goals of the two producers are not always identical. Too often the egg producer is interested in purchasing mature pullets at the lowest possible cost per bird regardless of its condition. For pullet growers to make a profit they must produce a bird at the lowest possible cost. With feed representing some 60 to 70% of the cost to produce a pullet, the obvious way for the pullet grower to reduce costs is to save on feed. While they may be able to save a small amount of feed by eliminating feed waste or by ensuring that house temperatures are optimum, the only way to save a substantial amount of feed is to place the pullets on a growing program such that feed consumption is reduced and/or cheaper diets are used. Because it is not possible to enhance the efficiency with which pullets convert feed into body weight gain, the net result is a smaller bird going into the laying house. If the birds have been on an increasing light pattern, they might well be mature, as judged from appearance, at the onset of production. However, such pullets must still grow before they reach their optimum weight and condition as a laying hen. Consequently, the egg producer will have to feed this pullet so as to attempt to bring the body weight up to normal if a profitable laying flock is to be obtained. If egg producers attempt to save on feed, the result will be underweight birds at peak egg production. This situation leads to smaller eggs, and often lower than normal peaks or birds dropping relatively quickly in production shortly past peak as discussed in the previous chapter.

It takes a certain amount of feed to produce a laying hen with optimum body size. If this feed is not consumed in the growing period, it must be fed in the laying house. Of course, one would have to be sure that the pullets are healthy and are not carrying an excess of body fat. However, the problem of excess body fat with today's modern type, early maturing pullet does not usually occur. Egg producers should also find out as much as possible about the pullets they are purchasing such as the type of feeding program they have been on, the health status of the flock, and the type of waterers used in rearing. With this type of information, they should be in a better position to turn the immature pullets into a profitable laying flock.

Over the last few years, it has become general practice to describe feeding programs for Leghorn birds according to the level of feed intake. It is well known that under normal environmental and management conditions, feed intake will vary with the egg production and/or age of bird, and this must be taken into account when formulating diets. While Leghorns may adjust intake according to diet energy levels, there is no evidence to suggest that such precision occurs with other nutrients.

The following daily intakes of nutrients are suggested under ideal management and environmental conditions (Table 4.5).

TABLE 4.5 Daily nutrient recommendations	
Crude protein	17 g
Metabolizable energy	280 kcal
Methionine	360 mg
Methionine + Cystine	640 mg
Lysine	720 mg
Calcium	3.5 g
Available phosphorus	0.4 g
Sodium	0.18g

However, as feed intake changes, specifications must be modified in order to maintain this intake of nutrients. Table 4.6 gives an example of the type of specification changes necessary for various levels of feed intake. As a generalization, these changes are something of a compromise, since it is often difficult and uneconomical to go to the extreme specifications necessary with very low levels of feed intake.

TABLE 4.6. Diet specifications and daily feed intake

	110g	100g	90g	80g	70g
Crude protein equivalent (%)	15.5	17.0	19.0	20.5	22.1
Metabolizable energy (kcal/kg)	2700	2800	2915	3025	3080
Calcium (%)	3.4	3.5	3.6	3.8	4.0
Available phosphorus (%)	.38	.40	.45	.50	.55
Amino acids (%)					
Lysine	.68	.72	.77	.84	.91
Methionine	.32	.36	.41	.47	.56
Methionine + cystine	.55	.64	.71	.80	.91
Tryptophan	.14	.15	.17	.18	.20

A knowledge of feed intake, and the factors that influence it, are therefore essential for any feed management program. To a degree, the energy level of the diet will influence feed intake, although one should not assume the precision of this mechanism to be perfect. In general, birds over consume energy with higher energy diets, and they will have difficulty maintaining normal energy intake when diets of less than 2500 kcal ME/kg are offered. In most instances under consumption rather than over-consumption is the problem, and so use of higher energy diets during situations such as heat stress will help to minimize energy insufficiency.

There is little doubt that body weight at maturity is a major factor influencing feed intake and so economic performance of laying hens. Body weight differences seen at maturity are maintained throughout lay almost regardless of nutrient profile of layer diets. It is therefore difficult to attain satisfactory nutrient intakes with small birds. Conversely, larger birds will tend to eat more, and this may become problematic in terms of the potential for obesity and/or too large an egg towards end of lay. Phase feeding of nutrients can overcome some of these problems, although a more simplistic long term solution is control over body weight at maturity. Under most economic conditions, heavier birds at maturity are ultimately most economical for table egg production in terms of egg revenue relative to feed costs.

This whole situation of feed management is also being confounded with earlier and higher peaks from today's strains of birds. Under these "ideal" situations it is again energy insufficiency that can cause problems during post peak production. Egg production curves that show a 5 - 8% reduction after peak are characteristic of birds with insufficient appetite caused by too small a pullet at maturity (Fig 3.1). The reduction in appetite is of concern relative to the adequacy of energy intake. Calculations of energy balance indicate a somewhat precarious balance around the time of peak egg numbers, indicating the need for stimulating feed intake *per se*, and the possibility of providing some labile energy reserves in the form of carcass energy (fat) stores. Tables 4.7 and 4.8 show such calculated values for Leghorn and brown egg strains respectively, and relate these to the required intake of a standard diet.

TABLE 4.7 Energy balance of Leghorn pullets during early egg production

	Theoretical Daily Energy Requirement (kcal ME per bird)					
Weeks Age	**Maintenance**	**+20% Activity allowance**	**Growth**	**Eggs**	**Total**	**Required intake of 17% CP, 2850 ME diet (g/d)**
16	111	133	43.7		177	62
17	114	137	43.7		181	64
18	118	142	43.7		186	65
19	122	146	43.7		190	67
20	125	150	43.7	5.0	199	70
21	129	155	43.7	10.0	209	73
22	132	158	43.7	24.0	227	80
23	136	163	43.7	42.6	250	88
24	139	167	43.7	60.9	272	95
25	141	169	43.7	77.8	269	94
26	142	170	21.9	84.9	277	97
27	146	175	21.9	87.1	284	100
28	147	176	21.9	88.6	286	100
29	149	179	21.9	89.6	291	102
30	150	180	21.9	90.7	293	103

The significance of energy intake as the limiting nutrient for egg production with modern strains of layer is shown in Fig 4.1. There is a dramatic response to energy intake from 184 - 312 kcal/bird/day, in the form of egg output. At very high energy intakes, there is little apparent response to protein intake over the range of 13 - 21 g/bird/day. Only when energy intake is limiting is there any measurable increase in egg numbers in response to increased protein intake. However, as will be detailed later (Fig 4.17) the converse applies in terms of egg size, when the bird shows a dramatic response to protein intake, and little response to energy intake.

TABLE 4.8 Energy balance of brown egg pullets during early egg production (kcal/bird)						
Weeks of age	Maintenance	+15% activity allowance	Growth	Eggs	Total	Required feed intake of 17% CP, 2850 ME diet (g/d)
16	128	147	80	-	227	80
17	131	151	80	-	231	81
18	136	156	80	-	236	83
19	140	161	80	2	243	85
20	144	166	80	8	254	89
21	148	170	70	61	301	106
22	152	175	70	66	309	108
23	156	179	55	83	315	111
24	160	184	40	94	320	112
25	162	185	40	95	320	112
26	164	186	40	96	322	113
27	166	191	38	97	326	114
28	170	196	34	98	328	115
29	171	197	34	99	330	116
30	172	198	32	100	330	116

Fig. 4.1 Egg production (18-66 weeks) in response to intakes of energy and protein.

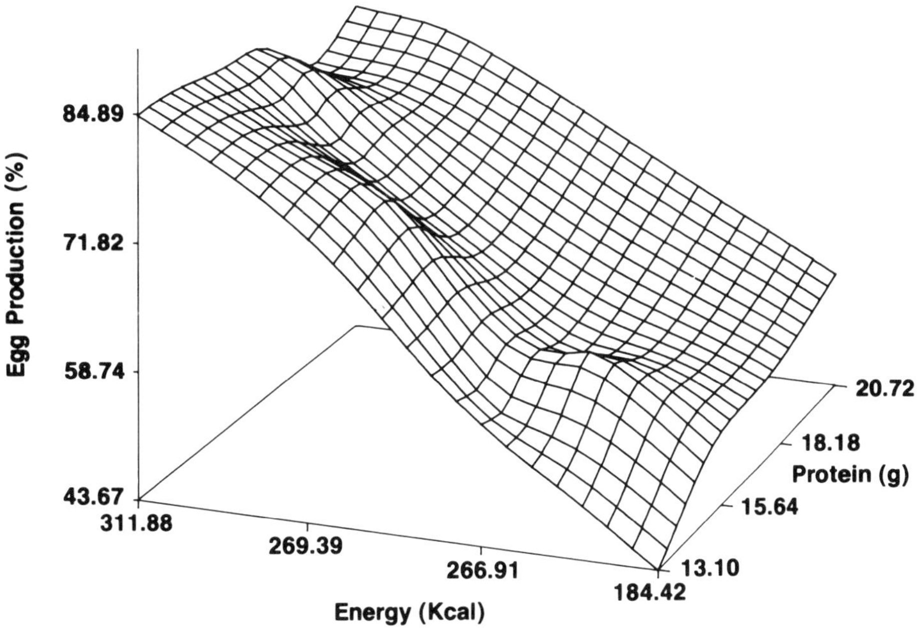

Many problems associated with reduced nutrient intake of white egg layers can therefore be overcome by ensuring optimum body weight and appetite of young laying pullets. Unfortunately, mean weight of the flock at this age is too often considered independently of flock uniformity. Pullets may be of "mean" body weight, yet be quite variable in weight, and often outside the accepted range of 85% of the flock being within ±10% of mean weight. The major problem with a non uniform flock is variability in age at first egg, and so variability in feed intake. If diets are tailored to feed intake, then late maturing smaller birds (with small appetites) will likely be underfed. Conversely large weight early maturing pullets with increased appetites may be overfed at this time. The consequence is often a delayed peak, and reduced overall egg production (Fig 4.2).

Fig. 4.2 Production curves for a uniform pullet flock (A) and a flock which is uneven (B)

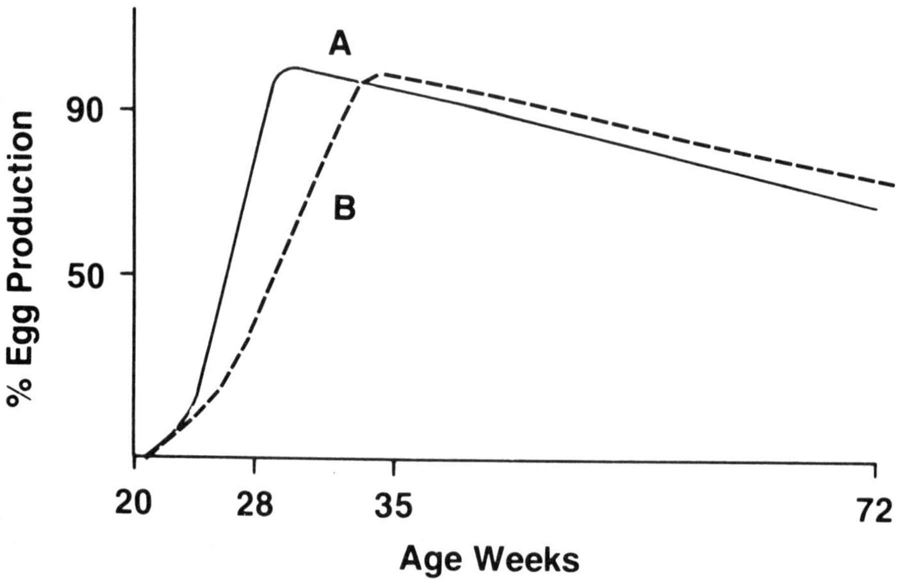

An argument against being overly concerned about uniformity, is that birds will adjust their intake according to nutrient (energy) needs, and so early maturing birds will eat more, and late maturing birds less, during the early phases of production. However if birds are given diets formulated on the basis of feed intake, this can lead to problems, the most serious of which is underfeeding of the larger early maturing bird. Another confounding factor, is that as birds mature within a flock, the % production realized on a daily basis does not reflect the number of birds laying at that time. As shown in Fig. 4.3, the proportion of laying birds always exceeds the % production, and this difference is most pronounced in early production. For example at about 40% production, there are in fact around 70% of the birds mature and requiring proportionally more nutrients than suggested by egg production alone. These differences as shown in Fig. 4.3 become more pronounced as flock uniformity declines, and so this makes tailoring nutrient needs to feed intake and performance characteristics that much more difficult to achieve.

Fig. 4.3 Comparison of number of birds producing eggs and actual egg production

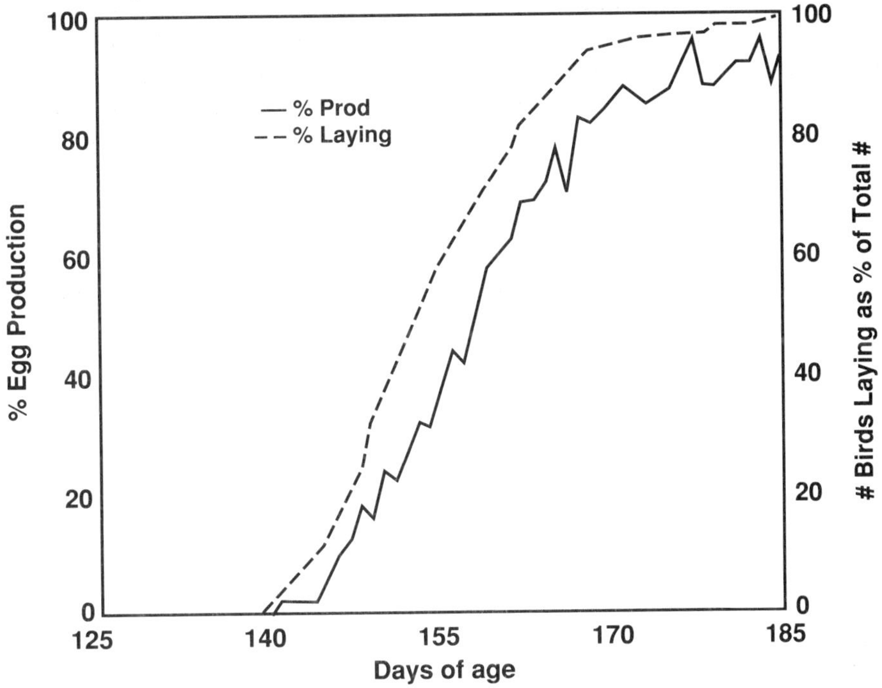

4.3 HEAT STRESS

An increasingly large proportion of the world's laying hens are kept in areas where heat stress is likely to be a major management factor at some stage during egg production. Basically the problem relates to birds not consuming enough feed at this time, although there are also some subtle changes in the bird's metabolism that affect both production and shell quality. While all types of poultry thrive in warm environments during the first few weeks of life, normal growth and development of older birds is often adversely affected. Obviously the bird's requirements for supplemental heat declines with age, because insulating feathers quickly develop and surface area in relation to body size is reduced. Heat stress is often used to describe bird status in hot environments, although it is obvious that more than just temperature is involved. Because birds must use evaporative cooling (as panting) to lose heat at high temperatures, humidity of inhaled air becomes critical. Thus high temperature and humidity together are much more stressful to birds than high temperature alone. Other environmental factors such as air speed and air movement also become important. It is also becoming clear that adaptation to heat stress can markedly influence bird response. For example, laying birds can tolerate constant environmental temperatures of 35°C and perform reasonably well. On the other hand, most birds are stressed at 35°C when fluctuating day/night temperatures are involved. In the following discussion, it is assumed that fluctuating conditions exist, since these are more common and certainly more stressful to the bird.

The main concern under hot weather conditions is the layer's ability to consume feed. As poultry house temperature increases, then less heat is required to maintain body temperature and the birds consume less feed. In this situation, "environmental" energy is replacing feed energy and is economical. However, the relationship between body heat production and house temperature is not linear, since at a certain critical temperature, the bird's energy demands are increased in order to initiate body cooling mechanisms. The following factors should be considered in attempting to accommodate heat stress.

a) Bird's response to heat stress

Figure 4.4 is a schematic representation of a heat stress effect. Minimal body heat production (and hence the most efficient situation) is seen at around 23°C. Below this temperature, (lower critical temperature) birds generally have to generate more body heat in order to keep warm.

Fig. 4.4 Environmental temperature and body heat production

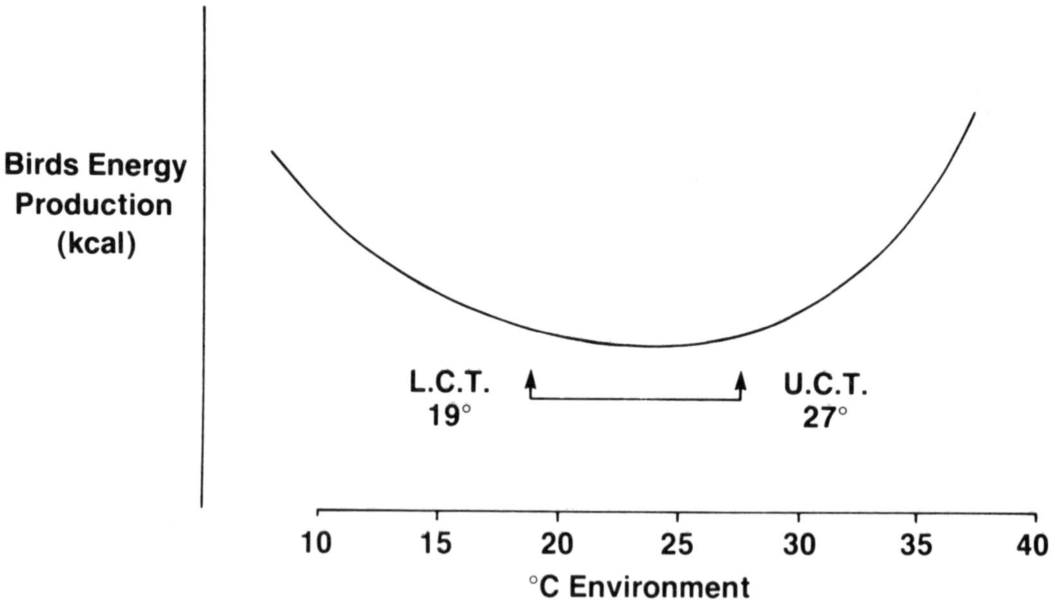

However, there is only a narrow range of efficiency between 19 - 27°C, over which heat production is minimal. Above 27°C, birds start to use more energy in an attempt to stay cool. For example, at 27°C, birds will start to dilate certain blood vessels in order to get more blood to the comb, wattles, feet etc. in an attempt to increase cooling capacity. More easily observed by the egg producer is the characteristic panting and wing drooping that occurs at slightly higher temperatures. These activities at high environmental temperatures mean that the bird has an increased, rather than decreased, demand for energy. Unfortunately, the situation is not as clear cut as depicted in Fig 4.4, and this is likely the reason behind the variability seen in flock response to various environmental conditions. Rather than lower and upper critical temperature being rigidly fixed under all conditions, heat production is likely to fluctuate in response to a number of very practical on-farm conditions.

Variation in response can be caused by such factors as a) increased feed intake, b) better feathering or c) increased bird activity. Such potential variability in bird response should be taken into account when interpreting the quantitative data discussed in Figs 4.5 - 4.6. The whole picture is further confused by the normal energy intake pattern of the bird (Fig. 4.5). The base line shown in Fig. 4.5 is a repeat of the temperature effect detailed in Fig. 4.4.

Fig. 4.5 Environmental temperature and energy balance

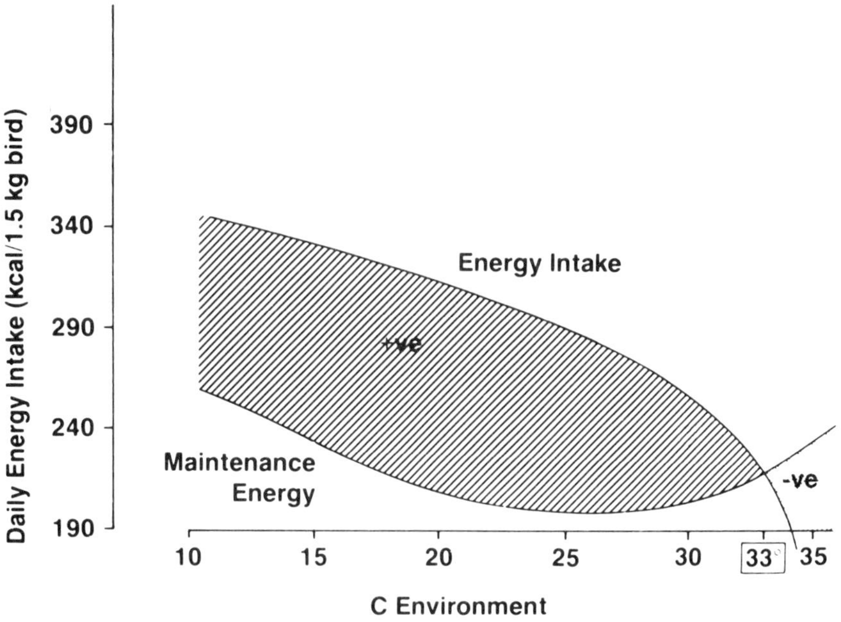

The upper line of Fig. 4.5 represents energy intake for a 1.5 kg layer. Again as environmental temperature increases, so energy (feed) intake declines. However above 27 - 28°C the decline becomes quite dramatic since the bird is changing its metabolic processes in response to the heat load, and actions such as panting, etc. adversely influence the feeding mechanisms in the brain and also reduce the time available for feeding. The shaded area between the lines in Fig. 4.5 represents the energy available for production. As we approach and exceed the critical 28°C mark, then energy available for production is dramatically reduced and around 33°C actually becomes negative.

If the shaded area (available energy), in Fig. 4.5 is itself plotted against temperature, then a clear pattern is seen with respect to potential for egg production (Fig 4.6). If we assume an average egg contains the equivalent of 80 kcal ME, then at 90% production, there is a daily need for around 70 kcal to meet needs for production alone. Our calculations from Fig. 4.5 indicate total available energy at 90 kcal per day, and so we have a small positive difference that will likely go for growth or increased body weight. At 28°C, there is energy available only for eggs and none for growth. Above 28°C, available energy cannot meet energy demands for 90% egg production. Either egg production must be decreased, or other energy sources used. The bird's body reserves (fat and muscle) could therefore be used at this time. These figures are not fixed and will likely vary with such factors as air speed, feathering etc. as previously detailed. However, for most flocks, these types of reactions, as depicted in Fig. 4.6 are likely to occur at ± 2°C of the values shown. In this scenario the bird is in negative energy balance at 33°C (Fig 4.6).

Fig. 4.6 Environmental temperature and energy balance.

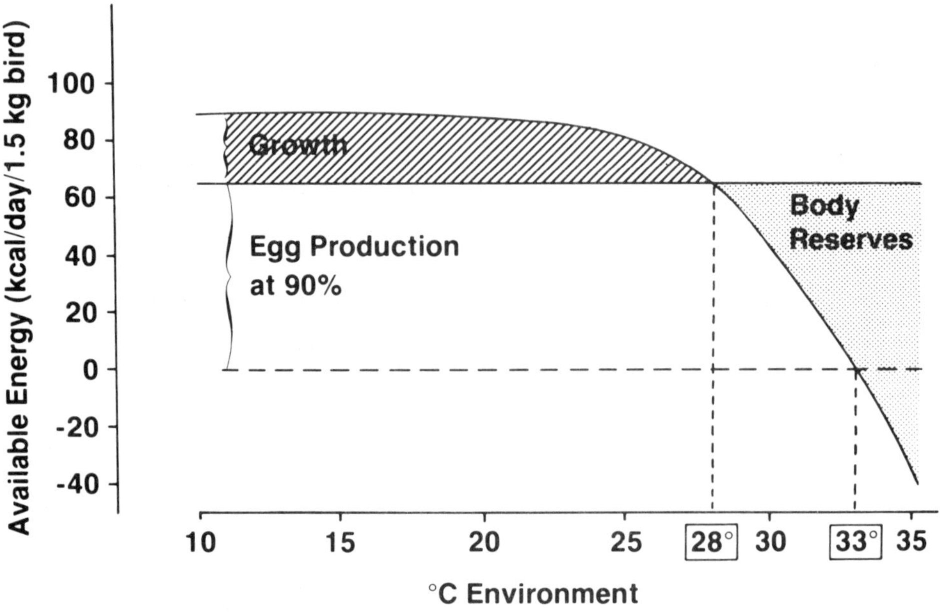

Various equations have been developed to relate energy intake to environmental temperature. For example, the equations given by NRC (1994) is ME(kcal/day) =

$$W^{.75} (173 - 1.95T) + 5.5\ \Delta W + 2.07EE$$

where W = body weight, kg; T = °C, ΔW = weight gain per day, g; and EE = daily egg mass, g. Solving this equation for environmental temperatures of 10 - 34°C, shows an almost linear relationship for a 1.3 kg bird producing 50 g egg mass per day and gaining weight at 2 g per day (Fig. 4.7).

Fig 4.7 ME intake of layers predicted between 10 – 34°C. (NRC, 1994)

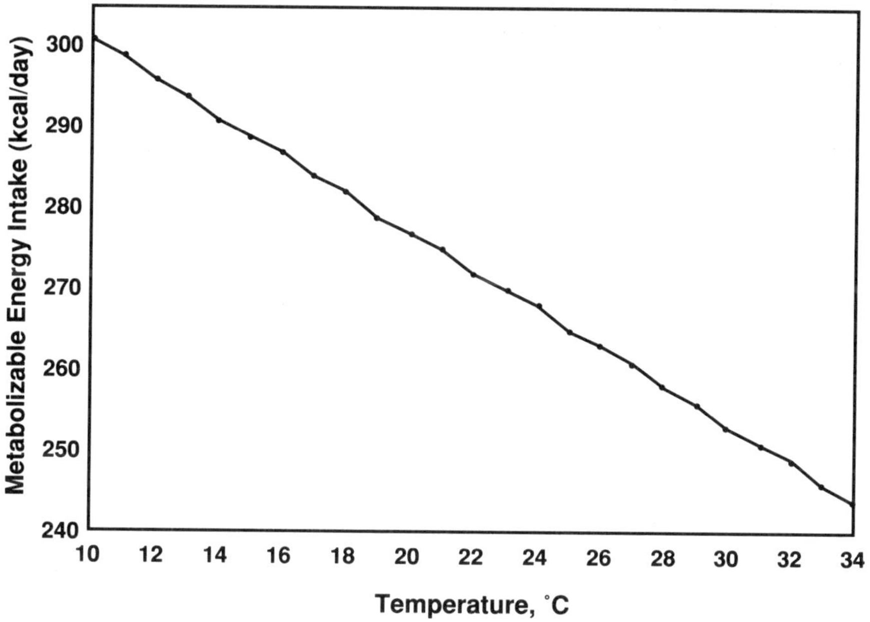

Fig. 4.8 ME intake of layers with 60, 75 or 90% feather cover at 10 – 34°C (Peguri and Coon, 1995)

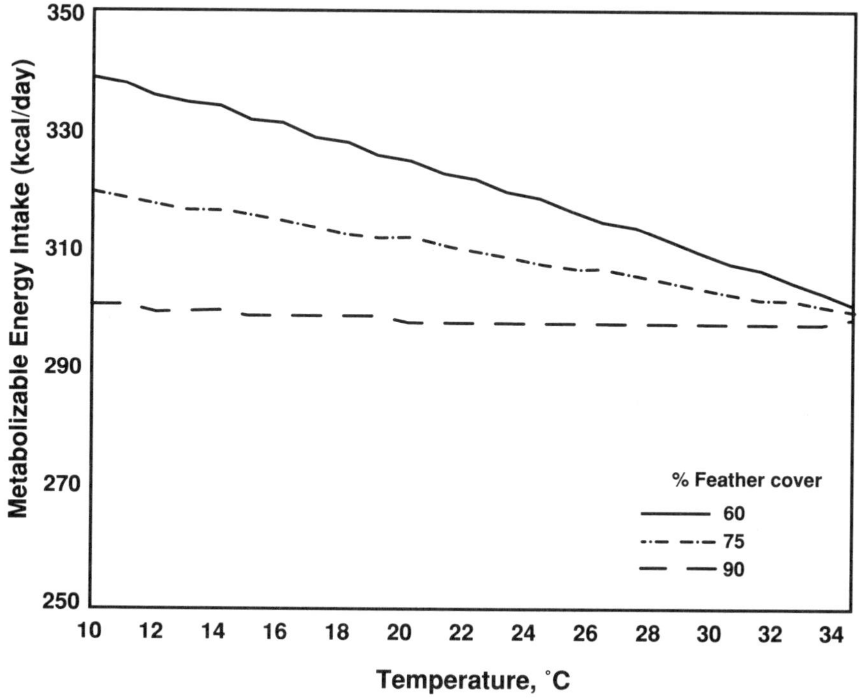

A major factor affecting this energy intake response to environmental temperature is feather cover, which represents insulating capacity for the bird. Coon and co-workers have developed equations that take into account degree of feathering. This equation is solved in Fig. 4.8 for birds having 90, 75 or 60% feather cover. As expected, at low environmental temperatures, feather cover has a major effect on feed intake, while at 34°C which is close to body temperature, there is no effect of feather cover.

b) Energy balance

Our main concern during heat stress therefore is availability of energy for egg production. Optimizing such energy availability may be approached by either:

(i) Increasing diet energy specifications

(ii) Stimulating feed intake

or (iii) Considering body energy reserves

(i) It is well known that birds consume less feed as the energy level of the feed increases. This is because the bird attempts to maintain a given energy intake each day. However, the mechanism is by no means perfect, and as energy level is increased, the expected decline in feed intake is seldom achieved. This obviously leads to "overconsumption" of energy. Also, as environmental temperature increases, the mechanism seems less perfect. The following results are seen when diet energy level is increased from 2860 kcal ME/kg to 3450 kcal ME/kg (Table 4.9, Payne, 1967).

TABLE 4.9 Effect of diet energy level on metabolizable energy intake

	at 18°C		at 30°C	
Diet energy (kcal ME/kg)	**Feed per day (g)**	**Energy per day (kcal)**	**Feed per day (g)**	**Energy per day (kcal)**
2860	127	363	107	306
3060	118	360	104	320
3250	112	364	102	330
3450	106	365	101	350

(From Payne 1967)

At 18°C there is fairly good adjustment by the bird, such that feed intake is markedly reduced with high energy diets; this in an attempt to normalize energy intake. At high temperatures, birds adjust feed intake less perfectly, such that "overconsumption" of energy occurs. It is not suggested that these extremes of diet energy be used, rather that energy intake will be maximized with as high a diet energy level as is possible. In order to increase diet energy level, the use of supplemental fat should be considered. Dietary fat has the advantage of increasing palatability and also reducing the amount of heat increment that is produced during its utilization in the body.

(ii) Various methods can be tried to stimulate feed intake. Feeding more times each day usually encourages feeding activity. Feeding at cooler times of the day, if possible, is also a useful method of increasing nutrient intake. If artificial lights are used, it may be useful, under extreme environmental conditions, to consider a so-called midnight feeding, when temperature will hopefully be lower and birds are more inclined to eat. Again, where conditions are extreme, making the diet more palatable may be advantageous. Such practices as pouring vegetable oil, molasses, or even water, directly onto the feed in the troughs may encourage intake. Whenever high levels of fat are used in a diet, or used as a top dressing as described here, care must be taken to ensure that rancidity does not occur. This can best be achieved by insisting on the incorporation of quality antioxidants in the feed and that feed is not allowed to "cake' in tanks, augers or troughs. Fresh feed becomes critical under these conditions.

Diet texture can also be used to advantage. Crumbles tend to stimulate intake while a sudden change from large to small crumbles also has a transitory effect on stimulating intake. It is interesting to observe that a sudden change from small to large crumbles seems to have a negative effect on intake (Table 4.10).

TABLE 4.10 Effect of sudden change in feed particle size on feed intake (5-7d) following this change

	Regular crumbs	Only small crumbs <2.4 mm	Only large crumbs >2.4 mm
Feed (g/bird/day)	112[b]	124[a]	81[c]

[a-c] **means followed by different letters are significantly different**

(iii) It is now realized that correct pullet rearing programs are essential for optimum economic return in the layer house. This becomes very critical under hot weather conditions as shown in Fig. 4.6 as the bird may have to rely on its body reserves to supplement energy required to maintain egg production. Rearing programs designed to maximize growth were discussed in the previous sections. In general, the larger the body weight at maturity, the larger the body weight throughout lay, and hence the larger the potential energy reserve and the greater the feed intake (Table 4.11).

TABLE 4.11 Pullet size and energy intake

18 wk body wt (g)	25 wk body wt (g)	Daily energy consumption (18-25 wks, kcal)
1100	1400	247
1200	1500	254
1300	1600	263
1400	1700	273

It is not suggested that extremely fat pullets are desirable, but it is obvious that birds of optimum weight with a reasonable fat reserve will likely stand up better to heat stress situations. Pullets that are subjected to heat-stress and have less "available" energy than that required to sustain production, have no recourse but to reduce egg mass output in terms of egg weight and/or egg numbers.

c) Protein nutrition

In the past, it has been common practice to increase protein levels during heat stress conditions. This has been done on the basis of reduced feed intake, and hence protein levels have been adjusted upwards in attempting to maintain intakes of around 17g crude protein/bird/day. It is now realized that such adjustments may be harmful. When any nutrient is metabolized in the body, the processes are not 100% efficient and as a result some heat is produced. Unfortunately, protein is the most inefficiently utilized nutrient in this regard and so proportionately more heat is evolved during metabolism of amino acids. The last thing that a heat stressed bird needs is additional waste heat being generated in the body. This extra heat production may well overload heat dissipation mechanisms (panting, blood circulation). We are therefore faced with a difficult problem of attempting to maintain "protein" intake in situations of reduced feed intake, yet we know that more crude protein may be detrimental. The answer to the problem is not to increase crude protein, but rather to increase the levels of essential amino acids. By feeding synthetic amino acids, we can therefore maintain the intake of these essential nutrients without loading up the body systems with excess crude protein (nitrogen). General recommendations are, therefore, to increase the use of synthetic methionine and lysine so as to maintain daily intakes of approximately 360 and 720 mg respectively.

d) Minerals and vitamins

Calcium level should be adjusted according to anticipated level of feed intake, such that birds consume 3.5 g per day. Under extreme conditions, this may be difficult since, as previously indicated, high energy diets are also desirable and these are difficult to achieve with the increased use of limestone or oyster shell. Table 4.12 shows the diet specifications needed to maintain intakes of Ca, P and vitamin D_3, all of which are critical for eggshell quality.

TABLE 4.12 Diet nutrient levels needed to maintain constant intake at varying levels of feed intake

Feed intake (g/d)	Av P (%)	Ca (%)	Vit D_3 (IU/kg)
70	0.60	5.35	4700
80	0.52	4.68	4125
90	0.47	4.17	3660
100	0.42	3.75	3300
110	0.38	3.41	3000

Because it is also necessary to increase the energy level of the diet when feed intake is low, then it is counterproductive to add high levels of limestone and phosphates which effectively dilute the feed of all nutrients other than Ca and phosphorus. The problem of potential calcium deficiency is most often met by top dressing feed with oystershell or large particle limestone. The situation for phosphorus is more complex, and in fact it may be deleterious to use the high levels shown in Table 4.12. In practice phosphorus levels are seldom increased to these extreme levels unless cage layer fatigue is an ongoing problem. The deficit of vitamin D_3 is best met with use of D_3 supplements in the drinking water.

There seems to be some benefit to adding sodium bicarbonate to the diet or drinking water. However, this must be done with care so as not to impose too high a load of sodium on the bird, and so salt levels may have to be altered. This should be done with great caution, taking into account sodium intake from the drinking water, which can be quite high during heat stress conditions. There is also an indication of beneficial effects of increasing the potassium levels in the diet, although again, this must be accomplished only after careful calculation, since high levels can be detrimental to electrolyte balance. While few reports indicate any improvement in adding supplemental B vitamins during heat stress, there are variable reports of the beneficial effects with the fat soluble vitamins. Although not always conclusive, increasing the levels of vitamins A, D_3 and E have all been shown to be advantageous under certain conditions. While vitamin C (ascorbic acid) is not usually considered in poultry diets, there is evidence to support its use during hot weather conditions. Birds require vitamin C, but under most circumstances are able to synthesize enough in their own bodies. Under heat stress such production may be inadequate and/or impaired. Adding up to 200 mg vitamin C/kg diet has proven beneficial for layers in terms of maintaining production.

e) Electrolyte Balance

As environmental temperature increases, birds increase their respiration rate in an attempt to increase the rate of evaporative cooling. As birds pant however they tend to lose proportionally more CO_2 and so changes in acid-base balance can quickly develop. With mild through to severe alkalosis, blood pH may change from 7.2 through 7.5 to 7.7 in extreme situations. This change in blood pH, together with loss of bicarbonate ions can influence eggshell quality and general bird health and metabolism. Under such heat stress conditions, it is the availability of bicarbonate *per se* which seems to be the major factor influencing eggshell thickness, and in turn, this is governed by acid-base balance, kidney function and respiration rate.

Under normal conditions shell formation induces a renal acidosis related to the total resorption of filtered bicarbonate. At the same time, shell secretion induces a metabolic acidosis because the formation of insoluble $CaCO_3$ from HCO_3^- and Ca^{++} involves the liberation of H^+ ions. Such H^+ release would induce very acidic and physiologically destructive conditions, and is necessarily balanced by the bicarbonate buffer system in the fluid of the uterus. While a mild metabolic acidosis is therefore normal during shell synthesis, a more severe situation leads to reduced shell production because of intense competition for HCO_3^-, as either a buffer or a shell component. A severe metabolic acidosis can be induced by feeding products such as NH_4Cl, and this results in reduced shell strength. In this scenario, it is likely that NH_4^+ rather than Cl is problematic because formation of urea in the liver (from NH_4^+) again needs to be buffered with HCO_3^- ions, creating more competition with uterine bicarbonate metabolism. Conversely, feeding sodium bicarbonate, especially when Cl levels are minimized, may well improve shell thickness. Under commercial conditions, the need to produce base excess in order to buffer any diet electrolytes must be avoided. Likewise it is important that birds not be subjected to severe respiratory excess, as occurs at high temperatures, because this lowers blood bicarbonate levels, and in extreme cases, causes a metabolic acidosis. Under practical conditions, replacement of part of the supplemental dietary NaCl with $NaHCO_3$ may be beneficial for shell production.

Acclimatization to heat stress is a confounding factor, because temporary acute conditions are more problematic. For example, pullets grown to 31 wk under constant 35 vs 21°C conditions exhibit little difference in pattern of plasma electrolytes. If birds are allowed to acclimatize to high environmental temperatures, there is little correlation between plasma electrolytes and shell quality. Temporary acute heat stress and cyclic temperature conditions seem most stressful to the bird.

Prevention of electrolyte imbalance should obviously be approached through incorporation of appropriate cations and anions in diet formulations. However it must be accepted that diet is only one factor influencing potential imbalance, and so general bird management and welfare also become of prime importance. Electrolyte balance is most usually accommodated by consideration of Na+K-Cl balance in the diet, and under most dietary situations this seems a reasonable simplification. Electrolyte balance is usually expressed in terms of mEq of the various electrolytes, and for an individual electrolyte this is calculated as Mwt ÷ 1000. This unit is used on the basis that most minerals are present at a relatively low level in feeds. As an example calculation, the mEq for a diet containing 0.17% Na, 0.80%K and 0.22% Cl can be developed as follows:

Sodium Mwt = 23.0, ∴ Eq = 23g/kg, ∴ mEq = 23 mg/kg
Diet contains 0.17% Na ≡ 1700 mg/kg ≡ $\frac{1700}{23}$ mEq = 73.9 mEq.

Potassium Mwt = 39.1, ∴ Ea = 39.1g/kg, ∴ mEq = 39.1 mg/kg
Diet contains 0.80% K ≡ 8000 mg/kg ≡ $\frac{8,000}{39.1}$ mEq = 204.6 mEq

Chloride Mwt = 35.5, ∴ Eq = 35.5 g/kg, ∴ mEq = 35.5 mg/kg
Diet contains 0.22% Cl ≡ 2200 mg/kg, ≡ $\frac{2200}{35.5}$ mEq = 62.0 mEq

∴ overall diet balance becomes Na + K - Cl = 73.9 + 204.6 - 62.0 = 216.5 mEq.

A balance of around 250 mEq/kg is usual, and so for this diet there needs to be either an increase in Na or K level of the diet, or a decrease in Cl level.

Under practical conditions, electrolyte balance seems to be more problematic when chloride levels are high. On the other hand, use of $NaHCO_3$ to replace NaCl, as is sometimes recommended during heat stress, can lead to a deficiency of chloride. Changes in diet electrolyte balance most commonly occur when there is a major change in ingredient usage, and especially when animal protein sources replace soybean meal and vice versa. Table 4.13 outlines electrolyte content and electrolyte balance of some major feed ingredients.

Within the cereals, electrolyte balance for milo is low, while wheat is high relative to corn. Major differences occur in the protein-rich ingredients, and relative to soy, all sources are low in electrolyte balance. As shown in Table 4.13, this situation develops due to the very high potassium content of soybean meal. Careful consideration to electrolyte balance must therefore be given when changes are made in protein sources used in formulation. For example, the overall balance for a diet containing 60% milo and 25% soy is 210 mEq/kg, while for a diet containing 75% milo and 10% fish meal the balance is only 75 mEq/kg. The milo-fish diet would perhaps need to be supplemented with $NaHCO_3$.

TABLE 4.13 Electrolyte content of feed ingredients.				
	% of ingredient			
Ingredient	Na	K	Cl	Na+K-Cl(mEq)
Corn	0.05	0.38	0.04	108
Wheat	0.09	0.52	0.08	150
Milo	0.04	0.34	0.08	82
Soybean meal	0.05	2.61	0.05	675
Canola meal	0.09	1.47	0.05	400
Meat meal	0.55	1.23	0.90	300
Fish meal	0.47	0.72	0.55	230
Cottonseed meal	0.05	1.20	0.03	320

Assuming that heat stress cannot be tempered by normal management techniques, then electrolyte manipulation of the diet may be beneficial. However, the technique should be different for immature birds compared to egg layers. With adult female birds, there is need to maintain the bicarbonate buffer system as it relates to eggshell quality. As such, diet or water treatment with sodium bicarbonate may be beneficial again emphasizing the necessity to meet minimal chloride requirements. On the other hand, treatment of respiratory alkalosis in layers, with acidifiers such as NH_4Cl while relieving respiratory distress, may well result in reduced shell quality. For immature pullets treatment with electrolytes is often beneficial and there is less need for caution related to bicarbonate buffering. Up to 0.3% dietary NH_4Cl may improve the growth rate of heat stressed birds, although as detailed previously, it is not clear if this beneficial effect is via electrolyte balance/blood pH or simply via the indirect effect of stimulating water intake. Under commercial conditions, adding salt to the drinking water of young birds has been reported to alleviate bird distress and to stimulate growth.

f) Water

A nutritional factor often overlooked during heat stress is the metabolism of water. It is well known that birds in hot environments drink more water, yet this has not been capitalized upon to any degree. It would seem logical to provide nutrients in the water, because water intake is increased at times when feed intake is depressed. Unfortunately, we have only met with very limited success to date with this type of management. What does seem more advantageous is to cool the water of laying hens. In studies with small numbers of birds, we have shown a distinct advantage to cooling the drinking water of birds housed in very warm environments (Table 2.41). In a more large scale study in a commercial unit in California, Bell (personal communication) indicated improved feed intake and egg production in response to the cooling of water by just 5°C at an environmental temperature of 32°C (Table 4.14).

TABLE 4.14 Effect of water temperature on egg production (%)		
	Environmental temperature 32°C	
Age (wks)	Water at 32°C	Water at 27°C
25	64	74
26	74	79
27	77	86
28	76	84
29	88	93
Mean	76	83
Feed intake (g/b/d)	83g	90g
		Bell (1987)

Adding various nutrients to the water during heat stress seems however to have a negative effect on performance (Table 4.15).

TABLE 4.15 Nutrient supplementation via the drinking water for layers during heat stress			
		Control	Water treatment[1]
Feed intake	3 d Heat stress	57 g/b/d	40g/b/d
	3 d post-stress	70 g/b/d	44g/b/d
Egg production	3 d Heat stress	64%	61%
	3 d post-stress	59%	19%
[1]Ca, amino acids, vitamins and propionic acid			DeSchutter (1988)

Another factor to consider with water intake, is the possible effect of dissolved minerals and contaminants etc., the effects of which may be greatly accentuated with increased drinking activity. In terms of mineral content, sodium concentration is the one most likely to cause problems.

g) *Diet changes*

Discussion to date has centred on the potential of diet manipulation to alleviate heat stress. However diet change *per se* may be detrimental under certain conditions. It seems that when the bird is confronted with an acute heat stress situation, diet change may impose another stress which merely tips the delicate balance in the wrong direction. For example, it was recently reported that a diet change brought about by adding fat caused an immediate rise in body temperature, which can be disastrous to the bird, and this persisted for about 4 days. At the same time, the diet change had the desirable effect of stimulating energy intake. For this reason, it is suggested that under extreme heat stress of 36 - 40°C, that no diet change be implemented, since it could lead to death from heat prostration. Under these conditions, it would be useful to be able to prejudge the rise in environmental temperature and make the diet change earlier, when the bird is under "moderately" stressful conditions (28 - 35°C). However, even with short term heat stress situations it may be inadvisable to change the diet (Table 4.16).

TABLE 4.16 Effect of diet change on layer performance during heat stress

		21 weeks of age			33 weeks of age		
	Diet type	Egg prod (%)	Feed intake (g)	Shell *deformation (μm)	Egg prod (%)	Feed intake (g)	Shell *deformation (μm)
Pre-test 7d (18°C)	Control	82	86	21	92	101	24
Stress 3d (35°C)	Control	92	64^{a}	22^{b}	71	50^{a}	35^{b}
	High CP	90	36^{c}	24^{a}	56	20^{b}	41^{ab}
	High Energy	94	40^{c}	23^{ab}	60	27^{b}	46^{a}
	High Density	96	53^{b}	24^{a}	67	28^{b}	37^{b}
Post-stress 4d (18°C)	Control	84^{a}	76^{a}	26^{c}	77^{a}	84^{a}	30^{b}
	High CP	39^{c}	24^{b}	35^{ab}	45^{b}	61^{bc}	41^{a}
	High Energy	56^{b}	33^{b}	41^{a}	64^{a}	57^{c}	42^{a}
	High Density	69^{ab}	76^{a}	31^{bc}	67^{a}	73^{ab}	29^{b}

***μm deformation, where a higher value denotes weaker shell**
[a-c] means followed by different letters are significantly different

In these studies, birds were fed a control ration for 7 d at an environmental temperature of 18°C. A heat stress of 35°C was suddenly imposed, and birds offered the same control diet, or diets high in energy, protein or all nutrients (termed high density). Feed intake was depressed almost immediately in response to heat stress, although changes in egg production and shell quality were not seen until after the 3 d stress period. However during this post-stress period, birds showed a dramatic loss in egg numbers and shell quality. There was no instance of diet change alleviating the effects of heat stress, and in most instances, the situation was much worse. Under such conditions of short term heat stress, it is suggested that sudden diet change merely imposed an additional stress and was not beneficial to the bird.

h) General recommendations concerning heat stress

Under normal conditions birds should be fed so as to attain optimum daily intakes of essential nutrients. Regardless of environment, the correct decisions cannot be made without knowledge gained from the monitoring of feed intake, body weight and egg weight. With heat stress situations (28 - 40°C), the following points should be considered:

1. Never place underweight pullets in the laying house. They will always remain small with low feed intake and have little body fat reserve to sustain optimum egg production.

2. Increase the energy level of the diet (2850 kcal ME/kg minimum) ideally by incorporation of fats or oils. Limit the use of fibrous ingredients if possible.

3. Reduce crude protein component (17% CP maximum) while maintaining daily intakes of methionine (360 mg) and lysine (720 mg).

4. Increase mineral-vitamin premix in accord with anticipated change in feed intake. Maintain daily intakes of calcium (3.5 g) and available phosphorus (400 mg).

5. Where shell quality is a problem, consider the incorporation of sodium bicarbonate. At this time, monitor total sodium intake, and ensure adequate chloride levels in the diet.

6. Use supplemental vitamin C (150 g/tonne) when heat stress occurs.

7. Increase the number of feedings per day and try to feed at cooler times of the day.

8. Keep drinking water as cool as possible. Analyze sodium content of water so as to calculate "salt" contribution from the water.

9. Use crumbled feed or large particle mash feed if available.

10. Do not make any diet changes when sudden short term (3 - 5 d) heat stress occurs.

4.4 PHASE FEEDING

Phase feeding refers essentially to reductions in the protein and amino acid level of the diet as the bird progresses through a laying cycle. The concept of phase feeding is based on the fact that as birds get older their feed intake increases, while their egg production decreases. For this reason, it should be economical to reduce the nutrient concentration of the diet. At this time, it is pertinent to consider a conventional egg production curve of a layer, and superimpose both egg weight and daily egg mass output (Fig 4.9).

If nutrient density is to be reduced this should not occur immediately after peak egg numbers, but rather after peak egg mass has been achieved. There are two reasons for reducing the level of dietary protein and amino acids during the latter stages of egg production namely, to reduce feed costs and secondly, to reduce egg size. The advantages of the first point are readily apparent if protein costs are high, but the advantages of the second point are not so easily defined and will vary depending upon the price of eggs. When a producer is being paid a premium for extra large and jumbo eggs, there is no advantage to using a phase feeding program unless egg shell quality is a problem.

Fig. 4.9 Bird age: egg production, egg weight and egg mass.

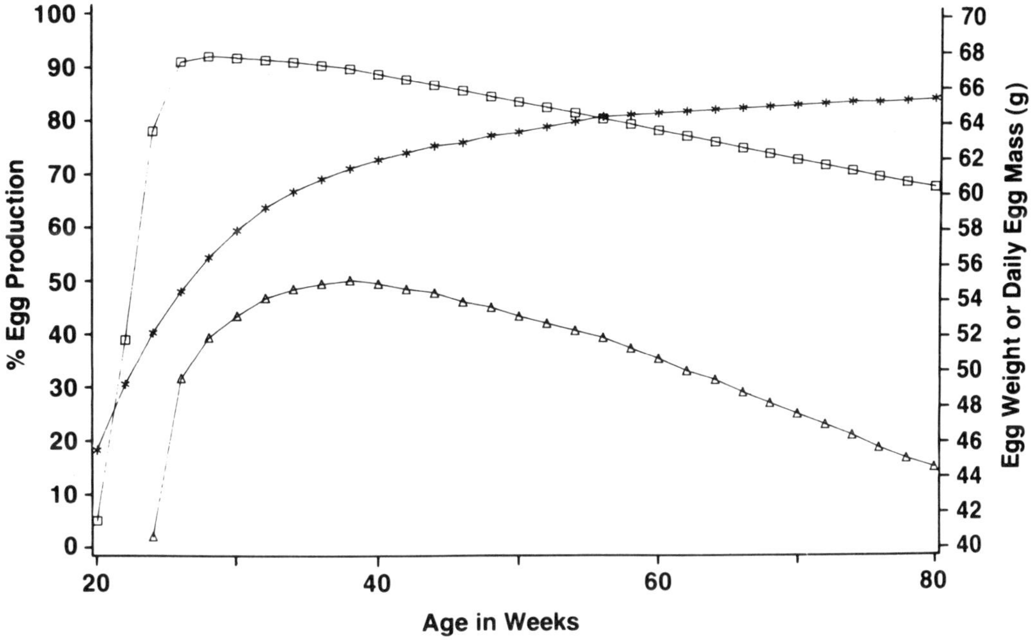

It is difficult to give specific recommendations as to the decrease in dietary protein level that can be made to reduce egg size without decreasing the level of production. The appropriate reduction in protein level will depend on the season of the year (effect of temperature on feed consumption), age and production of the bird, and energy level of the diet. Hence, it is necessary that every flock be considered on an individual basis before a decision is made to reduce the level of dietary protein. As a guide, it is recommended that protein intake be reduced from 17 g/day to 16 g/day after the birds have dropped to 80% production, and to 15 g/day after they have dropped to 70% production. With an average feed intake of 100 g/day, this would be equivalent to diets containing 17, 16 and 15% protein. It must be stressed that these values should be used only as a guide after all other factors have been properly considered. If a reduction in the level of protein is made and egg production drops, then the decrease in intake has been too severe and it should be immediately increased. If, on the other hand, production is held constant and egg size is not reduced then the decrease in protein intake has not been severe enough and it can be reduced still further. The amino acid to be considered in this exercise is methionine, since this is the amino acid that has the greatest effect on egg size. Phase feeding of phosphorus has also been recommended as a method of halting the decline in shell quality often seen with older birds. Using this technique, available phosphorus levels may be reduced from approximately 0.4% at peak production to slightly less than 0.3% at end of lay. Table 4.17 shows an example of phase feeding of protein, methionine and phosphorus, related to egg production, and independent of feed intake.

TABLE. 4.17 Phase feeding of major nutrients after peak egg mass, assuming constant daily feed intake at 100g

Bird characteristics		Diet levels (%)			
Age (wks)	Egg production (%)	Crude protein	Methionine	Calcium	Available phosphorus
<35	90	17.0	0.35	3.8	0.42
45	85	16.0	0.32	4.0	0.40
55	80	15.5	0.31	4.1	0.38
70	75	15.0	0.30	4.2	0.36
80	70	14.5	0.29	4.3	0.34

A major criticism of phase feeding is that birds do not actually lay "percentages" of an egg. For example if a flock of birds is producing at 75% production, does this mean that 100% of the flock is laying at 75% or is 75% of the flock laying at 100% production. If the latter is true, then the concept of phase feeding may be harmful. If a bird lays an egg on a specific day, it can be argued that its production is 100% for that day, and so its nutrient requirements are the same regardless of the age of bird. Alternatively it can be argued that many of the nutrients in an egg, and especially the yolk, accumulate over a number of days, and so this concept of 100% production, regardless of age, is misleading.

Advocates of phase feeding indicate that birds can be successfully managed by reducing protein/amino acid contents of the diet - others suggest that nutrient specifications are too high to start with initially, and that phase feeding merely accomplishes normalization of diet in relation to requirement. The bottom line is that environmental and management conditions vary from flock to flock, and certainly from season to season within a flock. For this reason, the basis of phase feeding must be an accurate assessment of the nutrient intake relative to requirement for production, growth and maintenance. Some years ago we tested the bird's response to phase feeding of protein as shown in Table 4.18.

Prior to describing research results, it should be emphasized that modern strains of birds with reduced appetites require 1 - 2% more dietary crude protein than used in this older study. The 16%, 14% and 12% protein diets were fed throughout the complete laying cycle, while the 16 -14 -12% treatment resulted in dietary protein levels being reduced by 2% every four months. The 14% protein diet resulted in 2 grams per day less protein consumption relative to the 16% protein diet, however egg size was reduced by one gram. Feeding the 12% protein diet resulted in a lower rate of production and a further decrease in egg size as compared to the 16% protein diet. However, protein intake was four grams per day less for the 12 vs 16% protein diet. The 16 - 14 -12% diets resulted in production and egg size equal to that seen with the 16% protein diet, however this was accomplished with a saving of 2 grams of protein per hen per day.

TABLE 4.18 Influence of dietary protein level on hen performance

Protein level (%)	Production H.D.B. (%)	Av. egg wt. (g)	Av. daily protein intake (g)
16	73	59.3	18.2
14	72	58.3	16.0
12	69	57.8	13.7
16-14-12*	72	59.6	15.7

*16, 14 and 12% diets each fed consecutively for 4 months

There seems to be an advantage to decreasing the available phosphorus, and increasing the calcium level over time. These phase-feeding changes aid in maintaining eggshell quality, especially when there is an attempt at tempering egg size through use of lower protein diets.

4.5 FORMULATION CHANGES AND FEED TEXTURE

With diets formulated to least cost ingredient input, it is often necessary to change ingredient concentrations, and depending upon economic circumstances, the computer invariably "asks" for major changes at certain times. In these situations, nutritionists are often reluctant to make major ingredient substitutions in consecutive diets, on the basis that such change may adversely affect feed intake and hence production. In a recent study birds were fed a range of diets over a 12 month cycle, with the situation of absolute least cost where major changes in ingredient use occurred in most months. Control birds were also fed least cost formulated diets, although in this situation major ingredient changes from month to month were not allowed, rather these changes occurred more gradually. Birds responded reasonably well to these changes and no major adverse effects were seen. However, a slight improvement in egg production and egg size with a conventional (least cost) system, where diet changes were tempered to prevent drastic swings in diet composition, somewhat negated savings in feed cost seen with absolute least cost. The economic situation in terms of egg return minus feed cost was in favour of conventional least cost, mainly due to a doubling of the mortality rate with major diet changes. It seems that while the absolute least cost diets are initially attractive in reducing feed cost, they offer little overall economic advantage and generally pose an additional economic risk.

The texture of diets for laying hens is perhaps subject to more variability than for any other class of poultry. In some countries very fine mashes are used, whereas crumbles are used in other areas. There is little doubt that any type of feed texture can be made to work physically, although bird response is not always the same. Our research data suggests that regardless of nutrient profile, layers prefer large particles of feed. When layers were offered a crumbled diet, they showed a marked preference for the largest size particles available. Smaller particles of feed only started to disappear later within a 24 h period, when all the large particles had been eaten. There was no disappearance of very fine particles < 0.6 mm, although this result may be confounded with the break down of large particles. Feed intake was increased when birds were suddenly presented with feed of small particle size, and intake temporarily declined when birds were offered only large size particles.

A criticism of mash diets is that they tend to separate out when used in long runs of feed trough, and especially where continuous chain feeders are used. From a survey of commercial flocks in Ontario, we found comparable physical separation of feed with both mash and crumbles (Table 4.19).

TABLE 4.19 Particle segregation and calcium analysis of feed collected from farms using either mash or crumbles (%)

Type of feed	Particle Size (mm)	At Feed Tank	Distance along feed trough (m)			
			+18	+36	+72	+108
Crumbles	>2.36	46.0	29.8	25.3	20.6	16.0
	>1.18	28.8	26.5	25.5	24.7	23.7
	>0.85	6.9	9.4	10.1	10.9	11.1
	>0.71	3.4	5.5	6.1	6.7	7.1
	>0.60	3.2	5.6	6.2	6.7	7.1
	<0.60	11.7	23.2	26.8	30.3	33.8
	% Calcium	3.5	4.3	4.5	4.7	4.5
Mash	>2.36	17.3	10.0	8.3	8.5	10.5
	>1.18	22.7	21.1	20.0	19.6	21.0
	>0.85	11.9	13.4	13.2	14.5	15.1
	>0.71	7.2	8.9	9.0	9.2	9.0
	>0.60	7.4	8.6	9.0	9.3	8.2
	<0.60	33.5	38.0	40.5	38.9	36.2
	% Calcium	4.0	4.9	5.3	5.6	5.0

In this study, feed samples were taken directly from the feed tank and then at points progressively further from the initial point of distribution within the feed trough. Particle and nutrient separation were seen with all farms (Table 4.19). With crumbled feed, particle size was dramatically reduced as feed travelled along the trough, although this was not associated with any major change in nutrients. Higher calcium levels *per se* in the trough rather than the tank relate to feed samples in the trough including all feed in front of the bird that included fine particles beneath the feeder chain. Particle separation was also seen with the mash feeds, although this was only during the first 18 m run of the feed trough.

4.6 NUTRITION AND SHELL QUALITY

The laying diets shown in Tables 4.2 and 4.3 contain all the calcium needed by the layer under most conditions. However, if egg shell quality is a problem during hot weather, or if the pullets have come into production at a fairly young age and have peaked very quickly, it may be advisable to increase the levels of calcium by at least 0.4%. Research has indicated that a marked improvement in shell quality can be obtained by feeding part of the dietary calcium as oyster shell or limestone chips. This is especially true if limestone flour rather than a granular source of limestone is used. The hen's requirement for calcium is relatively low, except at the time of the day when egg shell formation is taking place. Figure 4.10 shows the accretion of shell mineral over a 24 hour cycle. The greatest rate of shell deposition occurs in the dark phase, when birds are not actively eating feed. The source of calcium during this period then becomes residual feed in the digestive tract and the labile medullary bone reserve.

Fig. 4.10 Shell mineral deposition over a 24h ovulation cycle

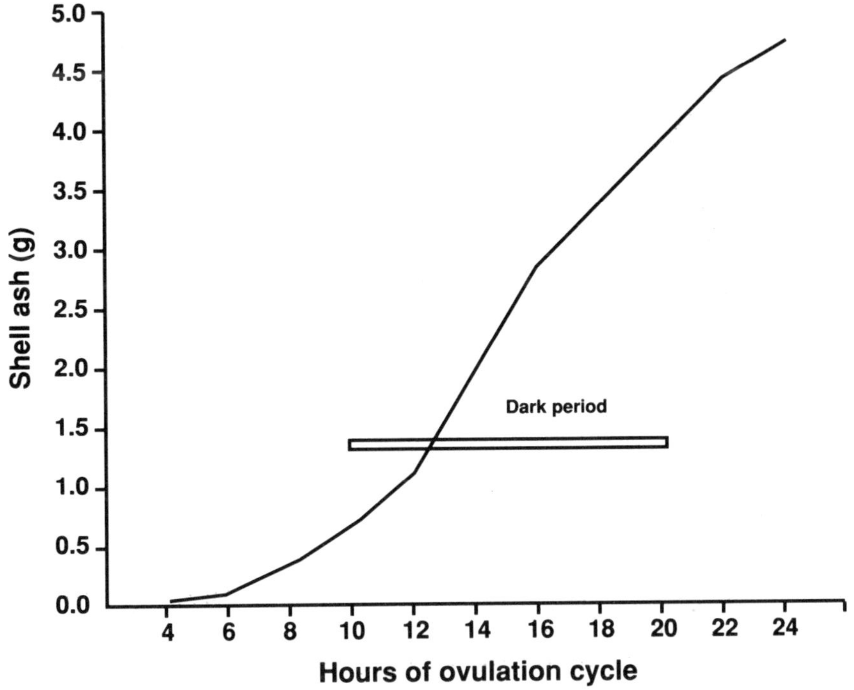

In the first 6 hours of the 24h laying cycle, there is virtually no shell deposition. This is the time of albumen and shell membrane secretion, and the time of redeposition of medullary bone. From 6 - 12 hr about 400 mg calcium are deposited, while the most active period is the 12 - 18 hr period when around 800 mg shell calcium accumulates. This is followed by a slower deposition of about 500 mg in the last 6 hr, for a total of around 1.7g shell calcium, depending upon egg size.

The voluntary intake of oyster shell or large particle calcium at various times during a normal 16 hour day is shown in Figure 4.11. With the control diet, the hen had no choice as to the time of day calcium was consumed. However, when given the simultaneous choice of diets providing energy, protein or calcium (E,P,Ca) the hen was able to select calcium at any time. Under these conditions, the hen consumed little or no calcium until the afternoon. This is the time of the day that the egg is usually in the shell gland, and the requirement for calcium should be higher at this time.

Fig. 4.11 Calcium intake on egg-forming days (Chah, 1972).

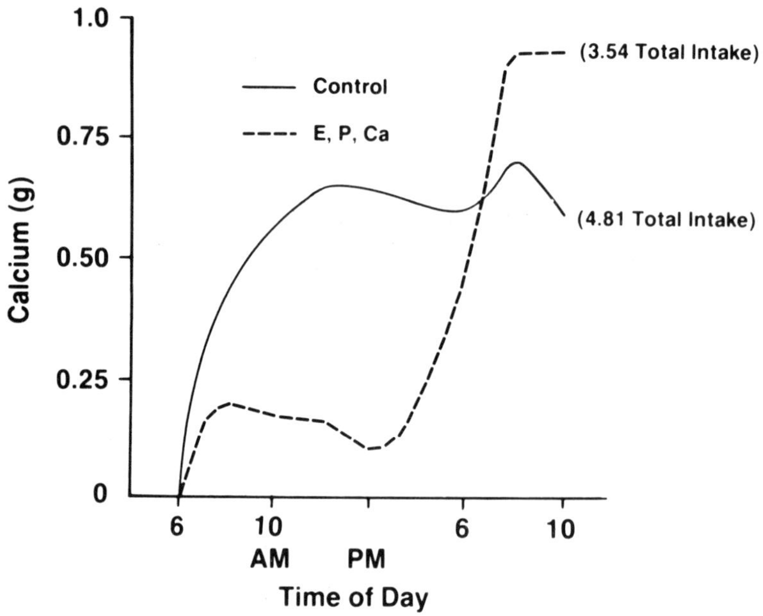

When feeding limestone chips or oyster shell, it is recommended that the diet contain 1 to 1.5% calcium and that the remainder be supplied by the supplemental source. The ideal time to feed this calcium supplement would be in the afternoon, since this is when the hen normally has a high calcium requirement. Since separate feeding of calcium is not very practical, the only apparent solution is to have the calcium supplement mixed in the feed. The hen has the opportunity of leaving the oyster shell or limestone chips until the latter part of the day when it is required. This type of feeding method is being used by a number of producers with very good results. The feeding of limestone or oyster shell on a continuous free choice basis, or on top of a diet containing the full calcium requirement, is not recommended. It has been shown that egg shells with chalky deposits and rough ends are probably a direct result of feeding too much calcium to laying hens. Feeding birds oyster shell *ad-libitum* can also result in the production of soft shelled eggs. This unusual circumstance is due to a deficiency of phosphorus. If too much calcium is ingested, it must be excreted, usually as soluble calcium phosphate. This can lead to a deficiency of phosphorus, which results in no medullary bone being redeposited between successive periods of calcification.

Calcium is the nutrient most often considered when shell quality problems occur, although it is realized that deficiencies of vitamin D_3 and phosphorus can also result in weaker shells. Vitamin D_3 is required for normal calcium absorption, and so if inadequate levels are fed, induced calcium deficiency quickly occurs. Results from our laboratory suggest that diets devoid of synthetic vitamin D_3 are quickly diagnosed, because there is a dramatic loss in shell weight (Fig 4.12). The same situation is seen in Fig. 4.13, when birds are fed deficient diets, and shell quality quickly deteriorates over 2 - 3 weeks. In this study, the basal diet was resupplemented with vitamin D_3 after 4 weeks, and there was rapid normalization of shell quality (Fig. 4.13).

Fig. 4.12 Decline in shell weight for hens fed diet devoid of Vitamin D_3 supplementation

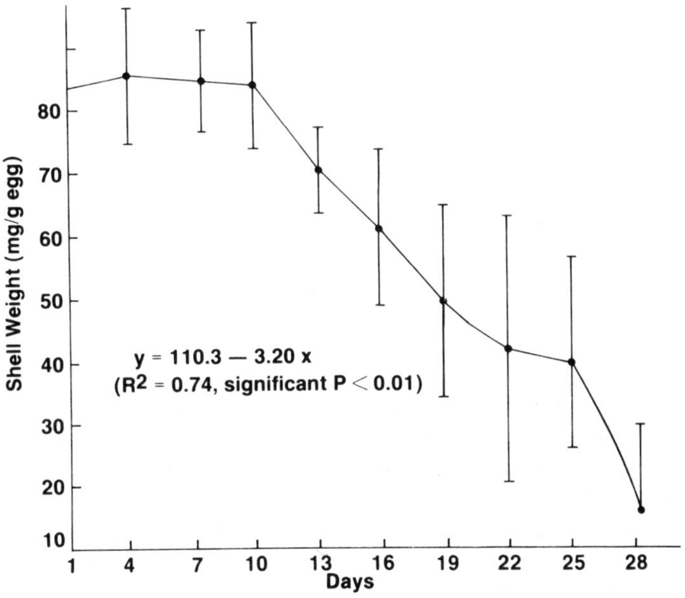

Regression Equation From 10-28 Days

However, a more serious problem occurs with sub-optimal levels of vitamin D_3, where changes in shell quality are more subtle but nevertheless of economic significance (Fig 4.13).

Fig. 4.13 Eggshell quality of layers fed various diets of vitamin D_3.

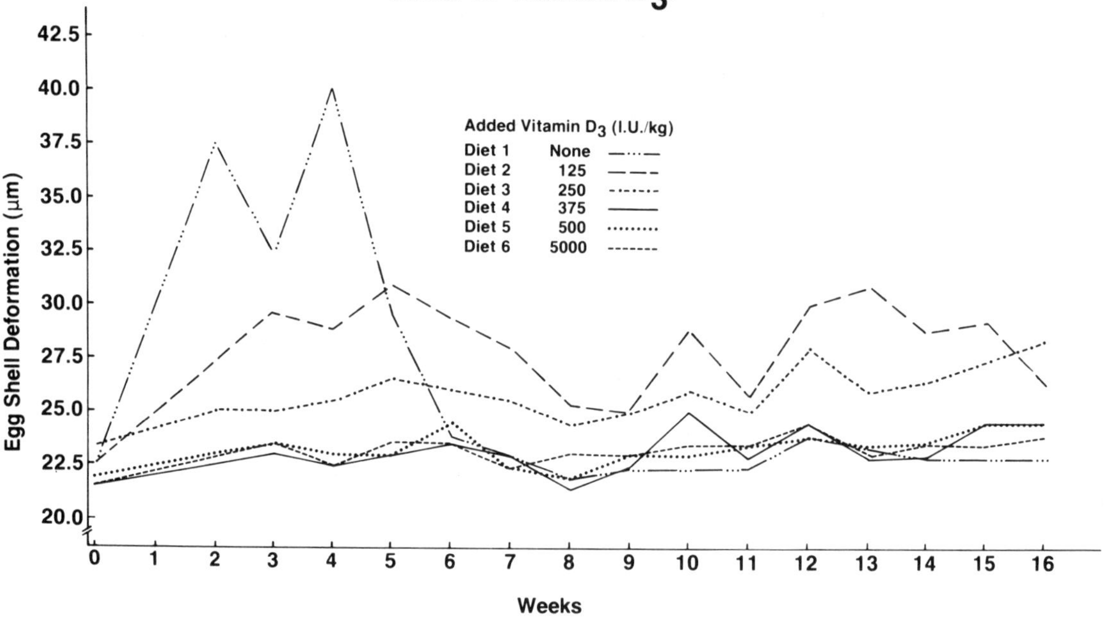

A major problem with deficiency of vitamin D_3 is that this nutrient is very difficult to assay in complete feeds. It is only at concentrations normally found in vitamin premixes, that meaningful assays can be carried out, and so if D_3 problems are suspected, access to the vitamin premix is usually essential. In addition to uncomplicated deficiencies of D_3, problems can arise due to the effect of certain mycotoxins. Compounds such as zearalenone, that are produced by *Fusarium* molds, have been shown to effectively tie up vitamin D_3, resulting in poor egg shell quality. Under these circumstances dosing birds with 300 IU D_3 per day, for three consecutive days, with water soluble D_3 may be advantageous.

Minimizing phosphorus levels is also advantageous in maintaining shell quality, especially under heat stress conditions. Because phosphorus is a very expensive nutrient, high inclusion levels are not usually encountered, yet limiting these within the range of 0.3 to 0.4%, depending upon flock conditions, seems ideal in terms of shell quality. Periodically, unaccountable reductions in shell quality occur and it is possible that some of these may be related to nutrition. As an example, vanadium contamination of phosphates causes an unusual shell structure, and certain weed seeds such as those of the lathyrus species, cause major disruptions of the shell gland.

Up to 10% reduction in eggshell thickness has been reported for layers fed saline drinking water, and a doubling in incidence of total shell defects seen with water containing 250 mg salt/litre. If a laying hen consumes 100g feed and 200 ml water per day, then water at 250 mg salt/ litre provides only 50 mg salt compared to intake from the feed of around 400 mg salt. The salt intake from saline water therefore seems minimal in relation to total intake, but nevertheless shell quality problems often occur under these conditions. It appears that saline water results in limiting the supply of bicarbonate ions to the shell gland, and that this is mediated via reduced activity of carbonic anhydrase enzyme in the mucosa of the shell gland. However it is still unclear why saline water has this effect, in the presence of overwhelmingly more salt as provided by the feed. There seems to be no effective method of correcting this loss of shell quality in established flocks, although for new flocks the adverse effect can be greatly reduced by adding 1g vitamin C/litre drinking water.

4.7 FEED INTAKE PATTERNS AND CHOICE FEEDING

Most laying hens are fed complete diets free choice, and so there is little fluctuation in intake on a daily basis. However, throughout the day, peaks in feed intake occur immediately after "lights-on", and again around 1 - 2 hr prior to "lights-off". The main peak in feed intake occurs some 2 hours before "lights-off" (Table 4.20).

TABLE 4.20 Feed intake and egg production of hens during a 16 hour day (6:00 a.m. to 10:00 p.m.)

Time	Feed intake (g)	% of the day's eggs laid
6-8 am	9.3	0
8-10 am	12.7	19.2
10-12 pm	14.7	44.4
12-2 pm	17.0	29.3
2-4 pm	15.3	5.1
4-6 pm	14.0	2.0
6-8 pm	17.1	0
8-10 pm	14.1	0

Chah (1972)

The peak in feed intake later in the day is undoubtedly an attempt by the bird to increase its calcium intake for impending shell calcification. If the bird is offered diets that allow self selection of nutrients, then some interesting patterns are seen. The intake of calcium, when separate calcium, energy and protein sources are offered, has already been detailed in Fig 4.11. Figs 4.14 and 4.15 show comparable intakes for energy and protein (Chah 1972). The interesting features of these nutrient intakes are the biphasic peaks, and the overall reduced intake relative to a conventional diet.

Fig. 4.14 Energy intake on egg forming days (Chah, 1972)

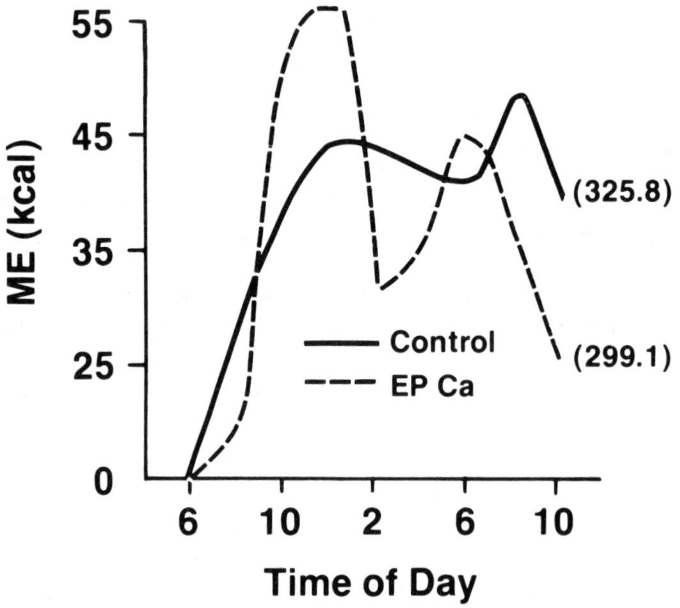

Fig. 4.15 Protein intake on egg forming days (Chah, 1972)

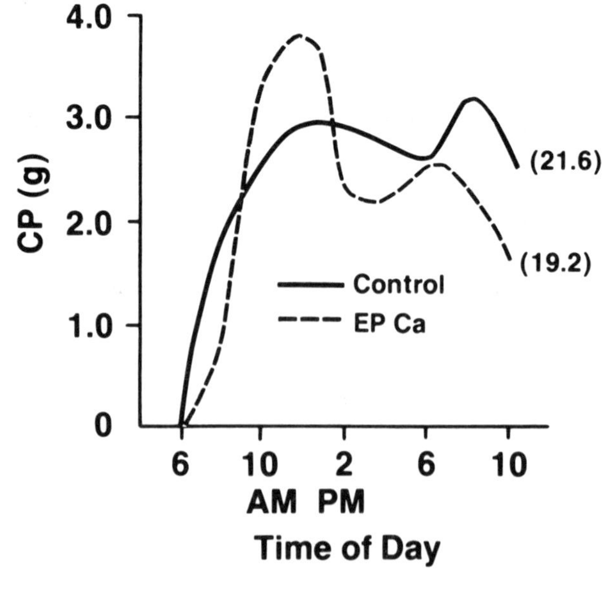

There was an increased intake of protein and energy in the morning about 5 - 7 hours after the lights came on at 6:00 a.m. Egg production, as a percent of the total eggs laid throughout the day, is shown in Table 4.20. As noted, and this is true for all flocks, the eggs for any one day are laid over a relatively short period of time (4 - 5 hours in this study). Egg production peaked at around 11 am and it was at this time that energy and protein intake also peaked. The hen will ovulate the next day's egg about 30 minutes post-laying. Consequentnly, there is a great deal of metabolic activity taking place and this may account for the high protein and energy intake at this time.

It is obvious from the feed consumption pattern that the hen has a daily cyclic rhythm rather than a constant intake throughout the day. However, the manner in which most hens are fed does not allow them to meet this cyclic need without consuming an excess of energy at certain times of the day. To this end the regular type of feeding program may be largely responsible for producing overweight birds. This idea leads to the concept of so called "choice-feeding".

In an attempt to utilize the information presented in the above studies, an experiment was set up in which laying hens were offered two diets in feed troughs that were partitioned down their length. Birds had a choice of diets with specifications shown in Table 4.21. The high calcium diet contained 13% calcium (or 34% limestone), and was pelleted in order to ensure that palatability was not a problem. Performance data of birds fed these diets is shown in Table 4.22.

TABLE 4.21 Diets fed to hens in a choice feeding test (% composition)

Ingredient	Control	Choice feeding	
Corn	64.0	69.9	20.0
Soybean (48% CP)	23.0	26.0	12.0
Barley	-	-	25.0
Animal-vegetable blend fat	1.0	1.0	1.0
Limestone	6.65	-	34.77
Calcium phosphate	2.0	2.0	-
Iodized salt	.25	.25	.25
Mineral:Vitamin Mix	.75	.75	.75
Methionine	.1	.1	.1
L-Lysine	.1	-	.1
Alpha floc cellulose	2.15	-	6.33
Calculated analysis			
Protein (%)	17.1	19.1	10.7
Energy (kcal ME/kg)	2794	3065	1740
Calcium (%)	3.0	0.47	13.1

TABLE 4.22 Performance of Leghorn hens offered a conventional diet or a choice feeding situation (20-40) weeks

Treatment	Egg Production (%)	Egg wt. (g)	Feed intake (g/bird/d)	Energy intake (kcal/d)	Wt. gain (g)	Carcass fat (%)
Control	79.2	60.8	118.4	331	209	50.6
Choice feeding	79.3	61.5	110.7	301	289	51.5

When offered a choice based on calcium, birds consumed about 6 - 7% less feed overall. This confirms the earlier work reported where birds had a choice of three diets. It is also of interest to note that there was increased weight gain with the free choice group even though egg size was similar and feed intake reduced. This situation suggests that the hen is obtaining more available energy by consuming the energy and protein at a different time of day relative to intake of calcium. It has been suggested that a high level of calcium in a laying diet does significantly reduce the metabolizable energy content of a diet. It can also be seen that body composition of end of lay birds from the two dietary treatments was similar indicating that the increased weight gain was not all as body fat but also partly as lean body mass.

Since conducting these studies, numerous other researchers have confirmed these findings, suggesting that the bird is capable of balancing its own diet when offered an array of ingredients. By closely tailoring individual nutrient intake to metabolic requirement at specific times of the day, there seems to be an overall saving in nutrient and feed intake. However, this improved feed efficiency must be considered in relation to the physical and mechanical problems related to simultaneously offering two or more diets to laying hens. A form of choice feeding has been attempted whereby layers are offered a mix of oyster shell, whole cereal grains and a pelleted protein supplement. Choice feeding of laying hens offers an interesting alternative to a complete laying diet. It may permit the hen to do a better job of meeting her daily cyclic requirement for nutrients as compared to using a complete or single diet with which she has to consume extra energy and protein in the afternoon when she has a specific appetite mainly for calcium.

4.8 DIET AND EGG SIZE

Increasing the hen's intake of balanced protein will result in an increase in egg size while feeding higher levels of protein at the onset of production may help to increase egg size more rapidly. For strains of birds that produce many extra large eggs during the latter part of their egg production cycle, lowering the level of dietary protein during this period will result in slightly smaller and more uniform eggs.

In these situations, when considering changes to the level of dietary protein, the energy content of the diet must also be taken into account. If diets are suboptimal in energy, little increase in egg size will be noted by increasing the level of protein since the hen will utilize protein to meet requirements for energy. Indeed one of the main factors limiting early egg size is that energy intake is suboptimal in many flocks that come into production at a young age and rise to peak production very rapidly. Where responses in egg size have been noted by adding tallow to laying diets, the rations (in most cases) have been low in energy and the increased energy has resulted in less protein being utilized for energy purposes. The effects of protein and energy on egg size are clearly demonstrated in Fig 4.16, which depicts the bird's response to a range of nutrient intakes. Unlike the situation with egg production, (Fig 4.1), there is an obvious relationship between increased egg size and increased protein intake. At low protein intakes (less than 14 - 15 g/d) there is an indication of reduced egg size when energy intake is increased.

Fig. 4.16 Egg weight (18-66 weeks) in response to intakes of energy and protein.

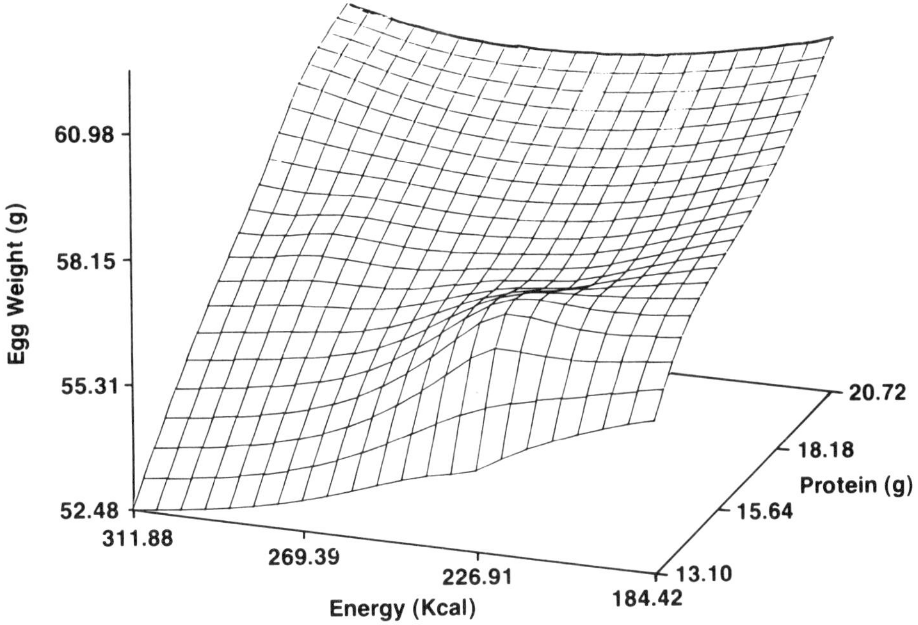

While it is fairly well established that levels of dietary crude protein and certain amino acids will influence egg size. It is not always clear if the responses recorded to crude protein *per se* can be totally attributed to intake of component amino acids. Table 4.23 shows a summary of 6 experiments reported by Waldroup where a range of methionine levels were tested, at 0.2% cystine, for various ages of bird.

Table 4.23 indicates that as methionine level of the diet is increased, there is an almost linear increase in egg size. As the bird progresses through a production cycle, the egg weight response to methionine changes slightly. In the first period, between 25 - 32 weeks, using 0.38 vs 0.23% methionine results in a 5.6% increase in egg size (Table 4.23). Comparable calculations for the other age periods show 7.3% improvement from 38 - 44 weeks, and 6.7% and 6.0% at 51 - 58 and 64 - 71 weeks respectively. The egg weight response to methionine, therefore closely follows the normal daily egg mass output of the laying hen. In another recent study Jensen shows a generalized curvilinear response in egg weight to graded levels of methionine (Fig. 4.17).

TABLE 4.23 Effect of methionine on egg weight - mean of 6 experiments

	% Diet Methionine[1]					
Bird age (wks)	**0.23**	**0.26**	**0.29**	**0.32**	**0.35**	**0.38**
25-32	49.8	51.0	51.9	52.1	52.0	52.6
38-44	53.2	55.0	56.4	56.3	56.3	57.1
51-58	56.2	57.9	59.6	59.2	59.2	60.0
64-71	56.8	59.4	59.5	59.5	59.5	60.2

[1](@ 0.2% cys). Adapted from Waldroup et al. (1995)

Fig. 4.17 Egg weight response to methionine
(Adapted from Calderon & Jensen, 1990)

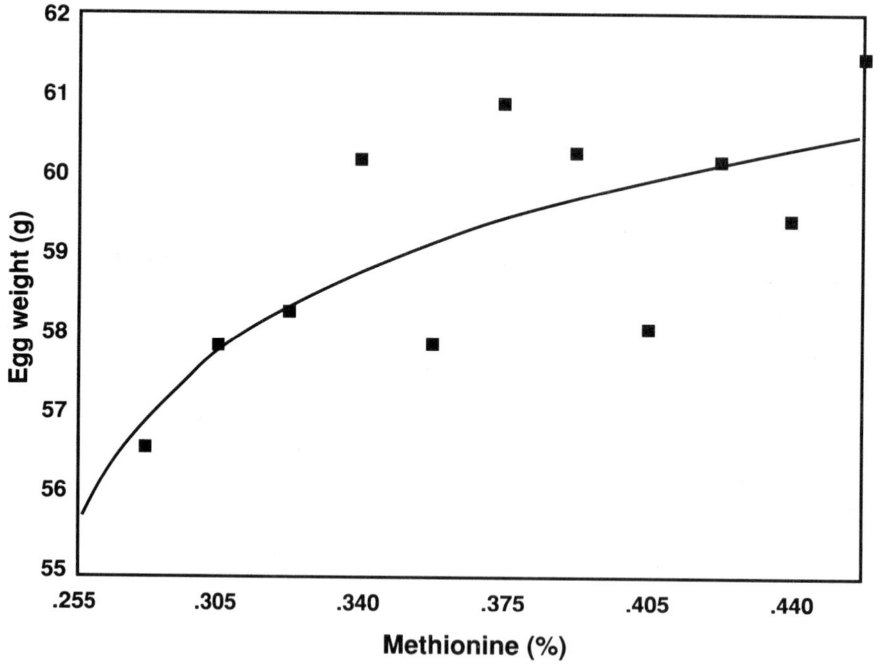

Over the last few years, there has been considerable research involving the source of methionine as it affects layer performance. When comparing DL-methionine with Alimet, Harms shows the classical response of egg weight to both methionine sources (Table 4.24).

TABLE 4.24 Effect of methionine source on layer performance[1]				
Diet methionine (%)	Exp #1, Egg weight (g)		Exp #2, Egg weight (g)	
	DL	Alimet	DL	Alimet
0.228 (basal)	54.5	54.5	51.5	51.5
0.256	56.2	55.3	53.2	52.7
0.254	56.8	56.8	55.1	56.2
0.311	57.6	57.2	55.9	55.7
0.366 - 378	58.0	57.5	57.0	56.8
[1]Mean 80% egg production			Harms and Russell 1994	

There has been a suggestion that L-methionine may in fact be superior to any other source. This product is not usually produced commercially, because routine manufacture of methionine produces a mixture of D- and L-methionine. This is the only amino acid where there is apparently 100% efficacy of the D-isomer. However most research data indicates no difference in potency of L- vs DL-methionine sources.

One role of methionine is as a methyl donor, and so the efficacy of methionine vs choline is often discussed. While choline can spare some methionine in a diet, it is obvious that there are severe limitations to this process, and this becomes most obvious when egg size, rather than simply egg production, is a major consideration. Data from Parsons clearly shows the advantage of using methionine over choline in terms of egg size, and that this effect becomes most critical as diet crude protein level is reduced (Table 4.25).

TABLE 4.25 Egg size with methionine vs choline (23-35 wks)			
Diet protein	Supplement	Egg prod. (%)	Egg weight (g)
16.0%	None	82.8	53.2
	0.1% methionine	84.0	56.6
	0.1% choline	82.4	54.0
14.0%	None	72.8	52.5
	0.1% methionine	84.5	54.9
	0.1% choline	78.9	51.9
			Adapted from Parsons and Leeper (1984)

Attempts at reducing or tempering egg size later in the production cycle by phase feeding of protein or methionine have met with only limited success, likely due to the fact that producers are reluctant to use very low protein diets. Our studies indicate that protein levels around 13% and less are necessary to bring about a meaningful reduction in egg size (Table 4.26). However with protein levels much less than this, loss in egg numbers often occurs.

TABLE 4.26 Effect of reducing dietary protein level on egg size of 60 week old layers (Av. for 2, 28-day periods)

Dietary protein level (%)	Egg production (%)	Av. feed intake per day (g)	Egg wt. (g)	Daily Egg mass (g)	Av. protein intake per day (g)
17	78.8	114	64.8	51.0	19.4
15	77.5	109	64.3	49.7	16.4
13	78.3	107	62.2	49.1	13.9
11	72.7	108	61.7	45.1	11.9
9	54.3	99	58.2	36.1	8.9

All diets 2800 kcal ME/kg

Methionine levels can also be adjusted in an attempt to control late cycle egg size. Results of Peterson show the tempering of egg size with reduced methionine levels (Table 4.27).

TABLE 4.27 Methionine and late cycle egg size (g)

Daily methionine intake (mg/day)	Exp. 1 38-62 wk	Exp. 2 38-70 wk	Exp. 3 78-102 wk
300	60.1[a]	63.7[a]	66.3[a]
285	60.3[a]	63.1[b]	65.5[b]
270	59.1[ab]	62.0[c]	64.0[c]
255	58.5[b]	62.0[c]	63.9[c]
Average egg prod (%)	86	80	75

Adapted from Peterson et al. (1983)

However, these results are often difficult to achieve under commercial conditions because reduction in diet methionine levels often leads to loss in egg numbers and body weight. Phase feeding of amino acids must, therefore, be monitored very closely. As stated at the outset of this section, mature body weight is the main determinant of egg size, and this applies particularly to late-cycle performance. The best way to control late cycle egg size is through manipulation of body-weight at time of light stimulation. Larger birds at maturity will produce much larger late cycle eggs and vice versa. There is an obvious balance necessary between trying to reduce late cycle egg size without unduly reducing early cycle egg size.

Waldroup recently outlined estimates of methionine and methionine + cystine requirements for both egg production and for egg weight/mass (Table 4.28).

TABLE 4.28 Estimated methionine and methionine + cystine requirements (mg/day)

	Bird age (wks)	Egg #	Egg weight	Egg mass
Methionine	25-32	364[b]	356[b]	369[b]
	38-45	362[b]	380[a]	373[b]
	51-58	384[a]	364[a]	402[a]
	64-71	374[ab]	357[b]	378[b]
Methionine + cystine	25-32	608[b]	610[ab]	617[b]
	38-45	619[b]	636[a]	627[b]
	51-58	680[a]	621[ab]	691[a]
	64-71	690[a]	601[b]	676[a]

Adapted from Waldroup et al. (1995)

During peak egg mass output (38 - 45 weeks) the methionine requirement for egg size is greater than for egg numbers, while the latter requirement peaks at 51 - 58 weeks of age. If these data are verified in subsequent studies, it suggests that we should be very careful at reducing methionine levels much before 60 weeks of age.

It is well known that linoleic acid also affects egg size. Although there is some controversy as to the level required for maximum egg size, it would appear that a level of 1% in the diet should be adequate. Results from a trial aimed at increasing early egg size with high levels of linoleic acid are shown in Table 4.29. In this study linoleic acid had no effect on egg size. It should be pointed out that this particular flock matured early and peaked very rapidly. It might well be that the pullets were producing maximum egg mass with respect to body size and thus could not use additional nutrients to advantage.

TABLE 4.29 Influence of linoleic acid intake on early egg size (average performance, 18-34 weeks)

Treatments	Egg Production (%)	Egg wt. (g)	Feed (g/d)	Linoleic acid intake (mg/day)
Carbohydrate	64.3	49.1	86.8	105[c]
10% Tallow	57.5	49.5	82.5	305[b]
10% Corn oil	60.1	49.9	88.6	4607[a]

All diets contained 15% protein
[a-c] means followed by different letters are significantly different

It seems unlikely that birds will respond to linoleic acid supplementation when corn-based diets are used. For example, in a recent study (Fig. 4.18), there is an apparent response in egg weight to adding safflower oil rather than tallow to a layer diet. However this wheat-based diet contained only 0.6% linoleic acid, and so a response to linoleic acid (in this case from safflower oil) is to be expected.

Fig. 4.18 Egg weight response over time to diet fats
(March and McMillan, 1990)

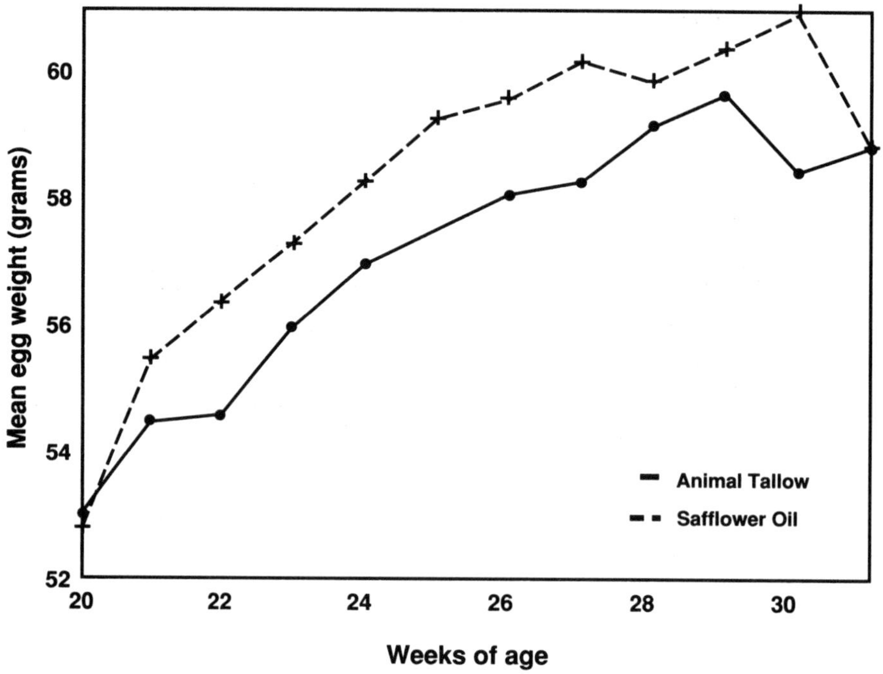

4.9 DIET AND EGG COMPOSITION

Prior to discussing the potential manipulation of egg composition through diet changes, it is perhaps pertinent to outline some basic data on normal egg composition. Tables 4.30 - 4.32. detail some chemical and physical characteristics of eggs.

TABLE 4.30 Major egg components (Romanoff and Romanoff, 1949)

		Albumen layer				
	Yolk	Outer	Middle	Inner	Chalaziferous	Shell
Weight (g)	18.7	7.6	18.9	5.5	0.9	6.2
Water (%)	48.7	88.8	87.6	86.4	84.3	1.6
Solids (%)	51.3	11.2	12.4	13.6	15.7	98.4
				Yolk	Albumen	Shell
Protein (%)				16.6	10.6	3.3
Carbohydrates (%)				1.0	0.9	-
Fat (%)				32.6	Trace	Trace
Minerals (%)				1.1	0.6	95.1

TABLE 4.31 Vitamin and mineral composition

Vitamins	Amount per 60 g egg (less shell)
A (I.U.)	300
D_3 (I.U.)	30
E (I.U.)	2
K (mg)	.02
B_1 (mg)	.06
B_2 (mg)	.18
B_6 (mg)	.20
B_{12} (mg)	.001
Pantothenic acid (mg)	1.2
Folacin (mg)	.008
Niacin (mg)	.06
Choline (mg)	350
Biotin (mg)	.01
Minerals (mg)	
Calcium	30
Phosphorus	130
Sodium	75
Chloride	100
Potassium	80
Magnesium	7
Manganese	2
Iron	1
Copper	.2
Zinc	1
Iodine	.02
Selenium	.01

TABLE 4.32 Egg components most often considered in human nutrition

Nutrient	Two eggs supply the following of an adult's daily requirement (%)
Protein	20
Energy	8
Calcium	10
Phosphorus	20
Iron	20
Vitamin A	25
Vitamin D_3	20
Thiamin	10
Riboflavin	30
Niacin	15

Yolk Color

Because corn is used in layer diets in many countries around the world, the consumer is most often receptive to eggs with a fair degree of yolk pigmentation. The nutritionist is therefore most often faced with the problem of maintaining a given yolk color, either through selection of ingredients and/or use of synthetic pigments. Fig 4.19 shows the effect of corn, and corn plus alfalfa meal on egg yolk color.

Fig. 4.19 Effect of diet on egg yolk color.

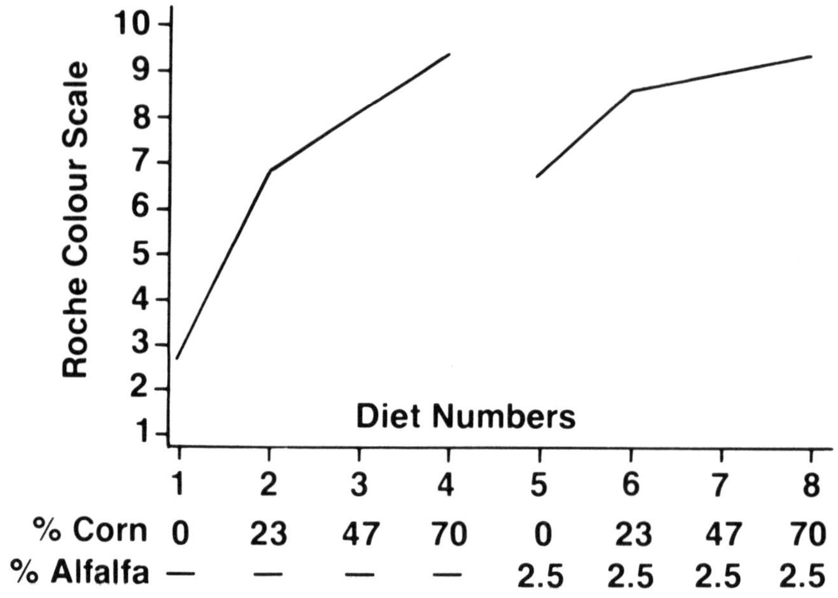

The basal ration employed contained only wheat as the cereal grain with corn replacing wheat for the various treatments. A Roche color fan number of 7 - 8 will usually be accepted for grade A eggs in most areas. Because corn and dehydrated alfalfa meal, or cereal grass, can vary in xanthophyll content throughout the year, one must vary the proportion of these ingredients in a ration if a constant yolk color is to be maintained. As a guide, it can be assumed that 1% of a dehydrated green feed will result in approximately the same yolk color as 20% of corn in a ration. Synthetic xanthophylls are now most commonly used to control yolk color. Approximately 7 to 8 g of xanthophyll per tonne of feed will usually give the desired yolk color. A level below 5 g per tonne of xanthophyll may result in egg yolks that are too pale to be readily accepted by the consumer (Fig 4.20).

Fig. 4.20 Roche color scale and dietary xanthophylls.

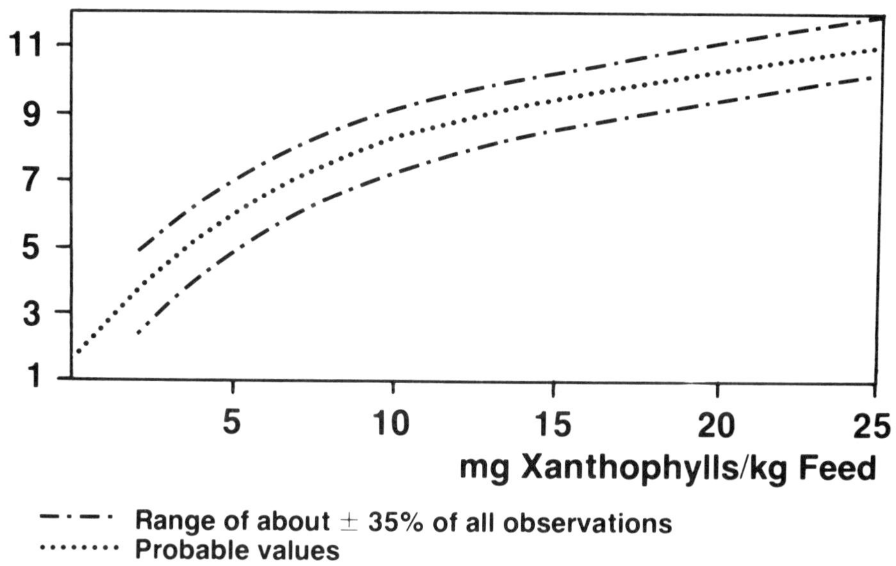

Carotenes have little pigmenting value for poultry, although the various hydroxylated carotenes are excellent pigments and the bird preferentially stores these in yolk, body fat, and its shanks. Yolk color is most easily determined in the field by use of the "Roche" fan, which provides a set of standard color shades from pale yellow to red/orange. The red/orange colors can be induced with products such as canthaxanthin, although usually this degree of coloring is unacceptable to most consumers. These pigments can be used in limited quantities as long as the diet has a base level of xanthophylls - otherwise the yolk color tends towards an objectionable red, rather than acceptable orange color. These red pigments also produce undesirable color in noodles made from egg yolk, and so care must be taken in selection of pigmenting agents in eggs destined for industrial uses.

There are a number of dietary and management factors which can reduce the effective deposition of xanthophyll in the yolk. Ingredients which are potential oxidizing agents, such as minerals, and certain fatty acids have been shown to reduce pigmentation. High levels of vitamin A, as sometimes used during water medication for various stress situations, have been shown to cause temporary loss in yolk pigmentation. High environmental temperatures, coccidiosis and aflatoxin contamination of feed are also implicated in production of pale colored yolks. Natural pigments in cereals tend to decline with prolonged storage, with up to 50% loss reported at elevated temperatures. Without blending of corn therefore, a slight natural loss in pigments is expected to cause subtle loss in yolk color throughout the year. Yolk color seems to be enhanced when high levels of vitamin E are used, and when the diet contains antioxidants.

Yolk Mottling

Egg yolk mottling continues to be a problem that appears sporadically in a number of flocks. Although the condition has been known for some time, there appears to be no definite evidence as to its cause or of ways to alleviate it. Diet has been implicated, but there is no real evidence that nutrition is a factor with the majority of mottling problems that appear. However, it is known that certain feed additives such as nicarbazin can cause a mottling condition if they are inadvertently added to a laying diet. By and large most yolk mottling is reported in the spring of the year and seems to disappear during the summer or fall. However, whether the season of the year or the type of laying house management is a factor has not been proven. Table 4.33 shows the result of a test in which eggs held for various lengths of time and under different environmental conditions, were checked for severity of yolk mottling. It is evident that the majority of mottling appears during storage. Storing, even at ideal temperature for one week, can result in a marked increase in the condition. It has been suggested that the modern strains of birds are more prone to yolk mottling than were birds of some years ago, although the results in Table 4.34 do not confirm this assumption.

TABLE 4.33 Yolk mottling as influenced by temperature

	Haugh units	Yolk color index	Severity of mottling (%)
Fresh eggs	85.4	11.3	7.1
Eggs held 1 week at 12.5°C	70.8	10.9	45.6
Eggs held 2 weeks at 12.5°C	66.7	10.9	44.0
Eggs held 1 week at room temperature	-	-	47.6
Eggs held 2 days at 31.7°C	-	-	60.0

TABLE 4.34 Influence of strain of bird on yolk mottling

Strain	Mottling severity (%)
Jersey Black Giant	60.0
Rhode Island Red	36.4
Brown Leghorn	22.2
Columbian Rock	33.3
White Rock	47.2
Columbian Leghorn	47.4
White Leghorn A	42.4
White Leghorn B	33.3

Where the interior quality of eggs is a problem, it has been suggested that the cereal grain employed in the ration could be a contributing factor. Table 4.35 shows the results of a study utilizing diets with corn, wheat, or barley as the sole cereal grain. It can be seen that little if any difference in egg quality measurements were noted in response to the three cereal grains. As expected, the yolk color of eggs from birds fed the wheat and barley rations was lower than for birds fed corn. There is a great deal of evidence to indicate that the type of diet has little influence on the Haugh unit value or albumen height of eggs, unless the diets are high in vanadium. Consequently, if egg quality becomes a problem, and is not due to the normal decline that occurs with age, or as a result of high temperatures, a producer should look for something other than feed as the likely source of the problem.

TABLE 4.35 Effect of cereal grain on egg quality

	Egg Production HDB (%)	Egg weight (g)	Shell weight (g)	Haugh units		Roche Yolk color	
				fresh	held[1]	fresh	held
Corn	70.3	62.3	5.57	72.6	61.0	10.2	10.0
Wheat	69.2	61.5	5.46	75.3	62.8	4.0	3.4
Barley	70.1	62.6	5.68	76.1	64.8	3.8	3.3

[1]Eggs held for 7 days at 15°C

Albumen quality

Over the last few years, there has been increasing concern about the quality of the thin albumen in fresh eggs. Most measures of albumen quality, such as Haugh unit, only measure characteristics of the thick albumen, and so apparently "Grade A" eggs can have problems with the thin albumen. In certain birds, we see the area of this albumen to be as much as 120 sq. cm., compared to 60 - 70 sq. cm for a "normal" egg. These spreading albumens are especially problematic in the fast-food industry where eggs are prepared on flat-surface grills. We have tested various levels of protein and amino acids, and fed birds diets of vastly different acid base balance, and seen no effect on this phenomena. We have selected birds producing normal and "spreading" thin albumens, and after inseminating with semen from a common male, find their offspring to show similar characteristics. The current thin albumen problem therefore seems to be an inherited characteristic.

Cholesterol

Over the last few years, there has been considerable publicity regarding the association between heart disease in humans and cholesterol intake. Since the egg contains considerable quantities of cholesterol, it has obviously come under close scrutiny. Recent analytical results show the egg to contain much less cholesterol than stated in classical reference material. Our data show a large-grade egg to contain around 198 - 208 mg cholesterol. While this is about 25% less than general standards, it is still quite high in relation to most other foods. For this reason, there has been considerable interest in the development of so called low cholesterol eggs, and invariably layer diet modification is considered in these programs.

Eggs are high in cholesterol because of its importance in sustaining the developing embryo. Cholesterol has many varied uses in the developing embryo including its role as a structural component of cell membranes, and as a precursor for adrenal and sex hormones, vitamin D, and the bile acids. Young chicks do not have the enzymes necessary for cholesterol synthesis, which emphasizes the importance of cholesterol being deposited in the egg.

Factors that influence egg cholesterol content include the hen's body weight and her intake of energy and fat. Dietary fat *per se* does not seem to be a factor, although in most instances high fat diets imply that high energy diets are used. The influence of dietary energy and body weight of the hen on egg cholesterol is mediated through their effect on egg size. Reducing energy intake in order to achieve a measurable reduction in egg cholesterol concentration has the disadvantage of adversely affecting egg production.

Dietary fibre influences cholesterol metabolism by a possible combination of different processes. These include lowered cholesterol absorption, binding with bile salts in the intestinal tract, shortening the intestinal transit time, and increasing fecal sterol excretion. Alfalfa seems to be the most effective of all the fibre sources tested that concomitantly has the least harmful effects on egg size, egg production, and feed efficiency. Reduction in egg cholesterol achieved by such dietary manipulations are, however only marginal, with little evidence to suggest a commercially significant reduction.

The reasons behind the lack of success in attempts to lower egg cholesterol content relates to the physical structure of egg fat components. Egg cholesterol level is determined by the cholesterol content of individual yolk lipoprotein fractions, rather than by the bird's plasma cholesterol concentration. Given that most cholesterol in lipoproteins is associated with the surface layers, reduction in egg cholesterol content can therefore occur only when the lipoprotein particle size is increased. Such a scenario will reduce the contribution of surface cholesterol molecules relative to total fat. Unfortunately, an increase in lipoprotein particle size will tend to reduce the efficiency of the critical transport of bigger sized "molecules" through the follicle wall. It seems, therefore that meaningful reduction in egg cholesterol content will likely accrue only through the genetic manipulation of processes involved in lipoprotein synthesis and transport to the developing follicle.

Egg fatty acid profile

The other major factor concerning egg content and human health, is the saturated fatty acid profile of the egg. Fortunately altering fat saturation in the yolk is much easier than influencing cholesterol level. Long term this may be the best route of research in an attempt to develop so-called "healthier" eggs. By feeding predominantly unsaturated fatty acids to the bird, the fatty acid profile of the egg can be moved favourably towards increased unsaturated fat content. Of the unsaturated fatty acids consumed by humans, two are receiving particular attention with respect to cardiovascular disease. Eicosapentaenoic acid (20:5 w-3) is commonly found in fish oils, and is the fatty acid that is thought to confer some protection against coronary heart disease in native people subsisting on a fish diet. Feeding layers and broiler chickens a diet with high levels of red fish oils can influence eicosapentaenoic levels in eggs or meat, although obviously this is associated with undesirable fishy taint in the eggs and meat. More recently there has been interest in the association between linolenic acid (18:3 w-3) intake and coronary heart disease. Patients suffering from CHD seem to have lower levels of linolenic acid in their adipose tissue. Linolenic acid is a precursor of prostaglandin E, which is reported to be a coronary vasodilator, an inhibitor of free fatty acid release (as occurs during acute CHD) and is one of the most potent known inhibitors of platelet aggregation. Unfortunately, the diet of most humans is not well fortified with linolenic acid, which is most commonly found in plant tissues. However the chicken has the somewhat unique ability to divert increased quantities of linolenic acid into the egg when its diet contains high levels of this nutrient. Again, most layer diets do not contain high levels of linolenic acid, yet there is one potential ingredient which is high in this fatty acid and so deserves some attention. This ingredient is flax, and our preliminary studies indicate a tremendous increase in egg linolenic acid content when the bird is fed up to 30% dietary flax seed (Table 4.36).

TABLE 4.36 The effect of dietary flax on egg composition (%)

Fatty Acid	Pretest	10% Flax	30% Flax
16:0 (Palmitic)	31.7	29.2	26.8
18:0 (Stearic)	8.3	8.3	8.3
18:1 (Oleic)	42.2	37.3	33.5
18:2 (Linoleic) 18:2w-6	13.9	15.6	14.8
18:3 (Linolenic) 18:3w-3	0.3	6.2	13.3
20:4 (Arachidonic)	1.7	0.9	0.7
22:6 (Docosahexaenoic)	0.5	1.0	1.1

In addition to there being a major increase in linolenic acid content of the egg due to feeding flaxseed, these eggs also contain less total saturated fatty acids. Unfortunately, flax seeds have some undesirable side effects when fed at these levels, yet hopefully these can be overcome through research and development. The reason for the bird accumulating so much linolenic acid in the egg, rather than converting it to other fatty acids as occurs with mammals, is not fully understood. It is conceivable that the egg is being used as a dumping ground for this somewhat unusual compound. As medical research continues to unravel the relationship between linolenic acid and CHD, it is likely that this fatty acid will attract more attention in the future especially from those consumers at risk from this disease, as well as from the growing number of people concerned about their nutritional habits. The egg may be ideally suited to meet some of these needs, and feeding programs for layers could be manipulated accordingly.

Vitamin enrichment of eggs

The increasing consumer awareness concerning nutritive value of foods has precipitated interest in elevating the vitamin content of eggs. As a consequence, the interest in modifying egg vitamin levels now extends beyond production considerations to designing a high quality food for human consumption that is of consistent nutrient composition. Factors that influence the egg vitamin content include nutrition, age, and genetics of the laying hen. The influence of dietary vitamin intake on vitamin enrichment of the egg is very variable between different vitamins. For example, egg riboflavin level in yolk and albumen responds rapidly to the dietary level of this vitamin. Similarly, the egg content of vitamin B_{12} is almost exactly proportional to diet content over one to four times the NRC requirement of the hen. There does not seem to be a ceiling on vitamin B_{12} transfer to the eggs even when up to four times the NRC requirement is fed. There is, however, a ceiling on the amount of riboflavin that the laying hens can transfer to the egg when fed diets containing between two and four times their requirement as established by NRC. On the other hand, when laying hens were fed vitamin A at four times their requirement, the egg content of vitamin A responded only partially. This situation is attributed to the modulating effect and storage capacity of vitamin A in the liver.

The vitamin content of feed has the greatest and most widespread influence on egg vitamin content. Using data from studies that reported dietary vitamin level and feed intake on the one hand, and egg output, ***i.e.*** egg weight and production on the other, Naber calculated the efficiency of vitamin transfer into eggs as a function of intake (Table 4.37)

TABLE 4.37 Classification of vitamins by relative transfer efficiency from diet to egg

Transfer efficiency[1]	Vitamin
Very High (60 to 80%)	Vitamin A
High (40 to 50%)	Riboflavin Pantothenic acid Biotin Vitamin B_{12}
Medium (15 to 25%)	Vitamin D_3 Vitamin E
Low (5 to 10%)	Vitamin K_1 Thiamin Folacin

Adapted from Naber (1993)

[1] **When the vitamin is fed at one or two times the NRC dietary requirement**

The transfer efficiency of vitamin A at dietary levels equivalent to one or two times NRC requirement is very high (up to 80%) but this markedly declines when the dietary level is raised to four times requirement. The transfer of dietary vitamin B_{12} into eggs was as efficient as for riboflavin, pantothenic acid and biotin, i.e. about 50% with dietary levels at one to two times requirement. Unlike riboflavin, however, this level of transfer efficiency continued in the case of vitamin B_{12} even at very high dietary levels of up to 40 times the requirement.

4.10. DIET IMPLICATIONS WITH SOME GENERAL MANAGEMENT PROBLEMS

Hysteria

Although not widespread in commercial flocks, hysteria can be a very serious nuisance and economic cost factor if encountered in a flock. Hysteria is easy to distinguish from an ordinary flighty flock as the birds seem to lose all normal social behaviour and sense of direction and will mill and fly in every direction making unusual crying and squawking sounds. Birds often go into a molt, and so egg production declines. The condition of hysteria is more difficult to distinguish from flightiness in birds that are cage reared rather than floor reared. However, if one studies the flock for a period of time, differences can be seen.

The exact cause of the condition is unknown, and attempts to artificially induce it in flocks have failed. Many people believe it is related to nutritional or environmental factors, or to a combination of both. Hysteria is more often encountered in birds 12 to 18 weeks of age; however, it is also seen in older birds. The overcrowding of birds is thought to be a factor in triggering the condition. It is not a disease and is not contagious. Many drugs, feed supplements and management practices have been tried in an attempt to cure the condition with little or no success. Some people believe that it is a behavioral problem with the hens reacting to any noise or stimulus with which they are not accustomed. Why some flocks react differently to others is not known; however it is well known that small differences in stress can result in marked responses in a flock. A number of diet modifications have been tried in an attempt to alleviate hysteria. These include high levels of methionine (2 kg/tonne supplement) niacin (200 g/tonne supplement) or tryptophan (up to 5 kg/tonne supplement). The latter is thought to have a sedative-like effect by influencing brain neurotransmitters. However the response has been variable, and hysteria seems to return once tryptophan is withdrawn from the diet. Also until the price of tryptophan is reduced the treatment is prohibitively expensive.

Flighty or nervous flocks are often a problem for some producers. While various strains may be more prone to nervousness than others, by and large the strain of bird is seldom the cause of a nervous flock. Excessive light or the use of fluorescent light, crowding, or poor beak trimming seem to aggravate the condition. Also, birds that are underweight usually are more nervous and this is especially true with the earlier maturing strains. These birds, at peak production, are often as much as 100 g below the breeder's recommended body weight for their age and production. Often such birds are far more prone to feather picking, and the nervous behaviour associated with this vice.

"Pickouts and Blowouts"

In the past, pickout or blowout losses of 2 to 3% per month over several months after housing pullets were not uncommon. Such losses were usually the result of a number of factors working together rather than any single problem. In most cases, the problem was due to picking rather than any physical stress resulting in "blowouts". Some of the problems that can lead to pickouts or blowouts are as follows:

- *lights too bright (or sunlight streaming into buildings)*
- *pen temperature too high (poor ventilation)*
- *improper beak trimming*
- *pullets carrying an excess of body fat*
- *poor feathering at time of housing*
- *too early a light stimulation*
- *too high protein/amino acid in the diet causing early large egg size in relation to body and frame size*

The condition is usually accentuated in multiple bird cages and is a factor of floor space per bird rather than bird density. Frequently the incidence of picking has been shown to be higher in multiple bird cages where there is in excess of 460 sq. cm. of space per bird. When birds are more crowded they do not seem to be as aggressive. One of the most effective ways of avoiding a problem is to keep the light intensity low. Where rheostats are available, these should be adjusted to a sufficiently low level that picking or cannibalism is kept to a minimum. With early maturing pullets, or pullets that reach peak production two to three weeks earlier than was commonly found a few years ago, a mortality picture similar to those outlined above, is seldom encountered. However, what many producers are encountering is a pickout problem that starts around peak production and, although not high, (less than 0.5% per month) often remains with the flock throughout the entire laying period. While this type of problem is aggravated by bright lights, as well as crowding and poor beak trimming, it is felt that one of the main factors triggering the condition is low body weight. Even if pullets mature at body weights recommended by the breeder, many of them are up to 100 g lighter than standard at peak production. This, we suspect, is due to the fact that the pullet is coming into production with a minimum of body reserves; it also has a low feed consumption as it has been conditioned on a feed intake near to maintenance just prior to commencement of lay and thus hasn't been encouraged to develop a large appetite. The pullet is laying at a rapid rate and thus utilizes her body reserves (fat) in order to maintain egg mass production. This smaller body weight bird is often more nervous and so more prone to picking. Under these conditions, the nutritional management program of pullets outlined earlier in this chapter should be followed.

Prolapse can sometimes be made worse by feeding high protein/amino acid diets to small weight pullets in an attempt to increase early egg size. Coupled with an aggressive step up lighting program this often leads to more double yolk eggs and so greater incidence of prolapse and blowouts. Such pullets are often below standard weight at 12 - 14 weeks, and so any catch-up growth is largely as fat, which also accentuates the problem. Being underweight at 12 - 14 weeks usually means that they have reduced shank length, and because the long bones stop growing at this time, short shanks are often used as a diagnostic tool with prolapse problems in 22 - 24 week old pullets.

Fatty Liver Syndrome

The condition of fatty liver in poultry has been observed for quite some time. While it is true that an alteration in liver function can lead to a build-up of fat in the liver, many cases of fatty liver in laying hens can be traced to an imbalance in nutrient intake. Low protein, high energy diets and those in which there is an amino acid imbalance or deficiency can be major contributors to a fatty liver condition in layers. It is known that diets low in lipotrophic factors such as choline, methionine and vitamin B_{12} can result in fatty infiltration of the liver. However, these nutrients are seldom directly involved in most of the fatty liver problems reported from the field. Excessive feed intake (more specifically energy intake) is also a major cause of the condition. It is well known that laying hens will over consume energy, especially with high energy diets and this is particularly true of high producing hens. Pullets reared on a feeding program that tends to develop a large appetite or encourages "over-eating" (high fiber diets or skip-a-day feeding), are often more susceptible to the condition when offered a high energy diet on a free choice basis during lay. There is some information to suggest that daily fluctuations in temperature, perhaps affected by the season of the year, will stimulate hens to over-consume feed. Hence, it is important to attempt some type of feed restriction program if feed intake appears to be excessive.

When fatty liver is a problem, adding a mixture of so-called "lipotrophic factors" to the diet is often recommended. A typical addition may involve 60 mg $CuSO_4$, 500 mg choline, 3 μg vitamin B_{12} and 500 mg methionine per kg of diet. It should be emphasized that in many cases the addition of these nutrients will not cure the problem. Increasing the level of dietary protein by 1 to 2% seems to be one of the most effective ways of alleviating the condition. However, such treatments do not work in all cases. Another treatment that may prove effective is to increase the supplemental fat content of the diet. This apparently contradictory move is designed to offer the birds a greater proportion of energy as fat rather than carbohydrate. The idea behind this diet manipulation is that by reducing carbohydrate load there is less stress on the liver to synthesize new fat required for egg yolk production. By supplying more fat in the diet, the liver merely has to rearrange the fatty acid profile within fats, rather than synthesize new fat directly. For this treatment to be effective the energy level of the diet should not be changed, the recommendation merely being substitution of carbohydrate with fat. This concept may be the reasoning for apparent effectiveness of some other treatments for fatty liver syndrome. For example, substitution of barley or wheat for corn has been suggested, and this usually entails greater use of supplemental fat with these lower energy ingredients. Similarly, substitution of soybean meal with canola or sunflower meals usually means using more supplemental fat if energy level of the diet is to be maintained.

More recent evidence suggests that mortality is caused by eventual hemorrhaging of the liver, and that this is accentuated or caused by oxidative rancidity of the accumulated fat. On this basis we have seen a response to adding various antioxidants, such as ethoxyquin and vitamin E. Adding ethoxyquin at 150 mg/kg diet and extra vitamin E up to 50 - 60 IU/kg has been shown to reduce the incidence of hemorrhage mortality.

Experience has shown that it is difficult to increase production in a flock once the condition is established. Thus it is important that a proper program be followed to prevent the development of fatty livers. In some cases, the cause of the trouble can be traced back to pullets coming into the laying house carrying an excess of body fat. If these birds are then subjected to a laying diet in which the balance of protein and energy is slightly suboptimal for a particular strain of bird, a buildup of fat in the liver may occur. Also the feeding of crumbles or pellets in the laying house may aggravate this condition since the hen may overeat such rations if offered on a free choice basis. The results of an experiment studying effects of the level of dietary protein on percent liver fat are shown in Table 4.38.

TABLE 4.38 Influence of dietary protein on percent liver fat

Dietary protein level (%)	Egg prod (HDB) (%)	Feed (g/d)	Liver fat (dry weight basis) (%)
13	76.4	108	49.3
15	77.0	107	40.2
17	78.0	107	38.2

These birds were all laying at a good rate and no fatty liver syndrome problems were reported. As can be noted, all birds had livers high in fat. This is perfectly normal for a good laying bird and thus should not be confused with the fatty liver syndrome where liver hemorrhage is the condition that usually kills the hen.

Recent information suggests that a condition similar to the so called fatty liver syndrome may be caused by certain types of molds or mold toxins. Although no definite relationship has been established to date between molds and fatty livers, care should be taken to ensure that molds are not a factor contributing to poor flock performance. From time to time, canola meal has been implicated with the fatty liver syndrome. While it is true that there were earlier reports with some of the high glucosinolate rapeseed meals triggering such a condition, there is no evidence to suggest that the new canola varieties are a factor in the fatty liver condition. Hemorrhage due to rapeseed is usually not associated with excess fat infiltration of the liver.

Unaccountable Egg Production Slumps

From time to time producers complain of unaccountable slumps in egg production. Rather than passing it off as just one of those unexplained hazards of the egg production business, a grower should try to track down the cause to prevent future occurrences. Egg production slumps generally fall into one of three main categories:

- false slumps
- those caused by environment, management or nutrition
- result of disease

False slumps are often the result of poor record keeping and this is the first place to look if egg production drops unexpectedly for any one day. Another common cause of false production drops may be called "clutch synchronization". Hens lay in clutches, then rest a day or two while preparing for the next clutch. In large operations, several thousand hens may reach the end of a clutch at the same time and production can vary several percentage points from day to day. The following environmental or management factors can often lead to slumps in production:

- changes in temperature, pressure and drafts within a house
- failure to supply sufficient water either through a breakdown of the water system or insufficient water depth in open drinkers
- changes in the consistency of the feed, or having insufficient feed depth in a trough, resulting in decreased feed consumption
- a sudden disturbance of the hens either by noise, unfamiliar visitors or pets
- mite infestation
- any change in routine, whether it is caretaker, time of feeding etc.

Nutrition can obviously play a role in production slumps, because a deficiency of most nutrients cannot be tolerated by the bird for very long. It is impossible to detail the effect of specific nutrients, since most of the 30 - 40 nutrients that are required by the bird, can be involved. Many nutrient deficiencies, such as energy, protein, salt, calcium etc. will have an effect within days, whereas others such as trace minerals and some fat soluble vitamins (except D_3) will only cause problems if deficiencies occur for a prolonged period ($>$ 7 d). The production curve as shown in Fig. 4.21, is typical of a chronic nutrient deficiency or imbalance.

Fig. 4.21 Egg production and nutrient deficiency

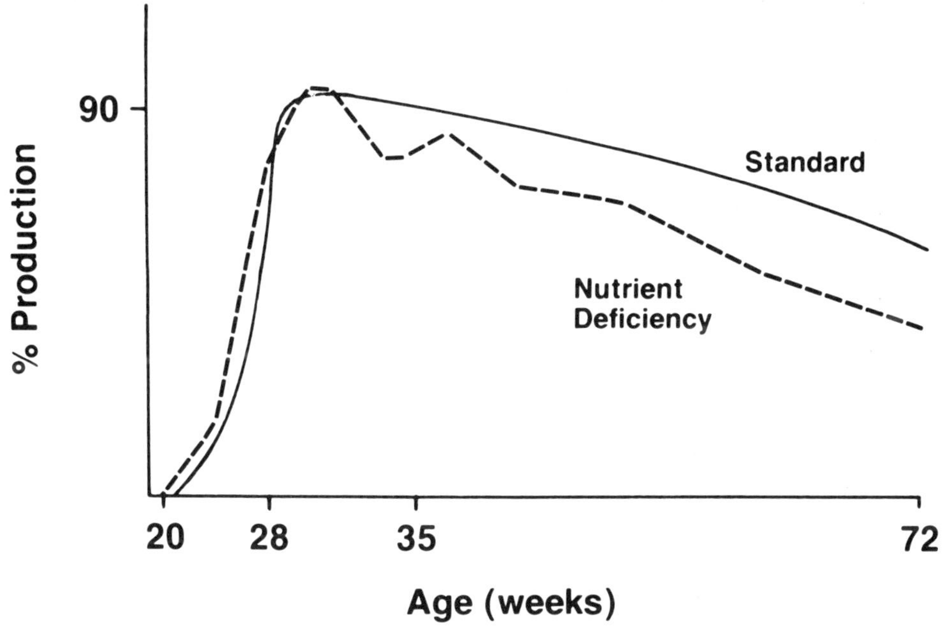

Molting programs

Molting programs are designed to prolong the productive life of layers and breeders. It is commonly practised for egg type layers and turkey breeders, and is also being tried with broiler breeders in some areas. The aim of a molting program is not necessarily to induce a feather molt, but rather to shut down the reproductive system for a period of time. When birds resume laying, egg shell quality is improved, although egg size will also be increased. A number of molting programs have been used over the years involving feed withdrawal, diets low in sodium or calcium, or diets with high levels of zinc.

Feed/water withdrawal usually involves a prolonged molt period, whereas low sodium and calcium and high zinc diets have been tried as a means of inducing a short "pause" in lay, rather than a distinct molt. In general these "pause" methods have not proven as reliable as the classical feed/water withdrawal. It is difficult to quantitate the economic advantage of a molting program. For example, if egg prices are high, then a short molt program may be beneficial, whereas with low egg prices, especially for large and extra large grades, there is no advantage to a very quick return to production. The criteria used in developing a molting program should be body weight and mortality. Mortality should not increase substantially during a molt and should not exceed 0.2% per day. If mortality is higher, then the duration and/or degree of feed restriction should be reconsidered.

Perhaps the most important criterion is body weight. Ideally birds start their second laying cycle at the same weight as when they were juvenile pullets. This means that the degree of feed restriction must be sufficient to cause weight loss equivalent to that gained by the bird during the first cycle. In practice, this is often very difficult. For example a flock of birds entering their first cycle at 1.3 kg may well be 1.4 kg entering their second cycle. In large part, this is the reason for increased egg size with second cycle birds. Weight loss should therefore be in proportion to bird weight gain during the first cycle. For flocks that are very heavy, more severe restriction is required, and vice versa. The time of initiation of a forced molt is also quite variable. In California, where most of the molting work has been pioneered by Dr. Don Bell, it is quite common to have two or three relatively short laying cycles - in this situation the first molt period may start at 60 - 65 weeks of age. More commonly, a molt will be initiated when layers are 75 - 80 weeks of age, when egg shell quality is a problem.

The main advantage of a molting program, is improved shell quality during the second and third laying cycles. The longer the molt period, within reason, the better the persistency of egg production and shell quality during a second or third cycle. For many situations therefore, the old saying of "short molt, short second cycle" holds true. The following guidelines are recommended for molting of not only layers, but also broiler and turkey breeders (Table 4.39.).

With the type of programs shown in Table 4.39, one can expect birds to molt and to decline to near zero percent egg production. The lowest egg production will likely occur about 5 - 7 d after initiation of the program, and maximum feather loss will occur a week later than this. Programs should be adjusted depending upon individual flock circumstances. For example, under very hot weather conditions it would be inadvisable to withdraw water for extended periods of time. Feed allocation should be adjusted depending upon body weight, and so this becomes the main criterion during the molt period (Table 4.39). The bottom line is to bring birds into their second laying cycle with a body weight comparable to that used for the initial lay cycle. The second, or successive lay cycles, will be shorter than the first, since shell and internal egg quality will deteriorate more quickly. Also egg production after peak will likely decline by 2.5 - 3% per month rather than the 2 - 2.5% seen in the first cycle.

TABLE 4.39 Molting Programs

	Egg-type layers		Broiler breeders	Turkey breeders
	White	Brown		
1. LIGHT[3]				
0-2d	**None, 8 hr or natural daylength**			
2-35d	**8 hr or natural daylength**			
35-42d	15 hr	15 hr	8 hr or natural	8 hr or natural
42-56d	15 hr	15 hr	15 hr	8 hr or natural
56d	15 hr	15 hr	15 hr	15 hr
2. WATER				
0-2d	**None**			
2-3d	Ad-lib	Ad-lib	None	None
3-4d	Ad-lib	Ad-lib	Ad-lib	None
4d→	Ad-lib	Ad-lib	Ad-lib	Ad-lib
3. FEED				
0-8d[1]	**None**			
8-10d[1]	Cracked cereal: 20 g/b/d		None	None
10-12d	Cracked cereal: 30-40 g/b/d		Cracked cereal: 45 g/b/d	None
12-20d	Cracked cereal: 40-45 g/b/d		Cracked cereal 50 g/b/d	Cracked cereal 100 g/b/d
20-35d	Pullet developer: Ad-lib		Pullet developer: 60 g/b/d	Holding[2] ration
35-42d	Layer diet: Ad-lib		Pullet developer: 100 g/b/d	Holding[2] ration
42-49d			Breeder diet: 100 g/b/d, then according to egg production	Holding[2] ration
49-56d				Holding[2] ration
56d				Breeder diet: Ad-lib

[1]Early feed restriction depending upon body weight loss desired
[2]Restricted or low-nutrient dense ad-lib
[3]In open sided houses, give birds 23-24 hr light/day for 5 days prior to start of molting program.

TABLE 4.39 Molting Programs Cont.				
	Egg-type layers		Broiler Breeders	Turkey Breeders
	White	Brown		
4. BODY WT (kg)				
1st cycle sexual maturity	1.4	1.6	2.5	10.5
End of 1st laying cycle	1.7	1.9	3.5	11.5
End of 1st molt cycle	1.5	1.8	2.7	10.7
End of 2nd lay cycle	1.8	2.1	3.5	11.6
5. EGG QUALITY				
Haugh units				
End 1st cycle	71			
1 month 2nd cycle	85			
8 months 2nd cycle	71			
Acceptable shells (%)				
End 1st cycle	76			
1 month 2nd cycle	97			
8 months 2nd cycle	76			

Cage layer fatigue

As its name implies, cage layer fatigue (CLF) is a syndrome most commonly associated with laying hens held in cages, and so its first description in the mid 1950's coincides with commercial acceptance of this housing system. Apart from the cage environment, CLF also seems to need a high egg output to trigger the condition, and for this reason it has traditionally been most obvious in White Leghorns. At around the time of peak egg output, birds become lame, and are reluctant to stand in the cage. Because of the competitive nature of the cage environment, affected birds usually move to the back area of the cage, and death can occur due to dehydration/starvation because of their reluctance to drink or eat. The condition is rarely seen in litter floor managed birds and this leads to the assumption that exercise may be a factor. In fact removing CLF birds from the cage during the early stage of lameness and placing them on the floor, usually results in a complete recovery. However this practice is usually not possible in large commercial operations. In the 1960 - 70's up to 10% mortality was common, although now the incidence is considered problematic if 0.5% of the flock are affected. There is no good evidence to suggest an association of CLF to general bone breakage in layers, although the two conditions are often described as part of the same general syndrome.

If birds are identified early, they seem alert and are still producing eggs. The bones seem fragile, and there may be broken bones. Dead birds may be dehydrated or emaciated, simply due to the failure of these birds to eat or drink. The ribs may show some beading although the most obvious abnormality is a reduction in the density of the medullary bone trabeculae. Paralysis is often due to fractures of the fourth and fifth thoracic vertebrae causing compression and degeneration of the spinal cord. If birds are examined immediately after the paralysis is first observed, then there is invariably a partly shelled egg in the oviduct, and the ovary contains a rich hierarchy of yolks. If birds are examined some time after the onset of paralysis, then the ovary is often regressed, due to reduced nutrient intake.

CLF is obviously due to an inadequate supply of calcium available for shell calcification, and the bird's plundering of unconventional areas of its skeleton for such calcium. Because calcium metabolism is affected by the availability of other nutrients, the status of phosphorus and vitamin D_3 in the diet and their availability are also critical. Birds fed diets deficient in calcium, phosphorus, or vitamin D_3 will show cage layer fatigue assuming there is a high egg output.

Calcium level in the prelay period is often considered critical in preventative measures for CLF. Feeding low calcium (1%) grower diets for too long a period, or even 2% calcium prelay diets up to 5% egg production often leads to greater incidence of abnormal bone development. It has been suggested that the reappearance of CLF in some commercial flocks may be a result of too early a sexual maturity due to the genetic selection for this trait coupled with early light stimulation. Feeding a layer diet containing 3.5% Ca vs a grower diet at 1% Ca, as early as 14 wk of age, has proven beneficial in terms of an increase in the ash and calcium content of the tibiotarsus (Table 4.40).

TABLE 4.40 Diet calcium and bone characteristics of young layers in response to prelay diet calcium

Time of change to 3.5% Ca	Tibiotarsus Ash (%)	Tibiotarsus Ca (mg/g)
20 wk	53.5[c]	182[b]
18 wk	55.7[b]	187[b]
17 wk	59.3[a]	202[a]
16 wk	58.9[a]	199[a]
15 wk	58.4[a]	197[a]
14 wk	57.9[a]	196[a]

Adapted from Keshavarz (1989)

Feeding a high calcium diet far in advance of maturity seems unnecessary, and in fact, may be detrimental in terms of kidney urolithiasis. Changing from a low to a high calcium diet should coincide with the observation of secondary sexual characteristics, and especially comb development which usually precedes first oviposition by 14 - 16 d.

There have been surprisingly few reports on the effect of vitamin D_3 on CLF in young birds. It is assumed that D_3 deficiency will impair calcium utilization, although there are no reports of testing graded levels of this nutrient as a possible preventative treatment. The other major nutrient concerned with skeletal integrity is phosphorus, and as expected, phosphorus deficiency can accentuate effects of CLF. While P is not directly required for shell formation, it is essential for the replenishment of Ca, as $CaPO_4$, in medullary bone during successive periods of active bone calcification. Without adequate phosphorus in the diet, there is a failure to replenish the medullary Ca reserves, and this situation can accelerate or precipitate the onset of CLF and other skeletal problems. Low phosphorus intake is sometimes caused by the trend towards lower levels of diet phosphorus coupled with very low feed intake of pullets through early egg production.

Bone breakage in older hens

CLF may relate to bone breakage in older hens, although a definitive relationship has never been verified. It is suspected that like the situation of CLF with young birds, bone breakage in older birds results as a consequence of impaired calcification of the skeleton over time, again related to a high egg output coupled with the restricted activity within the cage environment. Few live birds have broken bones in the cage, the major problem occurring when these birds are removed from their cages and transported to processing units. Apart from the obvious welfare implications, broken bones prove problematic during the mechanical deboning of the muscles.

Adding more calcium to the diet of older layers does not seem to improve bone strength, and in certain situations can lead to excessive eggshell pimpling. Adding both calcium and phosphorus to the diet has given beneficial results in some instances, although results are quite variable. In young birds at least, adding 300 ppm flourine to the water has improved bone strength, although there are no reports of such treatment with end of lay birds. Moving birds from a cage to litter floor environment seems to be the only treatment that consistently improves bone strength. This factor indicates that exercise *per se* is an important factor in bone strength of caged birds, but does not really provide a practical solution to the problem at this time.

It is not currently known how to improve the bone integrity of older high producing hens, without adversely affecting other traits of economic significance. For example it has been shown that bone breaking strength in older birds can be increased by feeding high levels of vitamin D_3. Unfortunately this treatment also results in an excessive pimpling of the eggshells (as occurs with extra calcium) and these extra calcium deposits on the shell surface readily break off causing a leakage of the egg contents. It may be possible to improve the skeletal integrity of older birds by causing cessation of ovulation for some time prior to slaughter. Presumably the associated reduction in the drain of body calcium reserves would allow re-establishment of the integrity of the susceptible medullary bones. Currently such a feeding strategy is uneconomical, although consideration for bird welfare may provide the impetus for research in this area.

4.11 FEEDING FOR REDUCED N AND P EXCRETION

Poultry manure has always been a valuable source of fertilizer for pasture land and arable crops. However, large scale intensive units have led to the concentration of manure in relatively small areas. Because layer manure contains so much water, it is expensive to transport and there are severe limits on economical distance for disposal. A 1 million bird layer operation yearly produces manure containing about 750 tonnes of nitrogen and 150 tonnes of phosphorus, which are the main nutrients of concern for effective land use relative to potential for environmental contamination. Land used for corn production requires about 60 kg nitrogen and 25 kg P_2O_5 equivalents each year. This means that to effectively utilize the manure from a 1 million bird unit, it should be spread over 5,000 hectares, which represents the limiting need for nitrogen disposal (2,500 hectares for P_2O_5).

Most of the nitrogen in manure originates from crude protein, and to a lesser extent free amino acids and non-protein nitrogen. Most diets are formulated to a crude protein specification, simply in order to supply component amino acids. If diets are formulated solely on the basis of essential amino acids, with no crude protein minimum need, then diets will usually contain less protein, and consequently less nitrogen. As protein/nitrogen level of the diet is reduced, there are resultant declines in nitrogen intake and nitrogen excretion. For example, in a recent study at Guelph, Summers fed diets varying in crude protein level from 19 down to a low of 5% crude protein, while maintaining levels of critical amino acids (Fig. 4.22).

Fig. 4.22 Nitrogen intake and excretion of layers in relation to diet protein level.

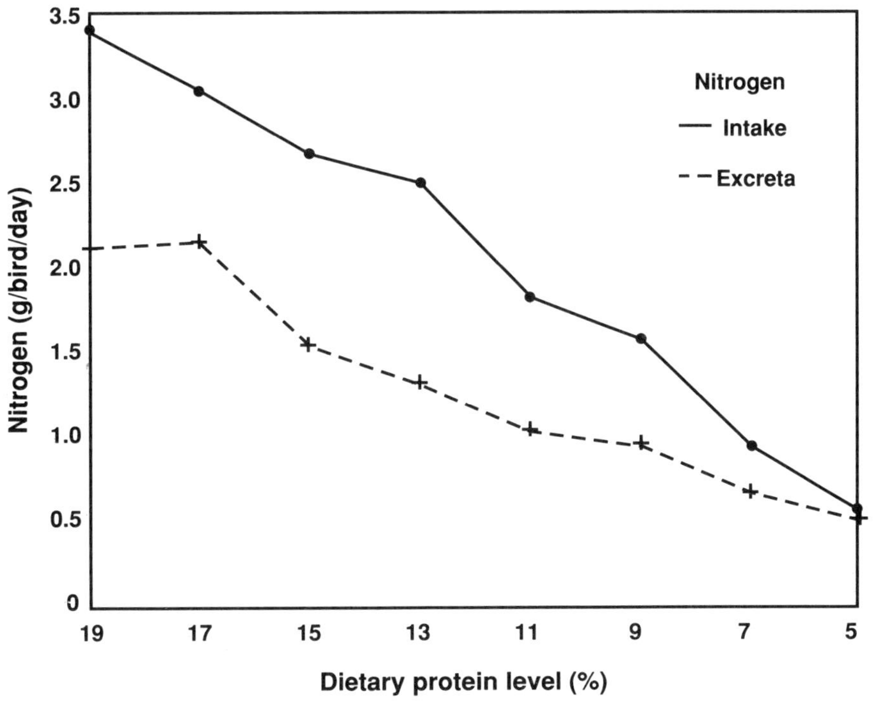

Layers do not seem to adjust their feed intake as the crude protein level of the diet is reduced, suggesting no appetite for protein *per se*. This is to be expected because any such appetite would likely be related to specific amino acids rather than protein *per se*. As protein intake declines, so does excreta nitrogen output (Fig 4.22). For example if the crude protein level of the diet is reduced from 17% down to 14%, then nitrogen excretion declines from around 2g/bird/day to about 1.5g/bird/day. For the 1 million bird unit referred to earlier, this means a reduction in manure nitrogen of about 200 tonnes per year, and in turn 1300 hectares less land are needed for manure disposal.

However, the overall economic picture is not quite as straightforward as depicted by land use alone. Unfortunately, as we make meaningful reductions in crude protein level, we often see a reduction in daily egg mass output. This latter situation may be due to incorrect assessment of amino acid needs and/or amino acid availability, or perhaps that the layer has need for some non-protein nitrogen in addition to amino acids. This loss in egg mass is often a consequence of reduced egg size, rather than reduced egg production. For the study depicted in Fig. 4.22, reducing crude protein from 17 to 13% resulted in a 2.3 g loss in average egg size.

Layers are often overfed phosphorus, and apart from contributing to reduced eggshell quality, this also results in increased phosphorus excretion in the manure. In some European countries there are now economic incentives to reduce manure phosphorus loading, and this has stimulated interest in phytase enzyme usage. Unfortunately much of the phosphorus found in plant tissue is complexed as phytate, which is a myo-inositol hexaphosphate (Fig. 4.23).

Fig. 4.23 Phytate structure♦

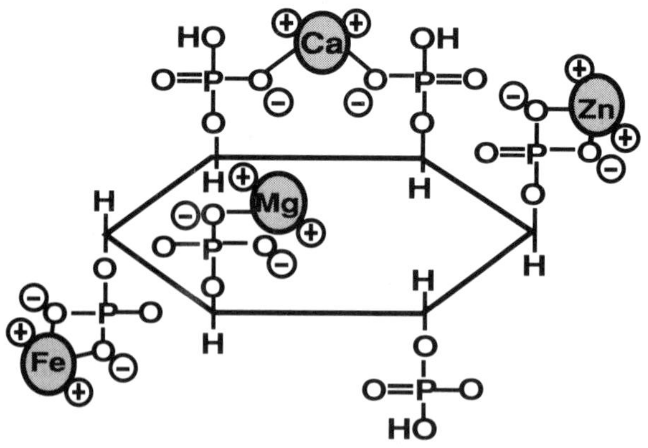

Each molecule of phytic acid therefore complexes six molecules of phosphorus and also variable quantities of calcium, zinc and other minerals. This so called "plant phosphorus" is available to varying degrees in different cereals and plant proteins, with availability of 5 - 60%. Usually, the younger the bird the lower the availability of phytate phosphorus. Birds do not produce a phytase enzyme, therefore adding a synthetic source to the diet can greatly increase phytic acid use, and consequently reduce the phosphorus appearing in the manure. If phytase is used in formulation there are a number of different approaches to account for increased phytate availability. Where few ingredients are used, then the available phosphorus level of these ingredients can be increased accordingly. Alternatively the specification for available phosphorus in the diet can be reduced or phytase enzyme can be included as an ingredient with a specification for providing available phosphorus, calcium (and possibly other minerals). Each 300 units of phytase activity is equivalent to about 1g of phosphorus as provided by sources such as dicalcium phosphate. Adding 300 units/kg feed therefore provides about 0.10% available phosphorus in the diet.

Currently there are two problems with using phytase, namely cost and degradation during pelleting. Most phytases are denatured at $\geq$ 65°C, and so any pelleting or feed processing involving high temperatures as used to control salmonella in feed, mean that phytase enzymes must be added as a liquid, post-processing. In many countries it is still economical to use phosphates rather than phytase enzyme. However this situation may change in the future as phytase costs are reduced, or there are economic incentives to reduced manure phosphorus levels. Future research may also indicate that phytase enzymes increase the availability of nutrients other than just phosphorus and calcium, because positively charged molecules, such as lysine can complex with the negatively charged phytate molecule.

CHAPTER 4

Selected References

Atteh, J.O. and S. Leeson 1985. *Response of laying hens to dietary saturated and unsaturated fatty acids in the presence of varying dietary calcium levels.* Poult. Sci. 64:520-528.

Calderon, V.M. and L.S. Jensen (1990). *The requirement for sulfur amino acid by laying hens as influenced by protein concentration.* Poult. Sci. 69:934-944.

Caston, L.J., E.J. Squires and S. Leeson 1994. *Hen performance, egg quality and the sensory evaluation of eggs from SCWL hens fed dietary flax.* Can. J. Anim. Sci. 74:347-353.

Chah, C.C. 1972. *A study of the hens nutrient intake as it relates to egg formation.* MSc. Thesis, Univ. Guelph.

Clunies, M. and S. Leeson 1995. *Effect of dietary calcium level on plasma proteins and calcium flux occuring during a 24h ovulatory cycle.* Can. J. Anim. Sci. 75:539-544.

Clunies, M., J. Emslie and S. Leeson 1992. *Effect of dietary calcium level on medullary bone reserve and shell weight of Leghorn hens.* Poult. Sci. 71:1348-1356.

DeSchutter, A. 1988. *The effect of nutrient intake on the performance of hens exposed to acute short-term heat stress.* MSc. Thesis, Univ. Guelph.

Harms, R.H. and G.B. Russell (1994). *A comparison of the bioavailability of DL-methionine and MHA for the commercial laying hen.* J. Appl. Poult. Res. 3:1-6.

Hoffmann-La Roche 1988. *Egg yolk pigmentation with carophyll.* 3rd Ed. Publ. F. Hoffmann-La Roche and Co. Ltd. Publ. 1218. Basle, Switzerland.

Keshavarz, K. 1989. *A balance between osteoporosis and nephritis.* Egg Industry. July p 22-25.

Leeson, S. 1993. *Potential of modifying poultry products.* J. Appl. Poult. Res. 2:380-385.

Leeson, S. and J.D. Summers 1983. *Performance of laying hens allowed self-selection of various nutrients.* Nutr. Rep. Int. 27:837-844.

Leeson, S., R.J. Julian and J.D. Summers 1986. *Influence of prelay and early-lay dietary calcium concentration on performance and bone integrity of Leghorn pullets.* Can. J. Anim. Sci. 66:1087-1096.

Naber, E.C. 1993. *Modifying vitamin composition of eggs: A review.* J. Appl. Poult. Res. 2:385-293.

Payne, C.G. 1967. In *Environmental Control of Poultry Production.* pp 40-54. Ed. T.C. Carter. Publ. Longmans, London.

Portella, F.J., L.J. Caston and S. Leeson 1988. *Apparent feed particle size preference by laying hens.* Can. J. Anim. Sci. 68:915-922.

Rao, K.S. and D.A. Roland, Sr. 1989. *Influence of dietary calcium level and particle size of calcium source on in vivo calcium solubilization by commercial Leghorns.* Poult. Sci. 68:1499-1505.

Roland, D.A., Sr. 1986. *Eggshell quality IV. (See also parts I-III).* World Poult. Sci. J. 42:166-171.

Roland, D.A. 1995. *The egg producers guide to optimum calcium and phosphorus nutrition.* Publ. Mallinckrodt Feed Ing.

Sell, J.L., S.E. Scheideler and B.E. Rahn 1987. *Influence of different phosphorus phase-feeding programs and dietary calcium level on performance and body phosphorus of laying hens.* Poult. Sci. 66:1524-1530.

Waldroup, P.W. and H.M. Hellwig 1995. *Methionine and total sulfur amino acid requirements influenced by stage of production.* J. Appl. Poult. Res. 3:1-6.

Zhang, B. and C.N. Coon 1994. *Nutrient modelling for laying hens.* J. Appl. Poult. Res. 3:416-431.

CHAPTER 5. FEEDING PROGRAMS FOR BROILERS

Page No.

BROILERS

5.1 DIET SPECIFICATIONS

The broiler chicken continues to show significant improvement in growth and overall feed efficiency as a standard market weight is achieved some 0.5 d earlier each year. In part, this is due to improved understanding of nutrient requirements, and the realization of continual change in proportion of nutrients directed towards growth vs maintenance. Perhaps one of the most striking features of the broiler to materialize over the last few years, is its ability to "adequately" respond to a range of diet situations. For example, varying the protein:energy of a diet seems to have less of an effect today than was recorded some 15 - 20 years ago.

In large part, the adaptability of the broiler chicken is due to its voracious appetite, and the fact that feed intake seems to be governed by both physical satiety as well as by cues related to specific nutrients. For example, varying the energy level of a broiler diet today has much less of an effect on feed intake, as expected on the basis of appetite being governed by energy requirement. This apparently subtle change in bird appetite has lead to increased variability in diet type and diet allocation used by commercial broiler growers. However, as will be discussed later, attempting to "cheapen" broiler diets through the use of lower protein/amino acid levels, while not having major effects on gross performance, leads to subtle changes in carcass composition. Feed programs may therefore vary depending upon the goals of the producer vs processor.

Another major change in broiler nutrition that has occurred over the last 5 years is the realization that maximizing nutrient intake is not always the most economical situation, at least for certain times in the grow-out period. A time of so-called "undernutrition", which slows down early growth rate appears to result in dramatic reduction in the incidence of metabolic disorders such as Sudden Death Syndrome and the various skeletal abnormalities. A period of slower initial growth, followed by "compensatory" growth is almost always associated with improved feed efficiency, because less feed is directed towards maintenance. Tables 5.1 and 5.2 show nutrient specifications for broilers and roaster birds respectively.

TABLE 5.1 Broiler diet specifications

	Pre-starter	Starter		Grower		Finisher/Withdrawal	
Approximate protein level (%)	**23**	**22**	**20**	**20**	**18**	**18**	**16**
Amino acids (% of diet)							
Arginine	1.40	1.20	1.10	1.05	0.95	0.90	0.85
Lysine	1.35	1.20	1.05	1.10	0.90	0.90	0.80
Methionine	0.52	0.48	0.42	0.44	0.38	0.37	0.36
Methionine+Cystine	0.95	0.82	0.75	0.73	0.65	0.64	0.61
Tryptophan	0.22	0.20	0.18	0.17	0.15	0.14	0.13
Histidine	0.42	0.40	0.35	0.32	0.30	0.28	0.27
Leucine	1.50	1.40	1.20	1.10	1.00	1.00	0.90
Isoleucine	0.85	0.75	0.60	0.55	0.50	0.47	0.45
Phenylalanine	0.80	0.75	0.65	0.60	0.55	0.53	0.50
Phenylalanine+tyrosine	1.50	1.40	1.20	1.10	1.00	1.00	0.90
Threonine	0.75	0.70	0.62	0.60	0.55	0.55	0.50
Valine	0.90	0.80	0.70	0.65	0.60	0.58	0.55
Metabolizable energy (kcal/kg)	3050	3050	2900	3150	3000	3200	3050
Calcium (%)	1.0	0.95	0.95	0.92	0.92	0.90	0.90
Available phosphorus (%)	0.45	0.42	0.42	0.40	0.40	0.38	0.38
Sodium	0.19	0.18	0.18	0.18	0.18	0.18	0.18
Vitamins (per kg of diet)							
Vitamin A (I.U.)	6500			5850			
Vitamin D_3 (I.U.)	3000			2700			
Choline equivalents (mg)	800			720			
Riboflavin (mg)	5.5			5.0			
Pantothenic acid (mg)	14.0			12.6			
Vitamin B_{12} (mg)	.013			0.012			
Folic acid (mg)	1.0			0.90			
Biotin (mg)	0.20			0.18			
Niacin (mg)	40.0			36			
Vitamin K (mg)	2.0			1.8			
Vitamin E (I.U.)	30.0			27			
Thiamin (mg)	4.0			3.6			
Pyridoxine (mg)	4.0			3.6			
Trace minerals (per kg of diet)							
Manganese (mg)	70						
Iron (mg)	80						
Copper (mg)	10						
Zinc (mg)	80						
Selenium (mg)	0.30						
Iodine (mg)	0.4						

Note: vitamin values in the first column span Pre-starter and Starter; those in the Grower column span Grower and Finisher/Withdrawal. Trace mineral values span all columns.

TABLE 5.2 Broiler roaster specifications

	Roaster Grower		Roaster Finisher	
Approximate protein level (%)	17	15	16	15
Amino acids (% of diet)				
Arginine	0.90	0.80	0.84	0.80
Lysine	0.90	0.75	0.80	0.72
Methionine	0.38	0.32	0.36	0.30
Methionine+Cystine	0.65	0.56	0.60	0.54
Tryptophan	0.15	0.11	0.14	0.10
Histidine	0.28	0.23	0.25	0.20
Leucine	0.95	0.87	0.90	0.85
Isoleucine	0.47	0.38	0.40	0.35
Phenylalanine	0.50	0.40	0.45	0.35
Phenylalanine+tyrosine	0.90	0.78	0.85	0.68
Threonine	0.50	0.45	0.47	0.42
Valine	0.50	0.45	0.50	0.45
Metabolizable energy (kcal/kg)	3175	3075	3200	3100
Calcium (%)	0.90	0.90	0.85	0.85
Available phosphorus (%)	0.38	0.38	0.35	0.35
Sodium	0.17	0.17	0.17	0.17
Vitamins (per kg of diet)				
Vitamin A (I.U.)	5850			
Vitamin D_3 (I.U.)	2700			
Choline equivalents (mg)	720			
Riboflavin (mg)	5.0			
Pantothenic acid (mg)	12.6			
Vitamin B_{12} (mg)	.012			
Folic acid (mg)	.90			
Biotin (mg)	.18			
Niacin (mg)	36			
Vitamin K (mg)	1.8			
Vitamin E (I.U.)	25			
Thiamin (mg)	3.6			
Pyridoxine (mg)	3.6			
Trace minerals (per kg of diet)				
Manganese (mg)	70			
Iron (mg)	80			
Copper (mg)	10			
Zinc (mg)	80			
Selenium (mg)	0.30			
Iodine (mg)	0.4			

5.2 FEEDING PROGRAMS

Tables 5.3 and 5.4 outline examples of diets that may be used for broilers and roaster birds. It is now more difficult to give general recommendations for allocation of these diets, since we now have the situation of meat birds being processed over a wide range of ages, and therefore requiring different nutrient inputs. For example, the allocation of a starter diet for a 35 d bird destined for the whole bird or cut up market may be quite different to that for a 56 d bird reared for the further processing market.

Following these example diets, is an attempt at an overview of feed allocation for males, females and mixed sex birds where appropriate. Diets are described only in terms of protein and energy. It is realized that all nutrients, and especially amino acids, are necessarily considered in describing diets, yet it is hoped that this simplified description will be sufficient to differentiate diet type (Table 5.5).

TABLE 5.3 Examples of broiler diets (kg)

	Starter		Grower			Finisher	
Corn	563	268	572	605	543	652	506
Wheat	-	200	-	-	174	-	270
Barley	-	200	-	-	-	-	-
Soybean meal (48%)	356	270	315	307	230	258	170
Meat Meal (50%)	-	-	10	-	-	-	-
Fish Meal (60%)	-	-	20	-	-	-	-
Fat	35	17	46	45	10	47	10
Ground limestone	17	16	13	15	15	15	15
Calcium phosphate (20% P)	15	15	10	15	15	15	15
Iodized salt	3	3	3	3	3	3	3
Vitamin:Mineral[1] premix	10	10	10	10	10	10	10
Methionine	1.2	1.0	0.8	0.6	0.8	0.6	1.0
Calculated analysis							
Crude protein (%)	22.0	22.0	21.8	20.0	18.0	18.0	16.1
Digestible Protein (%)	17.7	17.7	17.7	16.2	14.4	14.2	12.9
Crude Fat (%)	5.9	5.9	7.0	7.0	3.4	7.3	3.4
Metabolizable Energy	3060	3060	3145	3146	3022	3200	3050
(kcal/kg)	1.00	1.00	0.98	0.95	0.95	0.94	0.96
Calcium (%)	0.42	0.42	0.42	0.42	0.41	0.41	0.41
Av. Phosphorus (%)	0.17	0.17	0.18	0.17	0.17	0.17	0.18
Sodium (%)	0.48	0.48	0.46	0.40	0.38	0.37	0.37
Methionine (%)	0.82	0.82	0.80	0.71	0.65	0.64	0.61
Methionine & cystine (%)	0.31	0.31	0.30	0.28	0.25	0.25	0.22
Tryptophan (%)	1.25	1.25	1.27	1.10	0.93	0.96	0.78
Lysine (%)	0.94	0.94	0.94	0.86	0.75	0.78	0.65
Threonine (%)							

[1]**Use additional choline chloride if vitamin premix does not contain this vitamin**

TABLE 5.4 Examples of Roaster Diets (kg)

	Roaster Grower		Roaster Finisher	
Corn	600	648	529	643
Wheat	99	140	140	140
Barley	-	-	50	-
Soybean meal (48%)	220	160	190	160
Fat	40	10	50	15
Ground limestone	15	15	15	15
Calcium phosphate (20% P)	13	13	13	14
Iodized salt	3	3	3	3
Vitamin:Mineral[1] premix	10	10	10	10
Methionine	0.8	0.7	0.6	0.5
<u>Calculated analysis</u>				
Crude protein (%)	17.0	15.1	16.1	15.1
Digestible Protein (%)	13.8	12.0	12.9	11.9
Crude Fat (%)	6.5	3.8	7.4	4.2
Crude fiber (%)	2.4	2.5	2.7	2.5
Metabolizable Energy (kcal/kg)	3175	3083	3205	3100
Calcium (%)	0.90	0.91	0.91	0.92
Av. Phosphorus (%)	0.38	0.38	0.38	0.39
Sodium (%)	0.17	0.17	0.17	0.17
Methionine (%)	0.37	0.34	0.34	0.32
Methionine & cystine (%)	0.63	0.56	0.56	0.54
Tryptophan (%)	0.87	0.72	0.80	0.71
Lysine (%)	0.23	0.20	0.22	0.20
Threonine (%)	0.72	0.63	0.67	0.63

[1]Use additional choline chloride if vitamin premix does not contain this vitamin

TABLE 5.5 Feed allocation for regular type broiler diets (kg/bird)

	Starter	Grower	Finisher 1	Finisher 2	Total
Crude Protein (%)	22	20	18	16	
ME (kcal/kg)	3050	3150	3200	3200	
Bird Age/Type					
28d Cornish ♀	0.5	0.8	0.6		1.9
35d cut-up ♂	0.7	1.1	0.7		2.5
♀	0.6	0.9	0.8		2.3
Mixed sex	0.7	1.0	0.7		2.4
42d Whole bird ♂	0.7	1.9	1.3		3.9
♀	0.5	1.7	1.5		3.7
Mixed sex	0.6	1.8	1.4		3.8
49d Whole bird ♂	0.6	2.0	2.3		4.9
♀	0.5	2.0	2.0		4.5
Mixed sex	0.6	2.0	2.1		4.7
56d Whole bird ♂	0.5	2.0	2.3	1.1	5.9
60d Whole bird ♂	0.4	2.1	2.4	1.9	6.8
70d Roaster ♂	0.4	2.3	2.8	2.8	8.3

5.3 USE OF SPECIALTY FEEDS

Various types of feeding program are currently being considered by broiler producers and feed manufacturers, and these may be thought of as specialty feeds. These programs may involve low nutrient dense diets as a means of simply reducing feed cost, or diets of higher protein/amino acid content used in an attempt to reduce carcass fat content. Alternatively there is now interest in feed programs involving feed restriction or diet dilution.

a) Low-nutrient density program

By offering low protein, low energy diets (series shown in Tables 5.1 - 5.2) it is hoped to reduce feed costs and so make feeds more attractive to customers. However, it is obvious that the birds will necessarily consume more of these diets and that birds may also take longer to reach market weight. These two factors result in reduced feed efficiency. Surprisingly broiler chickens seem to perform quite reasonably with low nutrient dense diets, and in certain situations these may prove to be the most economical program. If diets of low energy level are fed, the broiler will eat more feed (Table 5.6).

In this study, only the energy level was changed and the broiler adjusted reasonably well in an attempt to maintain constant energy intake. Diet energy from 3300 - 2700 kcal ME/kg had no significant effect on body weight, and this suggests the bird is still eating for its energy need. Obviously these data on growth rate are confounded with the intake of all nutrients other than energy. For example, birds offered the diet with 2700 kcal ME/kg increased their protein intake in an attempt to meet energy needs. Using these same diets, but controlling feed intake at a constant level for all birds (Table 5.7) shows that energy intake *per se* is a critical factor in affecting growth rate.

TABLE 5.6 Performance of broilers fed diets of variable energy content

	Body weight (g)		Feed intake (g/bird)		
Diet ME (kcal/kg)	25d	49d	0 to 25d	25 to 49d	0 to 49d
3,300	1,025	2,812	1,468	3,003	4,471
3,100	1,039	2,780	1,481	3,620	5,101
2,900	977	2,740	1,497	3,709	5,206
2,700	989	2,752	1,658	3,927	5,586

TABLE 5.7 Performance of broilers given fixed quantities of feed

	Body weight (g)		Feed intake:body weight gain
Diet ME (kcal/kg)	25d	49d	0 - 49 d
3,300	825[a]	2,558[ab]	1.84[c]
3,100	818[a]	2,599[a]	1.82[c]
2,900	790[b]	2,439[b]	1.94[b]
2,700	764[b]	2,303[c]	2.05[a]

With low energy diets therefore, we can expect slightly reduced growth rate, because "normal" energy intake is rarely achieved, and this fact is the basis for programs aimed at reducing early growth rate. However, live body weight is often not the "end-point" of consideration for broiler production, since carcass weight and carcass composition are becoming of prime consideration. From the point of view of the processor or integrator, these cheaper diets may be less attractive. Carcass weight and meat yield are often reduced, and this is associated with increased deposition of carcass fat, especially in the abdominal region. Low protein diets are therefore less attractive when one considers feed cost/kg edible carcass or feed cost/kg edible meat. This consideration of carcass composition leads to development of diets that maximize lean meat yield.

Another concept for feeding broilers is true low nutrient dense diets, where all nutrient concentrations are reduced (in practice energy and protein/amino acids are most often the only nutrients changed in such a program). Examples of such diets are shown in Table 5.3. With this type of feeding program, one can expect slower growth and inferior feed efficiency, although this should not be associated with increased fat deposition. Depending upon local economic conditions and the price of corn and fat, this type of program can be very economical, and, as will be discussed later, is the basis of programs designed to induce compensatory growth.

b) *Diets for improved meat yield*

Diet nutrients can be manipulated in order to advantageously influence the composition of the broiler and roaster carcass. As already discussed, this is most easily achieved by varying the quantity of protein/amino acids in relation to energy and is accomplished by changing the energy or protein level of the diet. At a given protein level, leaner carcasses can be expected as the energy level of the diet is reduced. As shown in Fig 5.1 this effect of reducing the energy level is due to reduced carcass fat content rather than any measurable increase in carcass protein content.

Such diet manipulation is often not acceptable because feed efficiency is reduced with these low energy diets. A more attractive means of improving carcass quality is to increase the protein/amino acid level of the diet while maintaining standard energy levels (Fig 5.2).

The protein values shown in Fig 5.2, are obviously extreme in terms of commercial formulation. As protein level of the diet is increased, birds are leaner, although again this is due mainly to reduced fat deposition. There is an indication of reduced protein accretion with very low protein diets in Fig 5.2, although over the range 20 - 36% CP, there is little change in carcass protein content. Unfortunately, the high protein/amino acid diets necessary to influence carcass composition do not seem to have any major beneficial effect on growth rate. The economics of such a program therefore depend on being able to realize a premium price for these leaner carcasses, or there being sufficiently less fat trim during further processing.

c) *Feed restriction and compensatory growth*

Most broiler chickens are given unlimited access to feed, or at most, have limited access during brief periods of darkness. It is generally assumed that the faster that birds reach market weight, the better the feed conversion since maintenance requirement should be reduced. While this is usually true, there may be some potential for modifying the growth pattern of the bird in favour of an even greater reduction in maintenance requirement. If broiler growth rate could be reduced during early life, and this is followed by compensatory growth so as to achieve the same market weight for age, then maintenance requirements must be reduced - implying improved feed efficiency. This concept raises the question of restricted feeding and/or reduced nutrient intake during early life. If it is accepted that feed conversion in its classical sense (digestibility, metabolizability, etc) has improved little over the years, then improvements that we continue to see in feed utilization must be associated with the reduction in maintenance requirement.

In addition to improving feed utilization, there is also interest in manipulating growth because of mortality associated with metabolic conditions. The broiler chicken shows exceptionally fast early growth rate when fed high nutrient dense diets without any form of restriction. Table 5.8 shows the weekly increase in body weight of one strain of commercial broiler, and the changes that have occurred since 1989.

Fig. 5.1 Body weight and carcass composition of male broilers fed diets varying in level of energy.

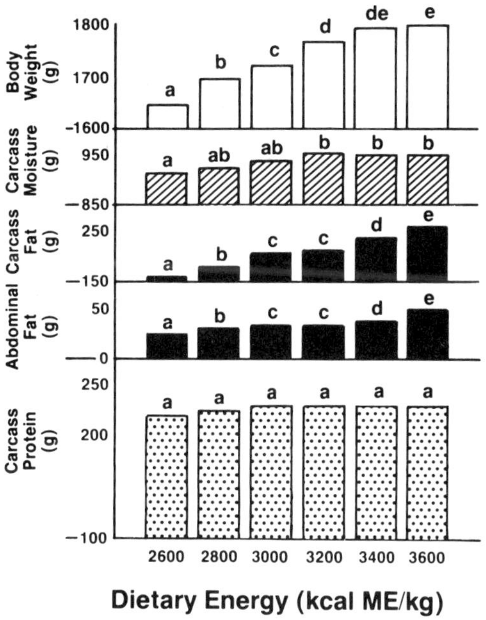

Fig. 5.2 Body weight and carcass composition for male broilers fed diets varying in protein level.

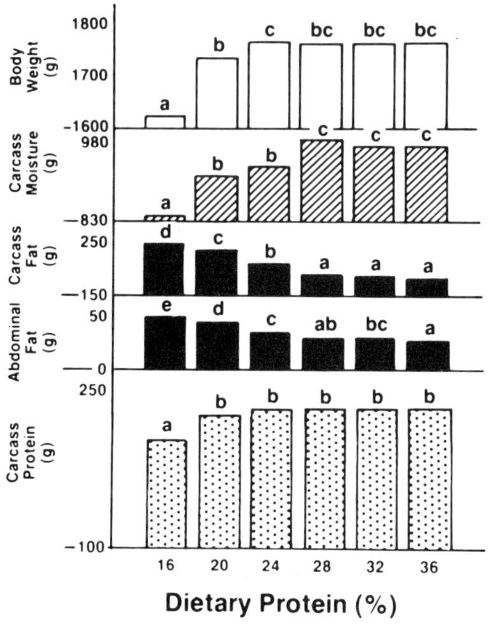

TABLE 5.8 Weekly increase in body weight of male broilers (%)			
Week	1989	1993	1996
1	243	343	242
2	140	141	142
3	81	84	88
4	60	53	75
5	50	40	47
6	30	29	32
7	23	23	22
8	18	18	18

Between 1989 and 1993, there was obviously selection for very rapid early growth rate, although this seems to have been reduced for the 1996 bird where there is greater emphasis on mid-stage growth. Overall there is greater body weight at market ages from 42d onwards, although the pattern of achieving this weight seems to have changed in the last few years (Table 5.9).

TABLE 5.9 Growth characteristics of the male broiler chicken					
	Body weight (g)			% Change	
Age (d)	1989	1993	1996	1993 vs 89	1996 vs 93
7	141	182	150	+29	-18
14	341	440	375	+29	-15
21	617	808	700	+31	-13
28	990	1235	1235	+25	0
35	1480	1734	1805	+17	+4
42	1970	2252	2380	+14	+6
49	2430	2769	2900	+14	+5
56	2869	3278	3430	+14	+5

The data in Tables 5.8. and 5.9. are specific to one strain of broiler, but do seem to be indicative of the industries' concern about the consequences of very rapid early growth rate. However, regardless of these "genetic" changes to the growth pattern, there is still potential to manipulate growth further through nutritional modification. It is difficult, or impossible, to achieve growth rates greater than standard, although the converse situation can easily be induced through some form of reduced nutrient intake. If growth rate is to be reduced then based on needs to optimize feed usage, such restriction must occur early in the grow out period (Table 5.10). As the bird gets older, a greater proportion of nutrients are used for maintenance, and less is used for growth. Therefore reducing nutrient intake in say the first 7 d, will have little affect on feed efficiency, because so little feed is going towards maintenance (Table 5.10). At 8 weeks of age, a feed restriction program would be more costly, because with say a 20% restriction there would likely be no growth, because 80% of nutrients must go towards maintenance. Early feed restriction programs therefore make sense from an energetic efficiency point-of-view, and also are the most advantageous in programs aimed at reducing the incidence of metabolic disorders.

TABLE 5.10 Proportion of nutrients for growth vs maintenance

	% Distribution	
Week	Maintenance	Growth
1	20	80
2	30	70
3	40	60
4	50	50
5	60	40
6	70	30
7	75	25
8	80	20

Fig. 5.3 is a schematic representation of various growth curves of broilers. Lines A, B and C represent three potential growth curves of broilers that reach approximately 2 kg at 42 days of age.

Fig. 5.3 Schematic representation of growth curves of broiler chickens.

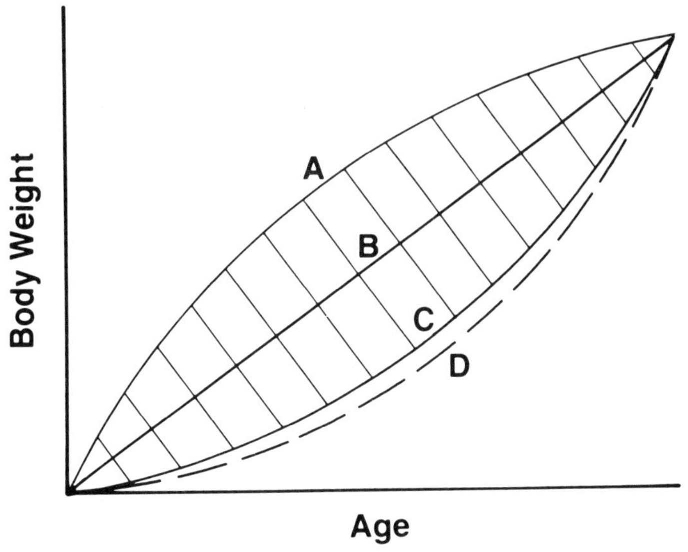

If birds grow at a uniform rate, growth will be as depicted by line B (Figure 5.3.), and this represents a biological ideal as far as minimizing stress is concerned ie: continual steady growth with no major periods of slow or rapid growth. However, few animals grow with such "statistical precision." Birds growing through lines A and C both reach 2 kg at 42 days, yet the routes they take are quite dissimilar. Bird A has more rapid initial growth, and subsequently, a slower growth approaching market weight. Bird C initially has a slower rate of growth, followed by an accelerated growth towards market weight. Bird C will likely exhibit a superior feed conversion because its maintenance requirement will be less. The reason for this reduced maintenance requirement is that at any specific age, prior to reaching market weight, the bird has a smaller body mass to maintain, and so will need less feed nutrients for this purpose. Smaller birds have proportionally higher maintenance requirements, but if C is sufficiently different to A (Figure 5.3), the absolute quantity of nutrients going towards maintenance will be less. Totalling these reduced maintenance needs, while still achieving desired body weight (Figure 5.3) must result in more feed directed to growth, and so improved feed efficiency. This theory assumes that carcass composition is unaltered.

Different strains of broiler chickens exhibit different growth patterns, yet it is possible to have two strains of birds reaching identical market weights, at identical ages, with significantly (and economically) different efficiencies of feed conversion. Detailed weighings of such birds throughout the growing period invariably shows differential growth patterns of the types shown in Figure 5.3 although obviously the effects are not quite as dramatic as depicted in this schematic. Fig 5.4 shows actual growth curves of birds exhibiting "normal" and compensatory-type growth. This situation is seen in many commercial flocks around the world, suggesting that it is not caused by isolated conditions of nutrition, disease, environment etc. that prevail at any one time.

Fig. 5.4 Compensatory growth-type curve exhibited by female broilers.

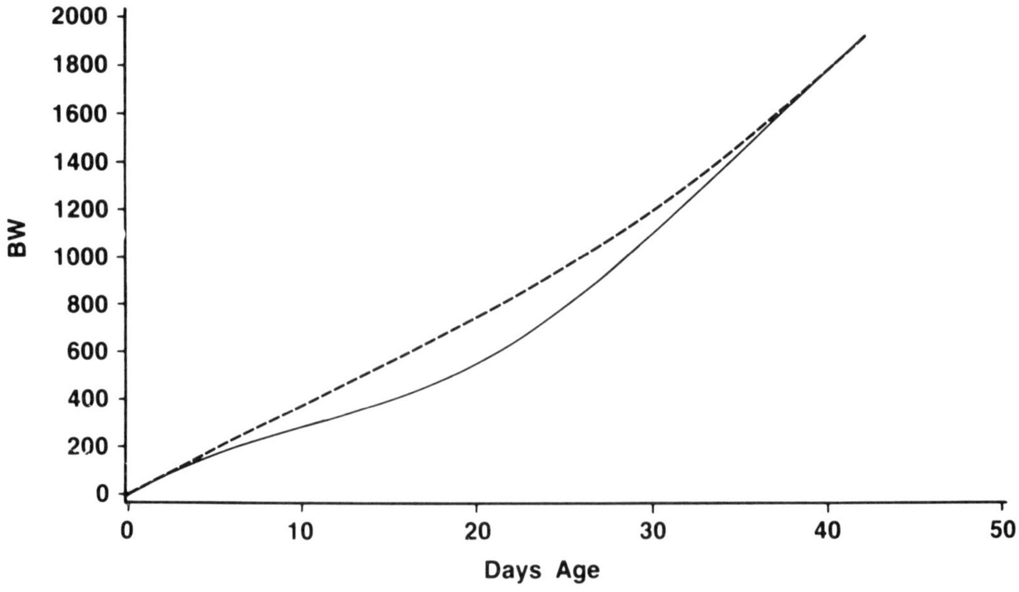

The more efficient growth curve depicted by line C in Figure 5.3, effectively capitalizes on compensatory growth. In addition to benefiting from reduced maintenance costs, these birds also exhibit slightly superior efficiency of converting feed into edible tissue during the period of compensation. This concept is well recognized for larger animals with more prolonged grow-out cycles, and was advocated for turkeys in the early 1970's (Auckland and Morris, 1971). Until recently, it was assumed that the broiler chicken simply did not have time to benefit from compensatory growth since this implies an initial period of slow or reduced growth. However, recent research findings suggest that the broiler is able to show compensation, even during a 6 week grow-out cycle.

The practical questions arising therefore are **a)** How do we convert a broiler that inherently has a growth curve depicted by A to that of C (Figure 5.3) and/or **b)** How do we further improve the efficiency of bird C to perhaps D (Figure 5.3)? The simple answer may be qualitative and/or quantitative feed restriction at some time during the early period of growth. There is little doubt that the broiler chicken can perform quite well on low protein starter diets, and that subsequent compensatory growth will likely occur. However, as previously discussed, such diets invariably result in increased carcass fat deposition. Similarly, offering birds reduced quantities of starter diet may initially seem attractive if carcass composition is disregarded.

In order to allow potential for compensatory growth, while maintaining carcass quality, some means of maintaining the correct balance of amino acids to energy must be achieved. This is best accommodated by physical feed restriction, or diet dilution, using conventional type diets. There is current interest in diet dilution of young broilers as a means of controlling fat deposition, because it is assumed that fat cell numbers increase most rapidly in the very young bird. Controlling fat cell growth at this age may therefore place an upper limit on subsequent fatness of the bird. Improvements in feed efficiency with such systems are claimed to be related to production of leaner birds, although such early qualitative feed restriction does imply compensatory growth. In some early studies, we fed broiler chickens conventional starter diets to 4 days of age and then the same diet diluted with up to 55% rice hulls from 4 - 11 days. After this time, the conventional starter was reintroduced, followed by regular grower and finisher diets. Table 5.11 indicates the amazing ability of the broiler chicken to compensate for this drastic reduction in nutrient intake from 6 - 11 days of age.

TABLE 5.11 Effect of diet dilution with rice hulls from 6-11 days of age, on compensatory growth of male broiler chickens

Treatment	Body weight (g)				Feed:gain		ME/kg gain
	21d	35d	42d	49d	21-35d	0-49d	0-49d
1. Control	733	1790	2390	2890	1.84	2.01	6.21
2. 50% dilution 6-11 d	677	1790	2380	2950	1.70	1.93	5.90

Adapted from Zubair and Leeson (1994)

Growth compensation was complete by 35 d of age, and this was associated with improvement in feed efficiency and a 5% improvement in energy efficiency. Such improvement in feed efficiency likely relates to a more favourable growth rate being induced (eg. C, Figure 5.3) and/or that birds utilize nutrients more efficiently during the period of compensation. Although not significantly different, there was also an indication that these birds deposit less fat, which is another factor that will improve feed efficiency. A practical problem with the type of diet dilution described in this study, is potential for wet litter conditions related to high fiber intake.

Diet dilution is not always a practical approach to reducing nutrient intake, because birds will compensate with increased feed intake, and diluents are very expensive per unit of energy provided. An alternative approach is to physically limit feed intake, and most experiments involve restriction down to the level of maintenance energy need, which is around 1.5 kcal/g $BWt^{.67}$. Fig. 5.5 shows the energy intake of broilers restricted to this level from 6 - 12 d of age.

Fig. 5.5 Energy intake of regular and feed restricted broilers (6 – 12d)

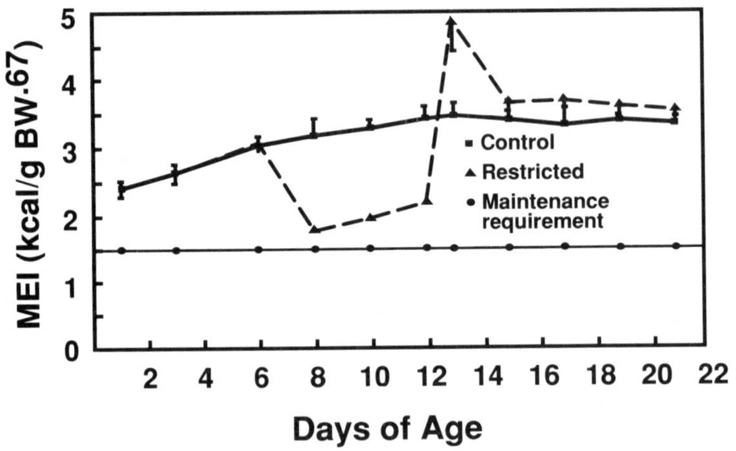

Maintenance energy need was slightly exceeded by the restriction program, although energy intake was obviously much less than that of the *ad lib* fed controls (Fig. 5.5.). A problem with such physical restriction programs occurs when birds are returned to ad-lib feeding, as occurred at 13 d in Fig. 5.5. Broilers greatly overcompensate in feed and energy intake, and take some 2 - 3d to normalize their eating pattern. This voracious appetite in the 24-36h following feed restriction can cause choking which in extreme cases is fatal. Return to full feeding should therefore be more gradual, over a 24 - 48h period. Following return to full feeding, and up to about 21 d (Fig. 5.5) previously restricted birds eat more feed and this is undoubtedly part of the "compensatory" effect. However this increased energy intake does contribute to fat cell hypertrophy (increased size rather than number of cells) and this is one reason why restricted-refed birds have similar overall body fat levels at market weight.

The reasons for improvement in feed efficiency with compensatory growth feeding programs are not entirely clear at this time. Based on classical studies with other animals, it has been suggested that birds have reduced maintenance energy needs, even during *ad lib* refeeding up to market weight. For the birds described in Fig. 5.5, we actually measured fasting metabolic rate of birds both during the actual underfeeding, and at 18d during subsequent *ad lib* refeeding (Fig. 5.6).

Fig. 5.6 FMR of birds full fed (—) or 50% restricted (---) from 6 – 11d of age

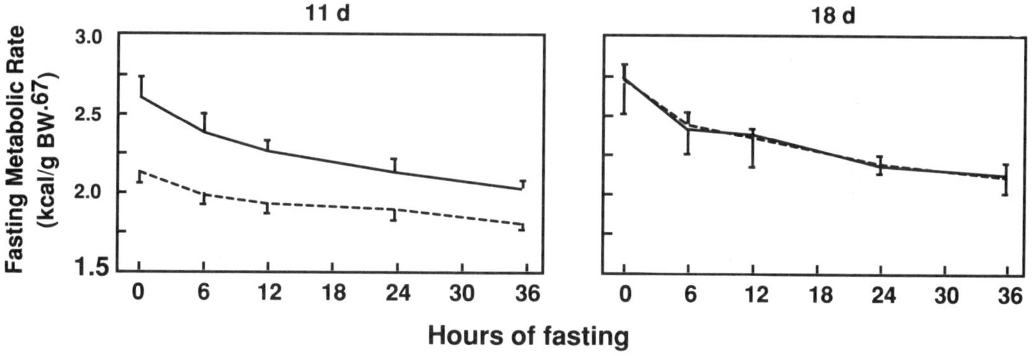

During the period of feed restriction, birds did exhibit a reduced FMR, even after all birds were fasted for 48 hr. However 6 days after return to *ad lib* feeding at 18d of age there is no difference in FMR for the two groups of birds (Fig 5.6). These data suggest there to be no "carry-over" of reduced maintenance needs per unit of body size, when birds are full fed, and so this is unlikely to account for overall improved feed efficiency seen with the compensatory type programs. A more likely reason for improved feed utilization, is that there is simply a reduction in overall maintenance energy needs associated with the bird being smaller for a significant part of the grow-out period. This hypothesis suggests that complete growth compensation should be delayed for as long as possible, and that ideally birds would not achieve normal weight-for-age until day of marketing.

Most research to date has not been able to duplicate the dramatic reductions in carcass fatness attributed to a compensatory growth program as originally described by Plavnik and co-workers. This may be due to other researchers not imposing a severe enough degree of undernutrition, since Plavnik used a feed allowance that accounted only for maintenance. In practice, this means an exceptionally low feed intake (10 - 12 g/bird/day for 4 - 11 d) which is difficult to calibrate with commercial equipment, and difficult to simulate with diet dilution. However, the potential benefits of a compensatory growth program seem exciting, and presumably any degree of modification of the growth curve will be beneficial.

The broiler chicken therefore appears able to benefit from a period of early "undernutrition," in that subsequent compensatory growth results in no overall loss of market weight, and should be associated with improved feed utilization. Depending on the method used to impose such undernutrition, there is potential for these birds to be as lean or leaner than conventionally fed birds. The only potential problem with this technique, is that it may shift the mortality/morbidity peak to later in the grow-out period. For example, leg problems and Sudden Death Syndrome are both related to fast growth rate, rather than body weight *per se*. With early undernutrition, we are shifting the time of most rapid growth (compensation period) to later in the cycle (3 - 5 weeks) and so growth related problems may be more prominent in this period. Such effects have not been recorded with experimental flocks, although research involving larger numbers of birds needs to be conducted. A period of compensatory growth may therefore be beneficial to grow-out of commercial broilers, and as long as good quality finisher diets are employed, there are interesting potential economic advantages to this technique. Even greater benefits may apply to roaster birds.

Another very practical problem with diet dilution or feed restriction, is deciding upon levels of anticoccidials and other pharmacological compounds. With diet dilution, birds will eat much more feed. If for example, feed intake is doubled due to a 50% dilution, should the level of anticoccidial be reduced by 50%? With 50% feed restriction on the other hand, does there need to be an increase in concentration of these additives? This general area needs careful consideration, and results may well vary with the chemical compounds under consideration due to potential toxicity at critical levels.

d) Lighting programs and feed intake

Many broiler chickens are grown under 23 or 24 h light each day, because it is thought that unlimited access to feed is required for maximum growth rate. However, there may be some potential for modifying the bird's feeding activity through lighting so as to improve the efficiency of feed utilization. Such programs involve periods of darkness for varying lengths of time throughout the day. The idea behind an intermittent program of, for example, 1 hour light: 3 hours darkness, repeated six times each day, is that birds will actively seek feed during the light period, and subsequently rest during the dark period. If enough feed has been consumed in the 1 hour of light, then feed efficiency should be improved, because birds will be quite docile during the dark period and so expend less energy on maintenance. After this period of darkness, the upper digestive tract should be empty, and the bird is again ready to eat. If this type of program is used, it is important to ensure that adequate feeder space is available such that all birds can eat during the light periods. These types of programs have been shown to improve feed efficiency to 49 d of age, by about 0.05 units. However such intermittent programs do not slow down growth rate and so do not help in reducing late cycle mortality associated with metabolic conditions.

More recently there has been interest in so called "step-down - step-up" programs for broiler chickens as advocated by Dr. Classen at the University of Saskatchewan. Broilers are subjected to very short days for a 1 - 2 week period, ostensibly in an attempt to reduce the incidence of leg problems and occurrence of SDS. In effect, the birds are likely to show reduced feed intake, and so this type of program fits in well with the concept of restricted feeding detailed in the previous section. An example of this type of lighting program being used commercially is shown in Table 5.12.

TABLE 5.12 Step-down lighting program for broilers

	Bird age (d)	Hours of light
Example 1	0-4	23
	4-10	8
	10-14	10
	14-18	14
	18-23	18
	23→Market	23
Example 2	0-4	23
	4-14	8
	14→Market	23
Example 3	0-4	23
	4-14	Natural daylength
	14-18	18
	18→Market	23

With very short periods of darkness, birds rarely eat in the dark, yet with the step-down programs it has been shown that birds will consume up to 30 - 40% of their feed in the dark period. This has caused problems for those producers using a high stocking density, since birds apparently clamber over one another during the dark period in an attempt to get to the feeder. This can result in increased scratching of the skin, and increased incidence of scabby-hip syndrome. At more liberal stocking densities, (14.4 birds/sq m.) such problems are less severe, and improved leg condition and lower incidence of early SDS is noticed. If normal market weight-for-age is desired, then such step-down, step-up programs must be ended, and birds returned to full lighting by at least 20d before expected market age. If reduced daylength is continued for too long a period, then reduced growth rate will have to be accepted. If mortality due to metabolic disorders is excessive (>8% in males to 45d age) then it may be economical to prolong the period of reduced daylength, and accept the delayed market age (see section on metabolic disorders).

Simple lighting programs involving a short daylength used from day-old to market age also seem to be beneficial in slowing down growth and reducing mortality (Table 5.13).

TABLE 5.13 Effect of short daylength on male broiler performance

Light Schedule	Body weight (g)				0-48d mortality (%)
	7d	21d	35d	48d	
23L:ID	138[a]	738[a]	1852[a]	2924	9.0
16L:8D	126[b]	684[b]	1798[b]	2912	3.0
14L:10D	121[bc]	641[c]	1727[c]	2850	3.5
Step down - step-up	115[c]	614[d]	1713[c]	2884	3.5

Adapted from Randon et al. (1993)

Shorter daylengths dramatically reduce mortality, and there is little difference in 48d body weight. However, as will be discussed in the following section on carcass composition, such slower growth rate may adversely affect breast meat yield.

5.4 CARCASS COMPOSITION

The composition of the broiler carcass is now receiving considerable attention with the poultry industry's major thrust in further processing. Just a few years ago the processor cut up whatever chicken was unsuitable for other markets, while today the trend is towards specialized carcass types and composition to meet specific demands for cut-up, deboning, and subsequent new product manufacture. While carcass portion yield is largely a factor of age and genetics, carcass composition can, to a large extent, be modified through diet choice. In general, diets high in energy produce fatter carcasses and vice versa. On the other hand, high protein diets produce leaner carcasses. The situation is a little more complex than this, since it is actually the balance of protein to energy that is important. If the bird consumes excess energy in relation to protein, a fatter bird develops, whereas a leaner bird can be produced by feeding larger quantities of protein in relation to energy. This manipulation of nutrients is sometimes referred to as changing the energy:protein ratio. Unfortunately, simple changes such as these are not economical, since the required degree of leanness in the carcass often only results from uneconomically high levels of protein.

When discussing the effect of diet protein or energy level on carcass composition, it is very important to appreciate the units of measurement. Often there is discussion about the effects of diet on percentage changes in composition, and in some situations the percentage of a component in the carcass changes simply because there has been a corresponding change in the level of another component. This situation is clearly shown in Tables 5.14 and 5.15.

TABLE 5.14 Proportional and absolute changes in carcass components in response to diet energy level				
	Carcass fat		Carcass Protein	
Diet energy (kcal/kg)[1]	Grams	Percentage	Grams	Percentage
2600	161[a]	37.5[a]	221	51.9[e]
2800	178[b]	39.3[b]	225	50.0[d]
3000	208[c]	42.4[c]	229	47.1[c]
3200	211[c]	42.6[c]	230	46.9[c]
3400	239[d]	45.6[d]	229	44.7[b]
3600	258[e]	47.9[e]	229	42.9[a]

[1]at constant protein

In this study, over the very wide range of energy levels used, there is the classic response for increase in % carcass fat and decrease in % carcass protein. However, only the actual quantity of fat changes, from 161 to 258 g as energy level increases, with no change in grams carcass protein. Producing a leaner carcass with low energy diets is therefore achieved by reducing fatness, rather than by increasing lean meat deposition.

Changing the protein level of the diet has the same basic effect on tissue deposition (Table 5.15).

TABLE 5.15 Proportional and absolute changes in carcass components in response to diet protein level				
	Carcass fat		Carcass Protein	
Diet protein level[1] (%)	Grams	Percentage	Grams	Percentage
16	252[d]	50.0[d]	202[a]	40.7[a]
20	237[c]	46.2[c]	227[b]	44.9[b]
24	210[b]	42.4[b]	233[b]	47.7[c]
28	189[a]	39.4[a]	233[b]	49.2[cd]
32	185[a]	39.2[a]	233[b]	50.3[d]
36	179[a]	38.3[a]	234[b]	50.7[d]

[1]at constant energy level

Except when very low protein diets are used (16%, Table 5.15) diet protein has no effect on the quantity of protein deposited in the carcass. Proportional changes in carcass protein are a consequence of less grams of fat being deposited as protein level increases. Consequently, producing "leaner" carcasses with higher protein diets is a consequence of there being less fat, rather than there being more protein.

Discussion on carcass composition invariably involves details of abdominal fat which is the leaf fat found in the vent area and that extends around the intestines and gizzard. Table 5.16 shows the distribution of this fat in the viscera and abdominal region.

TABLE 5.16 Abdominal fat (% distribution)

	Abdominal fat (%)	grams/49d male broiler
Pad	59.0	65
Gizzard	18.1	20
Viscera	17.4	19
Heart	3.6	4
Other	1.9	2

This fat is detrimental as far as consumer acceptance is concerned, and presents a waste water disposal problem in the processing plant. It is also of concern to the grower, since this wasteful fat has required a large amount of feed energy in its synthesis. This abdominal or leaf fat acts in the same way as the rest of the carcass fat to diet manipulation, and so a simple solution to reducing the quantity found is not possible at this time. It seems as though our present problem with carcass fat content is a result of the broiler's appetite. Over years of genetic selection for growth, we have in fact selected for appetite. Consequently, the bird that could eat the most feed, became the biggest bird in the pen and therefore was selected for the breeding program. In the future it will be necessary to continue selection of the biggest birds while at the same time ensuring that they are lean. This is currently a problem because our accuracy of judging fatness in live chickens is not too precise. The following tables detail the composition of broiler chickens. Tables 5.17 - 5.21 show the weight of broiler carcasses and the various portions and yields of deboned meat. Figures 5.7 - 5.9 show the fat, moisture and protein content of the carcass, breast and thigh portions.

TABLE 5.17 Body weight and offal characteristics of male broiler chickens

	35 days	42 days	49 days	56 days	70 days
Body weight (starved) g	1700	2200	2750	3200	4200
<u>% body weight</u>					
Total offal	14.0	13.1	11.2	10.8	7.5
Abdominal fat	1.8	2.0	2.9	3.0	3.6
Liver	2.4	2.3	2.1	2.0	1.8
Alimentary tract	8.0	7.0	6.0	5.5	5.0
Gizzard + proventriculus	2.2	2.0	1.9	1.7	1.6
Offal fat, % DM[1]	48.0	51.0	58.0	61.0	62.0
Offal CP[2], % DM	40.0	41.5	34.0	32.0	28.0
Offal DM, %	38.0	39.0	41.0	43.0	46.0
Offal fat, g	70.0	110.0	130.0	200.0	250.0
Offal CP, g	56.0	86.0	82.0	102.0	112.0

[1]Dry matter [2]Crude protein

TABLE 5.18 Live weight and carcass characteristics of high-yield male broiler chickens (g)

Weeks of age	Live Weight	Carcass Weight	Abdominal Fat Pad Weight	Wings	Drums	Thighs	Bone-in Breast	De-boned Breast
4	1224	818	20	83	120	143	240	151
5	1754	1237	32	128	164	210	372	245
6	2223	1596	52	155	208	260	498	320
7	2666	1982	64	192	269	321	626	409
8	3274	2500	86	234	335	412	812	542
9	3674	2731	111	273	398	499	905	596
10	4293	3320	121	305	440	594	1070	723

TABLE 5.19 Carcass weight and portion characteristics of high-yield male broiler chickens (% carcass weight)

Weeks of age	Carcass Weight (g)	Abdominal Fat Pad	Wings	Drums	Thighs	Bone-in Breast	De-boned Breast
4	818	2.5	10.2	14.8	17.6	29.4	18.5
5	1237	2.6	10.4	13.3	17.0	30.1	19.8
6	1596	3.3	9.74	13.1	16.3	31.2	20.1
7	1982	3.3	9.76	13.6	16.2	31.4	20.5
8	2500	3.4	9.44	13.5	16.5	32.5	21.6
9	2731	4.2	10.9	16.1	19.7	36.0	23.6
10	3320	3.7	9.2	13.3	17.9	32.2	21.8

TABLE 5.20 Live weight and carcass characteristics of high-yield female broiler chickens (g)

Week	Live Weight	Carcass Weight	Abdominal Fat Pad	Wings	Drums	Thighs	Bone-in Breast	De-boned Breast
4	1088	720	20	78	104	126	212	147
5	1582	1160	37	113	141	191	342	224
6	1910	1376	46	138	179	225	433	195
7	2382	1753	76	170	239	288	568	381
8	2730	1996	87	191	258	331	684	451
9	3125	2393	115	227	315	315	805	531
10	3546	2660	142	241	323	454	914	623

TABLE 5.21 Carcass weight and portion characteristics of high-yield female broiler chickens (% carcass weight)

Weeks of age	Carcass Weight (g)	Abdominal Fat Pad	Wings	Drums	Thighs	Bone-in Breast	De-boned Breast
4	720	2.8	10.8	14.4	17.5	29.5	20.4
5	1160	3.2	9.8	12.2	16.6	29.7	19.5
6	1376	3.4	10.7	13.4	16.4	31.5	21.5
7	1753	4.3	9.8	13.8	16.5	32.5	21.7
8	1996	4.3	9.6	13.0	16.6	34.2	22.6
9	2393	4.8	9.5	13.2	17.7	33.6	22.1
10	2660	5.4	9.1	12.2	17.1	34.3	23.4

Fig. 5.7 ***Effect of age on percentage moisture, protein and fat content of eviscerated carcass of male broilers: x observation means __ regression line.***

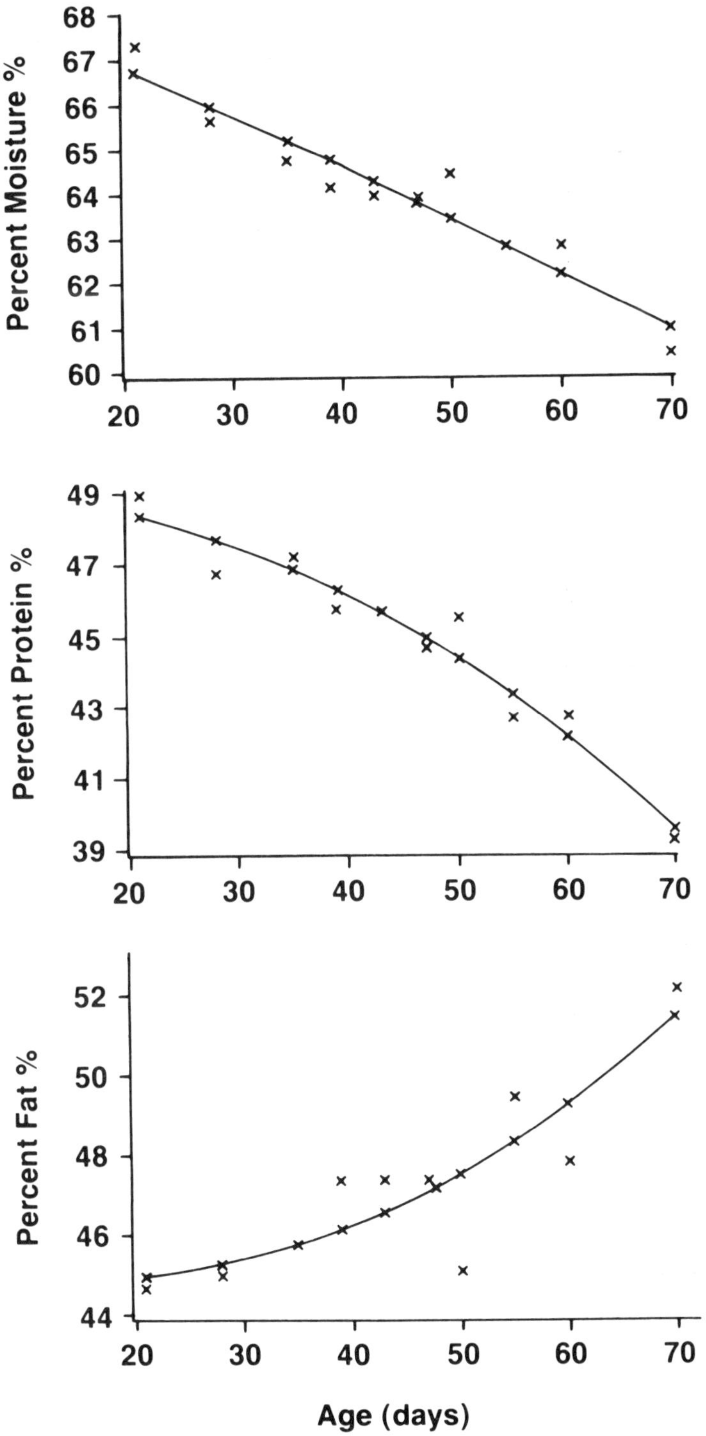

Fig. 5.8 ***Effect of age on percentage moisture, protein and fat content of male broilers breast portion (bone-in) x observations mean ___regression line.***

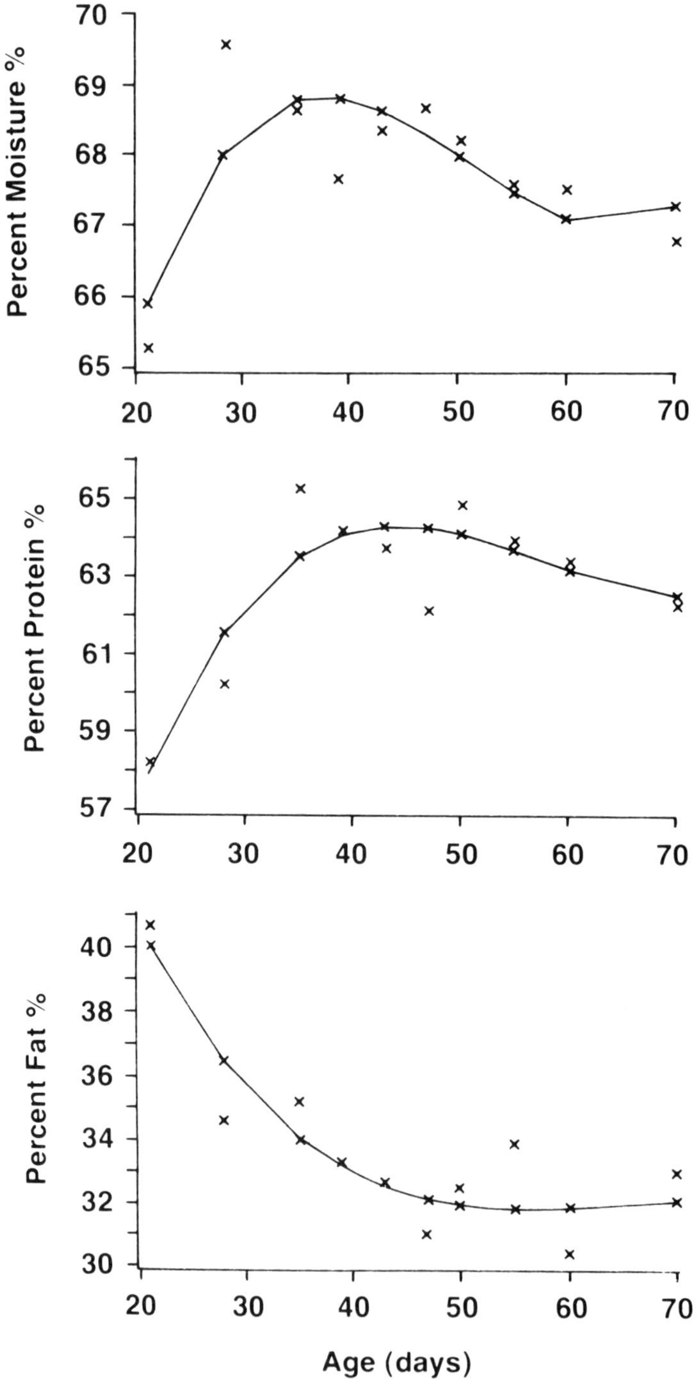

Fig. 5.9 *Effect of age on percentage moisture, protein and fat content of male broilers thigh portion (bone-in) x observations mean __ regression line.*

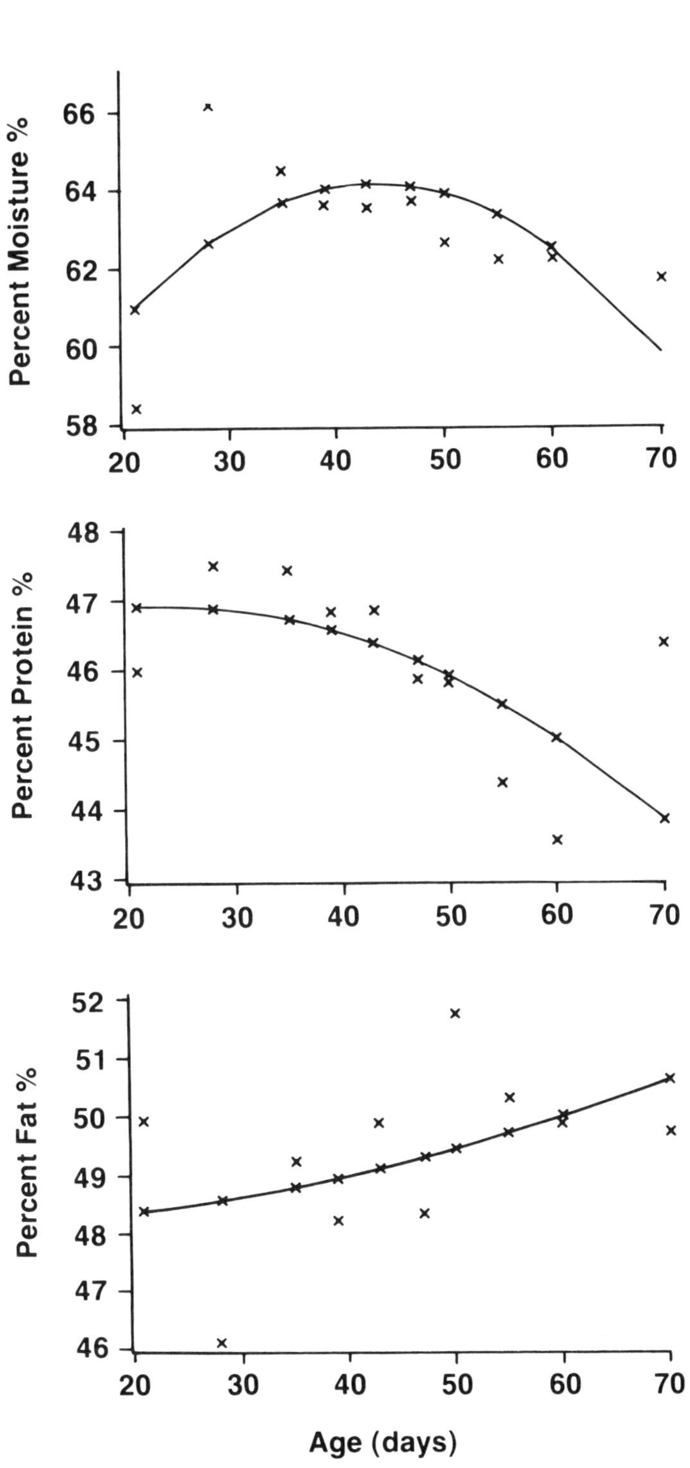

5.5. WITHDRAWAL/FINISHER DIETS

Over the last few years, there has been considerable interest in removing various nutrients from the last diet fed to broilers. Initially this related to certain vitamins and trace minerals, but now extends to protein/amino acids, various pharmaceutical additives and even energy level of the feed. It must be stressed that many such studies involve only growth rate and feed efficiency, and using these simple parameters it may sometimes be possible to reduce diet cost in the last 7 - 10 d of growth. However many nutrients influence carcass yield, bone integrity and immune status of the bird, and these factors often impact negatively on bird performance under stress conditions, and on overall economics of meat production. A truly "integrated" approach must be taken when considering deletion of nutrients from a withdrawal feed.

a) Protein/Energy dilution

We have carried out studies to observe the older broiler's response to diluting the withdrawal diet so as to provide less concentrated sources of energy, protein, or energy and protein. The idea behind this study was to see if the bird could be made to metabolize its fat depots to augment reduced energy intake, and so improve carcass characteristics. When energy level of the diet is reduced, by diluting out corn and fat with a mixture of indigestible fiber plus sand (so as to maintain normal bulk density), there is variable bird response according to time of feeding (Table 5.22).

The various diluted diets were fed from 35 - 49 d. At 42 d, growth rate was reduced with each increment of dilution, because birds failed to adjust their intake accordingly. However in the next 7 d, between 42 and 49 d of age, the birds ate correspondingly more of the diluted diets and this resulted in compensatory gain such that body weight was equalized by 49 d. In a second study, we diluted out both energy and protein/amino acids (Table 5.23).

Again, body weight was depressed by 42d, with compensation from 42 - 49 d. Overall, from 35 - 49 d, energy intake was not normalized through increased feed intake, although this seemed to be due mainly to low energy intake in the 35 - 42 d transitionary period. These data therefore suggest that it is possible to feed any nutrient density in the finisher period, as long as birds are given sufficient time for appetite compensation. However as discussed at the outset of this section, analyzing these birds further, shows adverse effects in various carcass traits (Table 5.24).

Even though body weight was not affected by diet dilution, there was a significant decline in carcass weight and yield of breast meat from these carcasses, at all levels of diet dilution. Unless diluted diets are fed for considerably longer periods of time, (and assuming the birds can adapt), then it seems uneconomical to dilute energy or protein in the withdrawal diet.

Although such diluted diets may only be of academic interest, there is an interesting pattern in the broiler's adaptation to nutrient dilution as shown in Table 5.25.

TABLE 5.22 Effect of energy dilution of finisher diet on growth of broilers

Diet energy	Body weight (g)		Feed intake (g/bird)		Energy intake (Mcal)
ME (kcal/kg)	42d	49d	35 to 42 d	42 to 29 d	35 to 49 d
3200	2,370	2,982	1,250	1,373	8.43
2950	2,395	2,998	1,301	1,401	8.00
2700	2,371	2,970	1.377	1,456	7.66
2450	2,331	2,913	1,371	1,585	7.24
2200	2,323	3,022	1,444	1,677	6.85
1950	2,277	2,946	1,482	1,946	6.65
Regression analysis	**	NS	**	**	**

Adapted from Leeson et al. 1996.

TABLE 5.23 Effect of dilution of energy and protein in the finisher diet on growth of broilers

Diet type		Body weight (g)		Feed intake (g/bird)		Energy intake (Mcal)
ME (kcal/kg)	CP (%)	42d	49d	35 to 42 d	42 to 49 d	35 to 49 d
3210	18.0	2,420	2,948	1,284	1,299	8.30
2890	16.2	2,367	2,921	1,318	1,445	7.99
2570	14.4	2,320	2,879	1,387	1,517	7.47
2250	12.6	2,263	2,913	1,429	1,844	7.36
1925	10.8	2,170	2,913	1,472	2,201	7.08
1605	9.0	2,218	2,892	1,685	2,610	6.90
Regression analysis		**	NS	**	**	**

TABLE 5.24 Effect of dilution of energy and protein in the finisher diet on carcass characteristics

Diet treatment		Carcass weight (g)		Breast Weight (g)	
ME (kcal/kg)	CP (%)	42d	49 d	42 d	49 d
3210	18.0	1,775	2,184	326	418
2890	16.2	1,719	2,107	320	404
2570	14.4	1,678	2,063	305	400
2250	12.6	1,654	2,088	205	402
1925	10.8	1,588	2,073	295	390
1605	9.0	1,540	2,038	285	378
Regression analysis		**	**	*	*

TABLE 5.25. Feed intake response to diet nutrient dilution from 35-49 d of age

Diet ME	Diet CP	Feed Intake		Feed Intake	
(kcal/kg)	(%)	35 - 42d (g)	% Adjustment	42 - 49d (g)	% Adjustment
3210	18.0	1280		1300	
2890	16.2	1320	+3	1450	+12
2570	14.4	1390	+9	1520	+17
2250	12.6	1430	+12	1840	+42
1925	10.8	1470	+15	2200	+69
1605	9.0	1690	+32	2610	+100

From 35 - 42d, the % adjustment of feed intake is insufficient to normalize nutrient intake. However from 42 - 49d, the % adjustment figures are amazingly accurate for that needed to normalize intake. For example, the 100% feed intake adjustment seen with the 1605 kcal/kg diet with 9% crude protein means that their energy and protein intake is identical to that of the control birds. This situation is convincing evidence that given time, the broiler is very efficient in controlling its nutrient intake.

b) Vitamins and trace minerals

Because it takes some time for vitamin or trace mineral deficiency symptoms to appear, in a bird previously fed a well-fortified diet, there has been interest in removing supplemental vitamins and minerals from the finisher diet (Table 5.26)

TABLE 5.26 Effect of removal of vitamins and trace minerals from 35 to 49 days of age

	35-49d weight gain (g)	F:G	Mortality (%)
Control	960	2.42	5.0
No supplements	990	2.39	4.7

Adapted from Skinner et al. 1992.

However in this same study, there were numerical losses in carcass yield (-1.2%) and increased incidence of leg disorders and TD score of cartilage plates. These authors conclude that under conditions of heat stress, mycotoxin challenge or other factors that adversely affect performance, that complete supplementation of minerals and vitamins seems advisable. This concept is confirmed in studies where birds are subjected to heat stress under controlled conditions (Table 5.27).

TABLE 5.27 Removal of vitamins and trace minerals from heat-stressed (24-35°C cycle) broilers

	35-49d weight gain (g)	F:G	Mortality (%)
Control	1280^a	2.66^a	9.6
No supplements	1240^b	2.86^b	13.2

Adapted from Teeter (1994)

5.6 FREE-CHOICE FEEDING PROGRAMS

The broiler chicken is able to select a relatively well balanced diet when offered an array of ingredients or a selection of diets differing in nutrient profile. Table 5.28 indicates the birds ability to adjust its feed intake when offered diets varying in energy content. In a recent study we offered male broilers a choice feeding situation involving diets differing only in energy content. The diets were offered in separate feeders, and the birds had free access to both feeds at all times (Table 5.28).

TABLE 5.28 Performance of male broilers given choice of two feeds differing only in energy content					
Diet choice (kcal/kg)		Body weight (g)		Nutrient Intake (0-49d)	
Basal	Test	25 d	49 d	Energy (Mcal)	Protein intake (g)
3,300 +	3,300	980	2,763	15.5	984
3,300 +	3,100	998	2,836	15.8	1,033
3,300 +	2,900	970	2,727	14.9	985
3,300 +	2,700	992	2,822	15.9	1,058

Birds performed remarkably well on the choice feeding program, with all birds consuming identical quantities of energy to 49 d of age.

A practical application of this feed selection ability of the bird, is to offer various ingredients as the choice "feed". In Europe this has been carried out mainly with wheat and a so-called grain balancer diet. The grain balancer is fortified with extra calcium, phosphorus, trace nutrients and certain amino acids, such that when the bird voluntarily consumes significant quantities of cereal, an approximately normal intake of most nutrients occurs. In a recent study, we investigated the ability of the broiler to select from either a cereal or a conventional broiler starter. The control birds were fed starter, grower and finisher to 49 d, while experimental diets involved the broiler starter + whole wheat or broiler starter + cracked corn from 7 d to 49 d (Table 5.29).

TABLE 5.29 Broiler performance with free-choice cereals					
	49 d body wt	Feed:Gain	Carcass wt	Breast wt	Breast yield
Treatment	(g)		(g)	(g)	(% Cwt)
Starter-Grow-Finisher	3030	1.93	2234[a]	480[a]	21.5[a]
Starter + Wheat	2922	1.99	2135[b]	420[b]	19.6[b]
Starter + Corn	2917	1.98	2139[b]	430[b]	20.0[b]

Body weight was comparable at 49 d, although there was a significant reduction in carcass weight and breast meat yield. As the birds got older, they ate less of the broiler starter and more of the cereal (Table 5.30).

TABLE 5.30 Feeding pattern of birds offered free-choice cereals				
Age	Wheat study		Corn study	
	% intake starter	% intake wheat	% intake starter	% intake corn
7-21d	84	16	87	13
21-35d	61	39	66	34
35-49d	60	40	57	43

Increasing the intake of cereal obviously reduced the intake of protein and amino acids, and this is the likely reason for loss of breast meat yield. Birds that are voluntarily consuming about 40% cereal from 35-49d are also receiving a diet with about 0.6% calcium and 0.24% available phosphorus, although there were no apparent skeletal problems in these birds. Diluting the diet with cereals obviously reduces the cost of the diet. Table 5.31 outlines the relative cost of producing the whole bird, the carcass and deboned breast.

TABLE 5.31 Relative feed cost with cereal feeding			
	Body weight	Carcass weight	Breast meat
Conventional diet	100	100	100
Free-choice wheat	91	92	100
Free-choice corn	90	90	96

There is a distinct saving in feed cost with free choice cereal feeding for production of live birds and also per unit of carcass weight. However because of associated reduction in breast meat yield, there is no cost advantage to this feeding technique for production systems relying heavily on further processing.

5.7 NUTRITION AND METABOLIC DISORDERS

a) Sudden Death Syndrome

Sudden Death Syndrome (SDS) has been recognized for over 25 years, although it has only been in the last 10 years or so that the occurrence in broiler chickens has been of significant economic importance. Also referred to as Acute Death Syndrome or "flip-overs", SDS is most common in males, and especially when growth rate is maximized. Mortality may start as early as 3 - 4 d, but most often peaks at around 3 - 4 weeks of age, with affected birds invariably being found dead on their back. Mortality may reach as high as 1.5 - 2% in mixed-sex flocks, and in male flocks the condition is often the major single cause of mortality, with death rates as high as 4% being quite common. The economic loss is therefore substantial. Confirmation of SDS by necropsy is difficult as no specific lesions are present. Birds are invariably well fleshed with a partially filled crop and gizzard. Julian (1987) suggests that since there are no gross or histopathological lesions, then there is no justification for including SDS with other diseases of the cardiovascular system. It is more likely of metabolic origin, with an imbalance of electrolytes resulting in left ventricular fibrillation.

There seems little doubt that any nutritional or management factors that influence growth rate will have a corresponding effect on SDS. Sudden Death Syndrome can virtually be eliminated with diets of low nutrient density (18% CP, 2400 kcal ME/kg) although these may not be economical in terms of general bird performance. Research data suggests that diets based on pure glucose as an energy source result in much higher incidence of SDS compared to birds fed starch or fat-based diets. It seems likely that some anomaly in electrolyte balance is involved in SDS, although this has not been clearly defined. In part, this is due to the fact that metabolic changes occur rapidly after death, and hence blood profiles taken from SDS birds will likely vary depending upon sampling time following mortality.

We have recently observed a high incidence (up to 100%) of SDS in birds dosed with lactic acid. Lactate is produced during periods of intense activity and when energy demands are high. After lactic acid dosing into the crop, diet content seems to influence timing of onset of SDS. With a predominantly glucose based diet, birds died within 1 minute, whereas with a starch diet onset of SDS did not occur until after 1 hour. In comparison to these values, birds fed a high fat diet did not show signs of SDS until at least 2 hours after dosing with lactic acid. It is interesting to observe that birds fed the glucose diet, where onset of SDS was rapid, showed the lowest plasma lactic acid levels, (Table 5.32). This apparent dichotomy may relate to more rapid uptake of lactate from the blood, by those birds fed glucose.

TABLE 5.32 Plasma lactate and diet composition

	Major diet energy source		
	Glucose	Starch	Fat
Blood lactate (mg/ml)	0.16	0.57	0.56

It is known that birds fed mash vs pelleted diets produce more lactic acid in the crop and so if lactate levels are a prime consideration in SDS, then pelleted diets would be advantageous. However, the pelleting of diets has been suggested as a causative factor in SDS, although this relationship may be simply due to consumption of pelleted feeds resulting in more rapid growth rate and therefore a higher incidence of SDS. High levels of dietary salt have also been implicated in SDS, although this may be an indirect consequence of increased incidence of ascites occurring with high diet salt levels. It is obvious that SDS is caused by rapid growth rate, a consequence of which is some alteration of metabolism and in particular acid-base balance. SDS can be reduced or eliminated by nutritional or management practices that reduce growth rate. Obviously such decisions will have to be based on local economic considerations. At this time, there is no indication of a single causative factor, and diet manipulation other than that related to reduced growth rate, is usually ineffective.

b) Skeletal Disorders

Abnormal skeletal development continues to be a major reason for mortality and/or downgrading in commercial meat birds. In North America this loss probably equates to some $30 M per year. Since there have been numerous studies on skeletal conditions, it is obvious that the aetiology is complex and not outwardly related to a single factor. The most common skeletal abnormalities seen in meat-type birds are tibial dyschondroplasia (TD) and field rickets. The fact that leg problems are more prevalent in broilers and turkeys than in egg-type birds, has led to the speculation of growth rate and/or body weight as causative factors. On this basis, one is faced with numerous reports of general nutritional factors influencing leg problems. For example, it has been suggested that energy restriction in the first few weeks of growth halves the number of leg problems in broilers, while reduced protein intake results in fewer leg abnormalities. Similarly, restricting access to feeder space also seems to result in fewer leg defects. However, most recent evidence suggests that body weight *per se* is not a major predisposing factor to leg problems. From experiments involving harnessing weights to the backs of broiler chickens and poults, it is concluded that severity of leg abnormality is independent of body weight, and that regular skeletal development is adequate to support loads far greater than normal body weight. There seems to be some disparity between the effects on skeletal development of (1) limiting the incidence by reducing the plane of nutrition and (2) failing to aggravate the problem by artificially increasing body weight. This apparent dichotomy suggests that it is the rate of growth, related to higher levels of nutrient intake, that is important in precipitating the condition.

We have recorded a higher incidence of leg problems when ionophore coccidiostats are used, and others have likewise indicated severe leg problems in turkeys inadvertently fed ionophores. However, it is also evident that leg problems commonly occur in broiler flocks fed non-ionophore medication, and hence the effect of these compounds may be part of complex antagonisms and be precipitated by certain management criteria. It is interesting to note that TD and other skeletal abnormalities are rare with broiler breeders. This fact is most often discussed relative to restricted feeding and reduced growth rate. A high incidence of leg problems is seen in broiler breeders fed broiler-type diets *ad libitum* and housed at broiler type stocking densities. This is important to breeding companies whose programmes involve a selection at broiler market weight following broiler-type growing conditions, and then require subsequent adequate reproduction in selected birds.

In addition to the confounding effect of genetics on skeletal development, there is also an indication of sex differentiation. In attempting to resolve this latter problem, researchers have shown that castrated turkeys have a higher incidence of leg abnormalities than do intact males or those treated with testosterone. It is suggested that androgens act to fuse the epiphyses and shafts of long bones. There may well be major sex differences in the hormonal control of skeletal development related to the balance of androgens:estrogens. However, the effect of androgens:estrogens *per se* on skeletal development in the relatively juvenile broiler of today is perhaps questionable due to the fact that little sex differentiation in tibiotarsal length is seen until after 5 weeks of age. The following general areas have been considered in relation to skeletal development.

i) Diet protein and amino acids

There has been recent interest to investigate the effect of diet protein on leg problems. From a more traditional point of view, leg defects seen with lower protein diets are attributed to a tempering of initial growth rate. However, our data suggests that leg problems today are related to more complex situations, and as such, nutrient interrelationships may be involved. In this context, it is known that excess protein produces a "stress" as evidenced by increased size of the adrenals. Diets high in protein can interfere with folic acid metabolism and in so doing, increase the incidence of leg problems. However, in recent studies involving folic acid deficient diets, we were unable to show an effect with 22 vs 30% crude protein diets. In studying factors influencing skeletal development in broiler breeders and Leghorns, we have shown that while early skeletal development was little influenced by mineral and vitamin fortification, shank and keel lengths could be increased by feeding diets of higher protein content (22 vs 16% CP). It is also conceivable that the ratio of amino acids:non-protein nitrogen may be of importance in the development of bone organic matrix. Evidence for this concept comes from experiments involving synthetic amino acids and purified diets. Under these conditions, the bird's nitrogen requirement for optimum organic matrix development is often greater than the apparent requirement for growth.

The wry-neck condition sometimes seen in broiler breeders, and especially males, may also be related to disrupted amino acid metabolism. While not directly a skeletal abnormality, the condition seems to be related to the metabolism of tryptophan or niacin. During incubation, wry-neck arises because of greater muscle pull on one side of the neck, which together with pressure from the amnion, causes the 'apparent' skeletal deformity.

ii) Feed ingredients

There is also current interest in the effect of feed ingredient type and source on leg problems. Much of the earlier work in this area centered around brewer's yeast and its apparent ability to reduce certain leg problems. Of more commercial importance perhaps are the reports that the source of soybean meal can have a marked effect on the occurrence of TD. In testing three samples of meal, two were reported to cause a high incidence of TD, while birds fed the other source of soybean meal exhibited a much lower incidence. Lowering the levels of supplemental chloride and increasing diet calcium concentration seemed to aggravate the condition regardless of soybean source. More recently Lui *et al.* (1992) showed that the chloride level of some soybean samples in China was causing severe TD. In this situation, the problem was partially resolved by adding magnesium to the diet, although this did cause some loss in growth rate (Table 5.33).

TABLE 5.33 Occurrence of TD in young broilers fed high levels of chloride in soybean meal

Diet minerals (g/kg)		Weight gain (g/d)	% TD
Chloride	Magnesium		
1.5	2.0	21	17
1.5	4.0	23	7
1.5	6.0	15	7
3.5	2.0	26	60
3.5	4.0	27	32
3.5	6.0	20	3

Adapted from Lui et al. (1992)

Comparable to the situation seen with rye, autoclaving of the soybean meal seemed to reduce subsequent leg problems. A correlation was also noted between leg problems and trypsin inhibitor level. Soybean source has also been implicated in the occurrence of foot pad problems with breeders. The exact cause of leg problems occurring with rye has not been fully elucidated, although impairment of vitamin D_3 absorption is often implied. The prompt increase in bone ash content resulting from removal of rye from the diet led workers to suggest the presence of a factor which interferes with nutrient absorption rather than there being some metabolic antagonism. In addition to the beneficial effect of autoclaving, as previously described with soybean meal, the feeding value of rye seems to be improved by both gamma irradiation and water extraction.

Just as similarities may exist between the deleterious effects of certain soy and rye samples on leg disorders, so similarities exist between sorghum and rapeseed meal. Tannins which are normal constituents of most rapeseeds and sorghums form complexes with metal ions. It has been proposed that perotic conditions related to feeding of rapeseed meal may be due to binding of minerals with tannin. However, in this situation, increasing the level of manganese in the diet does not alleviate the condition. It is possible that tannins are being absorbed and interfering with crosslinking in collagen. Other reports suggest that the deleterious effect seen with rapeseed meal may be due to the confounding effect of goitrogenic activity. A high incidence of perosis due to the addition of thiouracil is not corrected with manganese supplementation. Regardless of mode of action, it is obvious that a number of feedstuffs can aggravate certain skeletal abnormalities.

iii) Mycotoxins

It is realized that feedstuffs contaminated with certain mycotoxins can induce or aggravate skeletal problems. Oats contaminated with *Fusarium roseum* have been shown to cause TD. Aflatoxin and ochratoxin both decrease bone strength, and this may be related to vitamin D_3 metabolism. It is our experience that under such field conditions birds sometimes respond to water-soluble D_3 administered via the drinking water, regardless of the level and source of D_3 in the diet. Attempts at reducing leg problems by minimizing microbial contamination of the litter have met with varying success. Adding sorbic acid to the diet, or treating litter with potassium sorbate have resulted in improved leg condition only in isolated trials. A number of fungicides used in grain treatment can also themselves lead to leg problems. The presence of tetramethylthiuram significantly increases the incidence of TD, while tetramethylthiuram disulphide causes the 'classical' condition of irregular penetration of blood vessels into the cartilage.

iv) Electrolyte Balance

Since the initial observations that cation:anion balance can influence the occurrence of TD, there has been a wealth of generally corroborative data. It seems that in order to maintain acid-base balance, birds will regulate input and/or output and that this is best expressed in terms of three ions, namely Na^+, K^+, and Cl^-. High chloride levels induce TD, although since there are no major shifts in plasma ions with TD, it is concluded that the problem is not simply related to defective calcification. The occurrence of crooked legs seems to be greater when chicks are fed diets with a narrow range of cations:anions and the incidence of TD and bowed legs appears to increase with increase in anion content of the diet. There may be a relationship between ion balance and vitamin D_3 metabolism. Increasing the chloride content of the diet from 10 to 40 mEq/100 g was reported to markedly enhance cartilage abnormalities when the cation (Na^+, K^+) content of the diet was low. Thus with excess Cl^-, chicks become acidotic, although the condition can be corrected with dietary sodium and potassium carbonates, suggesting that if the diet is high in Cl^-, then it must be balanced with equimolar concentrations of Na^++K^+in the form of readily metabolizable anions. Workers from France have indicated that homogenates from acidotic chicks lose 50% of their capacity to synthesize 1,25-cholecalciferol to the active D_3 metabolite. This possibly infers a relationship between acid:base balance TD and vitamin D_3 metabolism.

v) Vitamins

Although impairment of D_3 metabolism is classically associated with rickets, there is again evidence to suggest that more complex systems may be involved in conditions such as field rickets. At certain times, a deficiency of D_3 will mimic both Ca and P deficiency situations. While Ca-deficient chicks are usually hypocalcemic and hyperphosphatemic, D_3 deficiency invariably results in hypocalcemia and hypophosphatemia. In the D_3-deficient chick a greater relative P deficiency is caused by parathyroid hormone. The time sequence involved in the recovery process of D_3 deficient chicks was recently investigated. Within 12 h of receiving an oral dose of D_3, distal portions of elongated epiphyseal vessels and adjacent maturing chondrocytes undergo necrosis. Within 72 h, calcification resumes and by 120 h necrotic cartilage has been removed and a normal growth plate restored. In situations of D_3 repletion, the skeleton seems to respond much more slowly than does the intestine, since the immediate effect of re-feeding D_3 is better 'absorption' of the diet Ca. There is also good evidence to suggest that D_3 is involved with collagen synthesis, where the maturation of collagen crosslinks seems D_3 dose related.

Controversy still exists with respect to the biopotency of D_3 sources and metabolites. Although bone growth is greater when chicks receive 1,25$(OH)_2$-D_3, high doses do seem to impair skeletal development. Similarly, while 24,25$(OH)_2$-D_3 corrects plasma Ca homeostasis, it exacerbates cartilage lesions due to apparent stimulation of chondrocyte proliferation with failure of mineralization. It was recently suggested that in terms of longitudinal growth rate of the tibia and its Ca and P contents, D_3 was more effective than 1,25$(OH)_2$-D_3, while 24,25$(OH)_2$-D_3 was intermediate. However, in subsequent studies this same group of researchers indicated the converse sequencing of 1,25;24,25;D_3 as being effective in promoting bone strength, leading to the suggestion that the degree of bone mineralization is not related to mechanical properties.

As previously mentioned, field rickets continues to be a problem especially with turkey poults. In studying a number of field outbreaks, Riddell at the University of Saskatchewan indicated adequate diet nutrient levels, even though in many instances the condition seemed to be diet related. This researcher concluded that inadequate feed distribution was the reason for the early (7 d) onset of rickets and also that water treatment with D_3 is usually an effective therapy. The exact involvement of vitamin metabolism in the aetiology of leg problems is certainly unclear at this time, and the general consensus is that 'complex deficiencies' are involved. We have seen remarkably similar gross leg problems in broilers fed diets deficient in a range of individual vitamins. In more detailed studies of these affected birds, we saw no clear trend to suggest a basis for diagnosis although vitamin deficiencies *per se* resulted in less of an effect on intercondyle depth than did manganese deficiency.

While most vitamins have been associated with leg problems, pyridoxine has perhaps received the most attention. There is overwhelming evidence to suggest that low levels lead to skeletal abnormalities and/or that supplementation reduces the incidence. It has been hypothesized that pyridoxine may exert its beneficial effect via involvement with zinc homeostasis and in particular the formation of picolinic acid which is involved in intestinal zinc absorption. There is an apparent synergism between zinc, B_6 and tryptophan involved in the prevention of leg weakness. The situation with pyridoxine is further complicated through the effect of diet protein as previously described with folic acid. Common to many other diet situations, pyridoxine deficiency manifests itself through epiphyseal lesions consisting of uneven invasion of irregular blood vessels into the maturing growth plate. Presumably the higher level of diet protein increases the metabolic requirement for pyridoxine through such processes as transamination and/or deamination. While deficiencies of many vitamins can, therefore, precipitate leg problems in broilers, there is also evidence to suggest that certain vitamin excesses may be detrimental. Very high levels of vitamin A in the diet increase the incidence of rickets, while impaired bone formation has been observed with excess dietary vitamin E. It must be pointed out, however, that all these reported effects of vitamin excess on bone metabolism relate to dietary levels grossly in excess of normal feeding levels and hence would only be practically encountered under unusual circumstances.

vi) Minerals

As with vitamins, deficiencies or excesses of a vast range of minerals can also influence bone development. The effect of abnormal levels and/or ratios of calcium:phosphorus are well documented. Confusion sometimes exists with respect to diagnosis of calcium or phosphorus deficiencies, and accurate on-farm diagnosis of phosphorus deficiency vs calcium excess is difficult, and immediate recommendations of diet change can be misleading prior to complete diet analysis. Identical lesions for the two conditions are seen suggesting that excess calcium forms insoluble $Ca_3(PO_4)_2$ in the intestine, thereby inducing phosphorus deficiency. Table 5.34 shows normal levels of minerals in bone ash, and so values which are much different to these are a cause for concern.

TABLE 5.34 Normal mineral content of bone ash

Calcium	37%
Phosphorus	18%
Magnesium	0.6%
Zinc	200 - 250 ppm
Copper	20 ppm
Manganese	3 - 5 ppm
Iron	400 - 500 ppm

The effect of manganese deficiency on the incidence of perosis is obviously well documented although some evidence suggests that interaction with iron may be a complicating factor. Administration of hydralazine, a manganese sequestering agent, causes leg defects very similar to those seen in classical manganese deficiency, and in fact successful Mn treatment has been recorded in these situations. Hydralazine blocks collagen secretion and this can be restored by administration of Fe^{2+} or Fe^{2+} with Mn^{2+}, but not by Mn^{2+} alone. The agent seems to block the synthesis of hydroxylysine and within this mechanism there seems to be a step requiring Fe^{2+}. Copper metabolism has always been suspect in studies of leg problems, since there are certain similarities between the cartilage of copper deficient birds and those with TD. However, attempts to correct TD with supplements of Cu have invariably proved disappointing. Solubilization studies indicate that dystrophic cartilage (TD) is not deficient in crosslinked collagen, a situation often seen with classical copper deficiency.

vii) Embryology

Because a number of skeletal disorders are seen in the first few days after hatching, it is possible that metabolic disorders are initiated during incubation. Skeletal mineralization starts at around the eighth day of incubation, and at this time the yolk serves as a source of calcium. Shell calcium is not utilized until about the 12th day of incubation, although during the course of embryonic development, the embryo will take up some 120 mg Ca from the shell. Culturing developing embryos in a medium deficient in calcium quickly results in gross skeletal abnormalities. There have been no reports linking breeder eggshell quality with bone formation in broiler offspring. Similarly there have been relatively few reports of the effect of breeder nutrition and management on skeletal development of the embryo. A marked reduction in the incidence of a twisted leg condition in broilers (17% vs 8%) was reported when their dams were fed diets containing 20% distiller's dried grains. There seems little doubt that more common leg problems, such as tibial dyschondroplasia (TD) are inherited to some degree and hence pedigree has a potentially confounding effect on studies of leg abnormalities. TD is related to a major sex linked gene, the recessive of which is associated with a higher incidence of TD. This situation suggests a large maternal component, and therefore female lines would greatly influence the expression. Whenever TD is seen in commercial broilers, consideration should be given to the 'environmental' conditions of female parent flocks.

c) Ascites

Ascites is quickly becoming one of the major causes of mortality/morbidity in broiler chickens. Once only seen at high elevations, ascites now causes problems in fast growing birds in most areas. Ascites is characterized by the accumulation of fluid in the abdomen, and hence the basis for the common phrase of "water-belly". Fluid in the abdomen is in fact plasma that has seeped from the liver, and this occurs as the end result of a cascade of events ultimately triggered by oxygen inadequacy within the bird. For whatever reason, the need to provide more oxygen to the tissues leads to increased heart stroke volume, and ultimately to hypertrophy of the right ventricle. Such heart hypertrophy, coupled with malfunction of the heart valve, leads to increased pressure in the venous supply, and so pressure build up in the liver, and often the characteristic fluid leakage.

Because of the relationship with oxygen demand, ascites is affected and/or precipitated by such factors as growth rate, altitude (hypoxia) and environmental temperature. Of these factors, hypoxia was the initial trigger some years ago, since the condition was first seen as a major problem in birds held at high altitude, where mortality in male broilers of 20 - 30% was not uncommon. Today ascites is commonly seen in fast growing lines of male broilers fed high nutrient dense diets at most altitudes and where the environment is cool/cold at least for part of each day. Mortality seen with ascites is dictated by the number of "stressors" involved and hence the efficacy of the cardio-pulmonary system to oxygenate tissues.

While growth rate *per se* is the major factor contributing to oxygen demand, the composition of growth is also influential, because oxygen need varies for metabolism of fats vs proteins. Oxygen need for nitrogen and protein metabolism is high in relation to that for fat, although it must be remembered that the chicken carcass actually contains little protein or nitrogen. The carcass does contain a great deal of muscle, but 80% of this is water. On the other hand, adipose tissue contains about 90% fat, and so its contribution to oxygen demand is proportionally quite high. Excess fatness in birds will therefore lead to significantly increased oxygen needs for metabolism.

Environmental temperature, and associated oxygen/energy demand is usually a factor in most cases of ascites. Keeping birds "warm", is perhaps the single most practical way of reducing the incidence of ascites. As environmental temperature changes, there is a change in the bird's oxygen requirement. If one considers the thermoneutral zone following the brooding period to be 20 - 26°C, then temperatures outside this range cause an increase in metabolic rate, and so increased need for oxygen. Low environmental temperatures are most problematic, since they are accompanied by an increase in feed intake with little reduction in growth rate. While there is an increased oxygen demand at high temperatures due to panting etc., this is usually accompanied by a reduced growth rate, and so overall reduced oxygen demand. Under commercial farm conditions, cold environmental conditions are probably the major contributing factor to ascites. For example, at 10 vs 26°C, the oxygen demand by the bird is almost doubled. This dramatic increase in oxygen need, coupled with the need to metabolize increased quantities of feed, often leads to ascites. It is interesting to note that birds maintained at high altitude under commercial conditions are often subjected to cool or cold night time temperatures.

Manipulation of diet composition and/or feed allocation system can have a major effect on the incidence of ascites. In most instances, such changes to the feeding program influence ascites via their effect on growth rate. However, there is also a concern about the levels of nutrients that influence electrolyte and water balance, the most notable being sodium. Feeding high levels of salt to broilers (>0.5%) does lead to increased fluid retention, although ascites invariably occurs with diets containing a vast range of salt, sodium and chloride concentrations.

Ascites may also develop secondary to rickets, because of poor rib structure. Julian *et al.* (1986) observed that broilers fed diets grossly deficient in phosphorus developed rickets as expected, but that these birds also displayed an increased respiration rate, increased arterial CO_2 and decreased arterial O_2. Most birds showed signs of cardio-pulmonary abnormalities and some birds died of hypoxia and some of RV failure with or without clinical signs of ascites. It was suggested that RV hypertrophy occurred in response to the pulmonary arterial hypertension due to hypoxia, which itself was caused by abnormal breathing due to poor rib strength and infolding.

Apart from obvious nutrient deficiencies, or excesses as in the situation with sodium, the major involvement of the feeding program as it affects ascites revolves around nutrient density and feed restriction. Ascites is more common when high energy diets are used, especially when these are pelleted. Dale and Villacres (1988) grew birds on high energy diets designed to promote rapid growth and likely to induce ascites. There was no correlation between 14 d body weight and propensity of ascites, although birds fed 3000 → 3100 kcal ME/kg rather than 2850 → 2950 kcal ME/kg had twice the incidence of ascites. Unfortunately Dale and Villacres (1988) did not show body weight data. In this study, the higher energy diets were produced essentially by increasing the level of supplemental fat. In a previous report, Dale and Villacres

(1986) support the concept that feed change *per se* is often the trigger to ascites, but that the condition is also seen in single diet feeding programs. In formulating diets of varying nutrient density and fat content, these workers show clear evidence of correlation between ascites and growth rate (Table 5.35). The highest incidence of ascites occurred when the highest energy level was fed, regardless of either energy:protein or fat content of the diet. These data (Table 5.35.) show little effect of added fat, although with both series of the energy:protein used, the greatest incidence of ascites occurred in the fastest growing birds.

TABLE 5.35 Effect of nutrient density and diet composition on incidence of ascites

Diet ME (kcal/kg)	Crude protein (%)	ME/CP	Diet Fat (%)	49d Body Wt (g)	Ascites mort. (%)
2950	23	128	0	1800	8.8
2950	23	128	4	1820	8.7
3100	24	128	4	1830	15.8
2950	21	140	0	1810	9.0
2950	21	140	4	1810	8.5
3100	22	140	4	1860	12.0

Adapted from Dale and Villacres (1986)

Because feeding program, nutrient density and growth rate are all intimately involved in affecting the severity of ascites, then there is invariably discussion on the possible advantages of feed restriction. Arce *et al.* (1992) carried out a series of interesting studies to record bird response to varying nutrient restriction programs. As pointed out by these authors, the goal of such programs is to reduce the incidence of ascites without adversely affecting economics of production. It is expected that nutrient restriction programs will reduce final weight-for-age to some degree, and obviously there is a balance between the degree of feed restriction and commercially acceptable growth characteristics. Arce *et al.* (1992) conducted studies in Mexico at 1940 m or at 2500 m elevation. Birds were either fed on a skip-a-day schedule for varying periods from 7 - 28 d or allowed access to the feeders for just 8 h per day (7 am - 3 pm). In all studies, control full fed birds were the heaviest and exhibited the greatest ascites mortality (40 vs 8 - 15%). As expected, the later in the grow out cycle that skip-a-day feeding is introduced, the greater the reduction in ascites mortality, although this is accompanied by greater reduction in body weight. By restricting the feed access time to 8 h per day, ascites mortality was greatly reduced and this seems a very practical system, especially where hand-feeding is practised, and feeders can easily be raised during the restriction period. Restricting feed access time with mechanical feeders is more complicated and may not be a practical solution. It appears as though ascites mortality can be reduced through limiting feed intake, and depending upon the timing of this restriction program, there will be about 100 - 200 g loss in weight-for-age. In most commercial operations, this will mean a 2 - 3 d delay in achieving 50 - 56 d market weight.

A 2 - 3 d delay in market age sounds quite a reasonable trade-off for a major reduction in ascites. However careful economic analysis must be carried out to determine the real cost of such programs. For example, suppose a flock of 100,000 male broilers is delayed in market age by just 1 day, then what saving in mortality do we need to "pay" for the extra feed cost of the entire flock?

- Extra feed cost for 100,000 broilers for 1 d delay
  ≃ 100 g/bird = 10 tonnes @ $250/t = $2,500

- The 1% mortality costs relate to feed eaten by ascitic birds and their chick price

1,000 chicks @ 25¢ =	$ 250
1,000 birds dying, consume average 3 kg feed =	$ 750
Total	$1,000

Therefore in this simple exercise, taking into account only feed and chick cost, there needs to be a reduction in late mortality of at least 2.5% in order to account for the extra cost of delaying the whole flock by 1 day.

Another factor of concern regarding the effect of the nutritional program on ascites, is the balance and the quality of the protein supply. It is well known that crude protein is supplied as a source of amino acids, and that birds have a minimal requirement for nitrogen *per se*. Excess nitrogen must be removed from the body, and this is an energy (oxygen) demanding process. There is a potential to reduce the oxygen demand through minimizing crude protein supply while maintaining essential amino acid levels in a diet. If we consider two diets providing the same level of available amino acids, but with 20 vs 24% crude protein, then there will be a need for birds to deaminate an extra 4% CP in the high-protein diet. If birds consume 130 g feed/d, this means an extra 5 g/d of protein for catabolism. Such protein catabolism will likely result in uric acid and fat synthesis, and these are calculated to need 2 and 1 litres of oxygen per day respectively. Therefore catabolism of an extra 5 g crude protein each day means a 3 litre increase in oxygen demand, which represents about an 8% increase relative to the bird's total requirements. There is an obvious incentive to minimize crude protein *per se*, because its catabolism merely imposes another stress on the oxygen demand of the bird.

If ascites mortality is sufficiently high, the following diet changes may be considered:

- Low energy feeds throughout the entire life cycle eg:

  starter 2850 kcal ME/kg
  grower 2950 kcal ME/kg
  finisher 3100 kcal ME/kg

- Use mash rather than pelleted feeds. Do not use too fine a mash diet, since this encourages feed wastage and causes dustiness at broiler level.

- Consider skip-a-day feeding from 7-20 d of age. Longer periods of restricted feeding may be necessary where ascites levels are very high. Water management becomes more critical with this system.

- Consider limit-time feeding, such that birds have access to feed for 8-10 hours each day. Extra care is needed in water management so as to prevent wet litter.

- Minimize toxin contamination of feed

- Limit sodium level to 0.20% of diet

5.8 SKIN AND FEATHER ABNORMALITIES

i) Feather growth

Although feathers are continuously being shed and regenerated, virtually all feather follicles arise during mid embryo development. The feather follicles appear in distinct tracts, and so it is normal to find large areas of the skin with no feather attachment. These follicle-less areas are however covered by feathers that grow out of adjacent feather tracts. Complete feather cover therefore relies on the normal shape and contouring of all feathers.

Table 5.36 shows normal feather composition, and deviations of greater than 15% from these values are cause for some concern.

TABLE 5.36 Average feather composition

	%
Crude Protein	90
Total amino acids	60
Methionine	0.7
Cystine	5.5
Arginine	7.1
Lysine	2.4
Threonine	4.2
Valine	6.5
Magnesium	0.2
Sodium	0.8
Iron	0.06
Copper	12 ppm
Zinc	10 ppm
Selenium	0.7 ppm

Normal feather growth, expressed as grams per bird, is shown in Fig. 5.10 for both male and female birds.

Most of the feather structure is keratin protein that is synthesized in the feather forming cells of the epidermis. The keratin structure is very rich in cystine, with each molecule containing around 8 half-cystine residues, and this is why methionine/TSAA levels are important for good feather structure. Marginal levels of methionine + cystine will cause abnormal feather growth and/or reduced feathering, although deficiencies of other amino acids will also cause feathering problems. With general amino acid inadequacy, the primary feathers have a characteristic spoonlike appearance that is caused by retention of an abnormally long sheath that covers the first 50% of the feather shaft. Deficiencies of many essential amino acids also cause abnormal curling of feathers away from the body. Interestingly, these same characteristics are seen with deficiencies of some of the B vitamins.

Fig. 5.10 Broiler feather growth

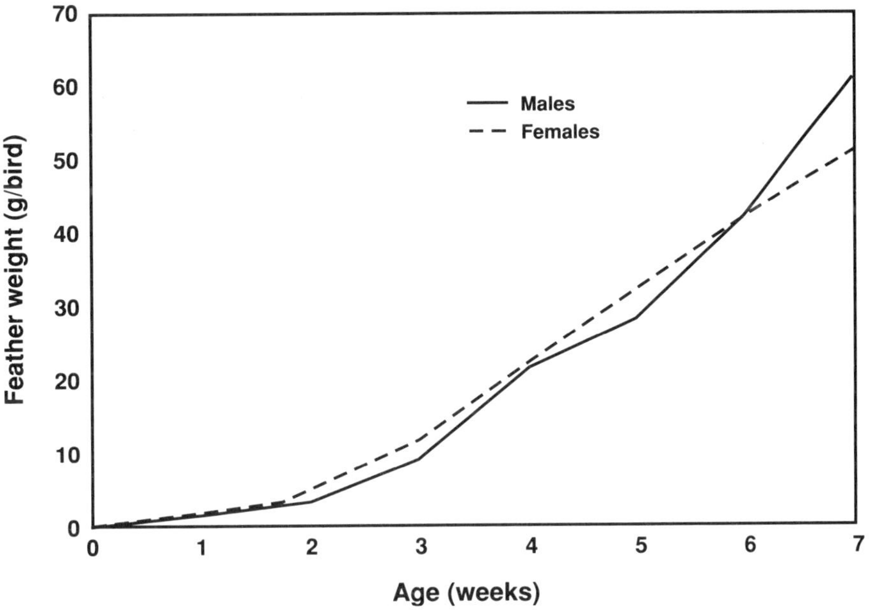

However methionine, and especially methionine + cystine levels come under closest scrutiny when abnormal feathering occurs. About 2% of dietary methionine and 25% of dietary cystine is needed for feather development. Because of the importance of methionine in maintaining normal growth and development, then this amino acid is not likely to be deficient in situations where growth is adequate and feathering is abnormal. More likely, if feathering problems relate to TSAA metabolism, then overestimation of digestibility/availability of cystine in feedstuffs is usually the major problem.

Bleeding of the feather pulp is characteristic of vitamin E/selenium deficiencies. Diets with much less than 40 ppm zinc have been shown to produce a characteristic frayed feathering especially in the fast growing primary feathers. A number of years ago there was concern about feathering in broilers fed various ionophone anticoccidials. However subsequent research failed to show any definitive relationship between these compounds and metabolism of sulphur containing amino acids. Feathering is however usually worse when ionophones are included in low protein diets (<18%CP equivalent) for very young broilers. Various *Fusarium* mycotoxins can affect feathering, the most significant being T-2 toxin.

Birds fed T-2 toxin (4 ppm) develop only sparse feathering, and the feathers that do develop tend to protrude from the bird at odd angles, leading to some areas of the skin being exposed. With T-2 toxin, most feathers are affected, unlike the situation with nutrient deficiencies that most characteristically first affect the primary feathers.

Feather growth is also affected by thyroid function, and so thyroid antagonists will delay normal feather growth. Poor feathering is sometimes seen at farms changing from corn to milo-based diets. While a number of diet situations may be involved in such a change, it is interesting to note that milo is very low in iodine content compared to other cereals.

Unfortunately, in most field cases of poor feathering there is no apparent dietary deficiency as determined by routine analyses or consideration of formulation consistency/changes. Often problems are isolated to particular flocks within a site where all flocks receive the same feed. These factors support the concept of poor feathering being caused by infectious agents (likely in the feather follicle itself) or factor(s) causing general malabsorption of nutrients. Because feathers are very fast growing, especially in the first 7 - 14 d of age, then their development is very sensitive to general availability of circulating nutrients.

ii) Oily bird syndrome (OBS)

As its name implies, birds with OBS have skin that is oily or greasy to the touch. Garrett (1975) was one of the first to document the condition, indicating the subcutaneous fat, especially in the back region, to be very oily. At that time, OBS was observed most frequently in older broilers and especially those fed high energy diets in the warmer summer months. The problem also seemed to relate to specific processing plants, where the occurrence was greater with the increased "stress" applied during processing, and especially plucking. Interestingly the condition is rarely seen in hand-plucked birds. The condition was also associated with increased water retention in the carcass, especially in regions of the carcass where skin "elasticity" had been affected. These pockets of water were most often seen in female birds. The problem is most noticeable in pockets of the skin that separate in the back region. Because the skin seems more prone to tearing, these pockets rupture and coat the surrounding skin with oil.

A general finding in situations of OBS is a change in the skin ultrastructure, such that either the layers of skin separate to allow the pockets of oil and/or chilled water to accumulate, or the skin tears more easily. Apparently, fat saturation is not a factor in OBS, rather there is some change in the integrity of the various layers of the skin because the skin from affected carcasses is easily separated and removed from the underlying musculature. The five collagenous layers beneath the epidermis seem less compact than normal, and the deepest layers contain the most fat cells.

While there is no real change in total skin thickness with OBS, its breaking strength seems to be reduced. Although males usually have thinner skin than do females, it is usually stronger possibly due to there being less subcutaneous fat. Males also exhibit more insoluble skin collagen, and so this may be important in reducing problems of solubilization and water uptake as often occurs with OBS carcasses in chill water tanks.

In more detailed studies of skin collagen in OBS birds, Ramshaw *et al.* (1986) studied changes in crosslinking. Interestingly, the electrophoretic pattern of collagen was similar in all birds from an OBS flock, and different to that seen in normal flocks, even though all birds in the OBS flock did not exhibit the typical greasy skin condition. These workers showed that the main collagenous layer in the skin of OBS birds was 30% weaker at 36°C than that from normal birds. When the temperature was increased to 90°C, then a 50% reduction in skin strength was observed. There was no real difference in collagen chains, although there appeared to be a difference in molecular crosslinking in OBS skin, suggesting that the major problem related to collagen crosslink maturation. In mammals at least, and in the formation of eggshell membranes, lysyl oxidase is thought to be the only enzyme involved in crosslink maturation of collagen and elastin, converting lysine and hydroxylysine into aldehydes. Lysyl oxidase is a copper metalloenzyme that requires pyridoxal phosphate as a co-factor, and copper deficiency is known to impair normal collagen crosslink structure. However it does not seem as though copper deficiency is the simple solution to this problem.

OBS occurs only in broilers grown in warm climates, and under experimental conditions the syndrome can only be duplicated by using warm growing conditions. At higher temperatures, birds carry more subcutaneous carcass fat, and so this may be the trigger mechanism. Because of the oily nature of the carcass, various diet ingredients and nutrient levels have come under investigation. Fat levels and sources in the diet have come under close scrutiny, although there does not seem to be a simple relationship. Higher levels of fat and/or energy in relation to the level of protein in the diet have caused more problems, although research results are inconsistent. Even though the bird's skin has an oily appearance, levels of unsaturated fatty acids do not correlate with OBS, and in fact more problems are seen in birds fed tallow.

If OBS occurs, then the only immediate practical solution is to modify the processing conditions, and in particular, scald temperature and pick time. Because the exact cause of impaired collagen crosslinking has not been identified, then other changes to the diet and/or environment are of questionable value. While fat levels in the diet do not seem to be a factor, there is an indication of more problems occurring with saturated fats such as tallow. The diet should contain adequate levels of copper and not contain excessive levels of zinc or vitamin A.

iii) Skin tearing

About 5% of downgrades at processing are due to skin tears. Most tears occur post-mortem, and so are related to scald water temperature and pick time. High temperatures with shorter pick time cause less tears than lower scald temperature with prolonged pick time. However, regardless of processing conditions, a proportion of carcasses have torn skin. Skin strength is greater in males vs females, and for both sexes it increases with age. Most problems are therefore encountered with carcasses from younger female birds. There is a genetic component, because different strains show differences in skin tearing, and in one study it was shown that skin from slow feathering strains was less elastic than that from fast feathering birds.

Skin strength is highly correlated with its collagen content, and so skin with greater collagen content is less prone to tearing. Any nutritional factor that influences skin collagen content will therefore indirectly affect susceptibility to tearing. The amino acid proline is a component of hydroxyproline which itself is responsible for the stability and rigidity of collagen. Zinc, copper and vitamin C all play a role in collagen synthesis and so deficiencies of any one of these nutrients results in less skin collagen production. However, gross deficiencies of these nutrients also cause poor growth rate, a characteristic that is not usually seen in situations of excessive skin tearing. Unfortunately there seems to be little benefit to increasing the dietary levels of these nutrients, or even increasing the level of proline in the diet.

A specific dietary situation involves the anticoccidial, halofuginone. When this product is fed at normal recommended levels, there is significant loss in skin thickness and skin strength, especially in female birds. In one study, using halofuginone at 3 ppm of the diet resulted in a 50% reduction in skin collagen content and 50% increase in the incidence of skin tears. Halofuginone seems to affect skin strength in female birds, more than it does with males, and because the female has an inherently weaker skin, this leads to the greater incidence of tearing. It has been shown that halofuginone interferes with the conversion of proline to hydroxyproline in the skin cells, and that this adverse effect cannot be corrected by adding more proline to the diet.

When skin tearing is a problem, then assuming that processing conditions have been scrutinized, the only potential nutritional factors are halofuginone and level of zinc, copper and Vitamin C. Skin tearing is more problematic in hot weather. This situation leads to recommendations of supplemental vitamin C, although birds under these conditions almost always carry more subcutaneous fat. Feeding higher levels of crude protein has also been shown to increase skin strength although the reason for this is not clear. More crude protein may provide more of the non-essential amino acid glycine which accounts for about 30% of the amino acids in collagen, or alternatively more protein *per se* may simply reduce carcass fatness.

CHAPTER 5

Selected References

Arce, J., M. Berger and C. Coello 1992. *Control of ascites syndrome by feed restriction techniques.* J. App. Poult. Res. 1:1-5.

Atteh, J.O., S.Leeson and R.J. Julian 1983. *Effects of dietary levels and types of fat on performance and mineral metabolism of broiler chicks.* Poult. Sci. 62:2403-2411.

Bartov, I. 1987. *Effect of early nutrition on fattening and growth of broiler chicks at 7 weeks of age.* Br. Poult. Sci. 28:507-518.

Cabel, M.C. and P.W. Waldroup 1990. *Effect of different nutrient restriction programs early in life on broiler performance and abdominal fat content.* Poult. Sci. 69:652-660.

Cherry, J.A., W.J. Swartworth and P.B. Siegel 1984. *Adipose cellularity studies in commercial broiler chicks.* Poult. Sci. 63:97-108.

Classen, H. 1988. *The role of photoperiod manipulation in broiler chicken management.* Canada Poultryman. 75(8):8-10.

Dale, N. and A. Villacres, 1986. *Nutrition influences ascites in broilers.* World Poultry Misset. April 1986, pp. 40.

Dale, N. and A. Villacres, 1988. *Relationship of two-week body weight to the incidence of ascites in broilers.* Avian Dis. 32:556-560.

Edwards, H.M., Jr. 1988. *Effect of dietary calcium, phosphorus, chloride, and zeolite on the development of tibial dyschondroplasia.* Poult. Sci. 67:1436-1446.

Fancher, B.I. and L.S. Jensen 1989. *Dietary protein level and essential amino acid content. Influence upon female broiler performance during the grower period.* Poult. Sci. 68:897-908.

Granot, I., I. Bartov, I. Plavnik, E. Wax, S. Hurwitz and M. Pines, 1991a. *Increased skin tearing in broilers and reduced collagen synthesis in skin in vivo and in vitro in response to the coccidiostat halofuginone.* Poult. Sci. 70: 1559-1563.

Howlider, M.A.R. and S.P. Rose 1988. *Effect of growth rate on the meat yields of broilers.* Br. Poult. Sci. 29:873.

Hickling, D., W. Guenter and M.E. Jackson 1990. *The effects of dietary methionine and lysine on broiler chicken performance and breast meat yield.* Can. J. Anim. Sci. 70:673-678.

Jackson, S., J.D. Summers and S. Leeson 1982. *The response of male broilers to varying levels of dietary protein and energy.* Nutr. Rep. Int. 25:601-612.

Jackson, S., J.D. Summers and S. Leeson 1982. *Effect of dietary protein and energy on broiler carcass composition and efficiency of nutrient utilization.* Poult. Sci. 61:2224-2231.

Jensen, L.S., C.L. Wyatt, and B.I. Fancher 1989. *Sulfur amino acid requirement of broiler chickens from 3 to 6 weeks of age.* Poult. Sci. 68:163-168.

Julian, R.J. 1993. *Ascites in poultry.* Avian Path. 22:419-454.

Julian, R.J., J. Summers and J.B. Wilson, 1986. *Right ventricular failure and ascites in broiler chickens caused by phosphorus-deficient diets.* Avian Dis. 30:453-459.

Leeson, S. 1993. *Potential of modifying poultry products.* J. Appl. Poultry Res. 2:380-385.

Leeson, S., and L.J. Caston 1993. *Production and carcass yield of broilers using free-choice cereal feeding.* J. Appl. Poultry Res. 2:253-258.

Leeson, S., L.J. Caston and J.D. Summers 1988. *Response of male and female broilers to diet protein.* Can. J. Anim. Sci. 68:881-890.

Leeson, S., L.J. Caston and J.D. Summers 1996. *Broiler response to diet energy.* Poult. Sci. 75:529-535.

Leeson, S. and J.D. Summers 1988. *Some nutritional implications of leg problems with poultry.* Br. Vet. J. 144:81-92.

Perrault, N. 1989. *Growth and development of male broilers.* M.Sc. Thesis, University of Guelph.

Plavnik, I., and S. Hurwitz 1985. *The performance of broiler chicks during and following a severe feed restriction at an early age.* Poult. Sci. 64:348-355.

Plavnik, I. and S. Hurwitz 1989. *Effect of dietary protein, energy, and feed pelleting on the response of chicks to early feed restriction.* Poult. Sci. 68:1118-1125.

Rosebrugh, R.W., N.C. Steele, J.P. McMurtry and I. Plavnik 1986. *Effect of early feed restriction in broilers II. Lipid metabolism.* Growth 50:217-227.

Summers, J.D., M. Bedford and D. Spratt 1987. *Sudden Death Syndrome: Is it a metabolic disease?* Feedstuffs, January 26th, p 58-63.

Summers, J.D. and S. Leeson 1984. *Influence of dietary protein and energy level on broiler performance and carcass composition.* Nutr. Rep. Int. 29:757-768.

Summers, J.D., S. Leeson and D. Spratt 1988. *Yield and composition of edible meat from male broilers as influenced by dietary protein level and amino acid supplementation.* Can. J. Anim. Sci. 68:241-248.

Teeter, R.G.. 1994. *Broiler nutrition strategy considerations involving vitamin fortification.* Proc. BASF Tech-Symp. Indianapolis. May 25.

Zubair, A.K., and S. Leeson 1994. *Effect of varying period of early nutrient restriction on growth compensation and carcass characteristics of male broilers.* Poult. Sci. 73:129-136, 1994.

Zubair, A.K., and S. Leeson, 1994. *Effect of early feed restriction and realimentation on heat production and changes in sizes of digestive organs of male broilers.* Poult. Sci. 73:529-538.

CHAPTER 6. BROILER BREEDERS

6.1 DIET SPECIFICATIONS

Most of the current advantages seen in growth and development of the broiler chicken pose a number of problems to those involved with management of the parent breeders. Growth rate and reproductive performance do not go hand-in-hand, and so it is obvious that breeding stock cannot be allowed to express their genetic potential for growth. As appetite and weight for age increase in commercial broilers, so nutrient restriction of young breeders must start at earlier ages and/or be of increasing severity at older ages. Our modern breeder hen at 20 weeks of age must be comparable in weight to their offspring at 6 weeks of age. It is therefore not too surprising that appetite control of parent flocks is becoming more challenging. Unlike most other classes of poultry, the absolute requirements of broiler breeders is influenced by both restricted feeding level and nutrient specifications of the diet(s). This obviously complicates any research program and is undoubtedly a factor related to the dearth of current research activity with breeders. However, this dual effect means that nutrient intake can be controlled much more closely, and so represents a great potential for matching intake to requirement.

Over the past few years, there has been the introduction of high meat-yield birds, and these breeders seem to have some unique characteristics that must be accounted for in a feeding program. High-yield females are often slightly later maturing (4 - 7 d) than conventional broiler breeders, and also have a longer feed clean-up time. In general, managers should not react too quickly in changing the feed allocation or diet as they normally would to circumstances arising with conventional breeders. High-yield roosters also pose some interesting new feed management problems, basically related to their aggressive behavior. Table 6.1 shows diet specifications for male and female breeders, and Table 6.2 gives some examples of diet formulations.

TABLE 6.1 Broiler breeder diet specifications

	Starter-Grower		Breeder (peak feed/bird/day, g)			Separate Male Feed
			165	155	140	(g/bird/d)
Approximate protein level (%)	18.0	16.0	15.5	16.0	16.5	12.0
Amino acids (% of diet)						
Arginine	.72	.62	.53	.58	.62	.48
Lysine	.90	.72	.60	.64	.68	.50
Methionine	.42	.38	.30	.33	.35	.23
Methionine + cystine	.74	.63	.53	.57	.60	.40
Tryptophan	.18	.14	.11	.12	.13	.10
Histidine	.19	.17	.14	.15	.16	.12
Leucine	.92	.84	.71	.75	.80	.60
Isoleucine	.75	.67	.51	.58	.62	.40
Phenylalanine	.62	.55	.41	.46	.49	.30
Phenylalanine + tyrosine	.95	.90	.65	.70	.79	.50
Threonine	.75	.70	.52	.59	.63	.40
Valine	.80	.70	.60	.62	.65	.50
Metabolizable energy (kcal/kg)	2850	2800	2850	2900	2950	2700
Calcium (%)	0.95	0.90	3.25	3.35	3.50	0.85
Available phosphorus (%)	0.42	0.40	0.39	0.40	.42	0.38
Sodium	0.18	0.18	0.18	0.18	0.19	0.18
Vitamins (per kg of diet)						
Vitamin A (I.U.)	7,000					
Vitamin D_3 (I.U.)	3,000					
Choline equivalents (mg)	800					
Riboflavin (mg)	10					
Pantothenic acid (mg)	14.0					
Vitamin B_{12} (mg)	.013					
Folic acid (mg)	1.0					
Biotin (mg)	0.20					
Niacin (mg)	40.0					
Vitamin K	3.0					
Vitamin E (I.U.)	25					
Thiamin (mg)	2.2					
Pyridoxine (mg)	2.5					
Trace minerals (per kg of diet)						
Manganese (mg)	80.0					
Iron (mg)	70.0					
Copper (mg)	10.0					
Zinc (mg)	80.0					
Selenium (mg)	0.3					
Iodine (mg)	1.0					

TABLE 6.2 Examples of broiler breeder diets (kg)

	Starter-Grower		Breeder			Separate Male Feed
Corn	320	453	470	575	668	431
Barley	409	305	250	120		280
Wheat shorts	-	-	-	-	-	200
Soybean meal (48%)	220	190	167	197	205	40
Fat	10	10	10	10	20	10
Limestone	14	15	77	71	70	16
Calcium phosphate	15	15	15	15	14	12
Salt	3	3	3	3	3	3
Vitamin:Mineral premix[1]	5	5	5	5	5	5
Methionine	1.4	1.0	.4	0.8	2.5	0.2
L-Lysine					1.3	0.5
Calculated analysis						
Crude protein (%)	18.1	16.5	15.0	16.0	16.1	12.3
Digestible protein (%)	14.2	12.4	11.2	13.0	12.8	8.2
Crude fat (%)	3.1	3.3	3.4	3.3	4.3	4.3
Crude fiber (%)	4.7	3.8	3.7	3.5	2.8	4.8
Metabolizable energy (kcal/kg)	2840	2820	2765	2850	2900	2760
Calcium (%)	0.95	0.95	3.3	3.3	3.3	0.92
Av. Phosphorus (%)	0.42	0.41	0.40	0.42	0.42	0.36
Sodium (%)	0.16	0.16	0.16	0.16	0.16	0.16
Methionine (%)	0.44	0.38	0.30	0.35	0.42	0.23
Methionine + cys (%)	0.74	0.63	0.53	0.60	0.68	0.40
Lysine (%)	0.93	0.80	0.73	0.81	0.90	0.50
Tryptophan (%)	0.27	0.23	0.21	0.23	0.24	0.17
Threonine (%)	0.73	0.66	0.62	0.67	0.72	0.46

[1]**Use additional choline if premix does not contain this vitamin**

6.2 PULLET REARING PROGRAMS

Both male and female breeders, regardless of commercial strain should be grown so as to optimize feed efficiency. Feed represents about 60% of the cost of producing pullets. However, this does not mean to say that the feeding program should be such that the "cheapest" possible pullet is produced. By and large, such approaches invariably lead to reduced adult performance. Optimizing feed utilization refers to the potential practice of manipulating growth rate in order to achieve a quality pullet at least cost.

It is generally recognized that each commercial strain of bird has an optimum weight at the time of approaching sexual maturity - for most strains, this is around 2.2 kg at 22 weeks of age. In fact, as will be discussed later, the best performing flocks are invariably slightly heavier at this age, regardless of strain. With the potential to influence nutrient intake with both diet modification and degree of feed restriction, then it is obvious that target weights can be achieved by various routes, and these will influence rearing (feed) costs. Over the years both qualitative and quantitative nutrient restriction programs have been studied.

a) Qualitative feed restriction

By modifying the "quality" of the diet, growth rate can be controlled. For example, diets that are deficient in protein or an essential nutrient should limit growth, and hopefully formulation can be adjusted to manipulate growth up to mature body size. In most instances, these programs have failed since all birds in a flock do not have identical nutrient requirements. For example, reducing the methionine content of the diet by 25% may well lead to a reduction in mean flock weight. Unfortunately, those birds with a high inherent methionine requirement will be very light weight, while those birds with an inherently low methionine requirement will be little affected by the diet and grow at a normal rate. Therefore, while mean flock weight can often be manipulated with qualitative feed restriction, uniformity of flock weight is usually very poor, often reaching 30 - 40% compared to 80% under ideal conditions (% of birds ± 15% of flock mean weight). For example, our studies with salt deficient diets indicated that mean flock weight could be quite accurately controlled by regulating the level of salt added to a corn-soybean meal based diet. Unfortunately, flock uniformity at 20 weeks was very low, and consequently many birds were over- or under-weight in the breeder house, and egg production and fertility were both impaired. Similar attempts at qualitative feed restriction have been made with manipulation of fatty acid and amino acid levels in the diet. In some countries, very low-energy high fibre diets are used for growing breeder pullets. It is difficult to fully control growth rate with low energy diets alone, because the breeder pullet is able to increase its feed intake quite dramatically in an attempt at normalizing its energy intake. However low-energy diets can be used to help reduce the stress of quantitative feed restriction programs. However low energy - high fiber diets are expensive to transport and store because of low bulk density, and do cause wetter litter and greater manure production. Total production costs per unit of energy consumed must be calculated with such programs.

b) Quantitative feed restriction

Some form of physical feed restriction is now universal with broiler breeders of all ages. With mechanical weigh-scales attached to automatic feeders, it is relatively easy to realize accurate feed allocation under commercial conditions. An example of a feed restriction type program is shown in Table 6.3.

TABLE 6.3 Skip-a-day feed restriction program for high-yield broiler breeders (diet at 2900 kcal/kg)

Age (weeks)	Pullets (g)	Cockerels (g)
1	Ad-lib	Ad-lib
2	Controlled 25/d	Controlled 30/d
3	Controlled 30/d	Controlled 40/d
4	70	80
5	80	90
6	90	100
7	100	110
8	105	115
9	110	120
10	115	125
11	120	130
12	125	135
13	130	140
14	135	145
15	145	150
16	155	160
17	165	170
18	175	180
19	185	190
20	195	200

Actual feed intake will obviously depend upon nutrient density and environmental conditions, yet these values can be used as guidelines. The concept of feeding to body weight, and the regulation of body weight will be discussed more fully in a subsequent section. Table 6.3 indicates a restricted feeding program for both pullets and cockerels to be initiated at 4 weeks of age. Prior to this, "controlled" feeding should be practised so as to acclimatize birds to a limited feed intake. Controlled feeding should be adjusted to ensure that birds are cleaning up their feed on a daily basis within 4 - 6 hours. Because different strains of birds have different growth characteristics, then initiation of controlled and restricted feeding must be flexible in order to control body weight. For strains with inherently fast initial growth rate, true restricted feeding on a daily-basis may be necessary as early as 7 - 10 d of age. For other strains, ad-lib feeding to 3 - 4 weeks is possible since they have a slow initial growth rate.

With the skip-a-day program shown in Table 6.3, birds are given these quantities of feed only every-other-day. The concept behind this program, is that with every-other-day feeding birds are offered a considerable quantity of feed and this is easier to distribute around the pen so that even the smallest most timid bird can get a chance to eat. The usual alternative to skip-a-day feeding, is feeding restricted quantities every day. For example at 11 weeks of age, pullets could be fed 60 g each day. The problem with every-day feeding, is that feed is eaten very quickly, and so all birds within a flock may not get adequate feed. With such small quantities of feed, and using slow-speed feed chains or auger delivery, it is not unheard of for birds to "keep-up" with feed delivery close to the feed hoppers, and reduce effective feeder space. With every-day feeding, birds may well consume their daily allocation within 30 minutes, and so adequate feeder space is essential with this type of program.

There are some real advantages to every-day feeding however. The previous statement that every-day feeding at 11 weeks means that birds should get 60 g/day since skip-a-day birds get 120 g every-other-day (Table 6.3) is in fact incorrect. In practice birds will become overweight if regular skip-a-day allowances are halved in order to calculate every-day allowances. Unfortunately this error is made by most breeding companies, since they invariably make this statement in their management guides. The reason for this error, is that birds will utilize their feed 10 - 15% more efficiently when fed every-day, rather than skip-a-day. When fed skip-a-day, the bird must store nutrients (body fat and some body protein) which must be utilized by the bird for growth and maintenance the following day when it has no access to feed. Unfortunately this deposition and re-utilization of nutrients is not 100% efficient. For example, in feeding dwarf breeders the same overall feed allocation either every-day or skip-a-day, the every-day birds were 6% heavier at 20 weeks. Similarly, in feeding regular breeders on these two programs, but using the same overall quantity of feed, we recorded weights of 1700 g vs 1500 g at 17 weeks in favour of every-day feeding. As shown in Fig 6.1, there are major differences in heat production (energy expenditure) of birds fed every-day vs skip-a-day, with the latter showing major differences on feed vs off-feed days. The mean difference in heat production for these birds was around 10% in favour of every-day feeding. These data suggest that if every-day feeding can be made to work under commercial conditions, then there should be a saving of up to 10% in feed required to produce a pullet of 2.3 kg body weight.

Fig. 6.1 Heat production of broiler pullets.

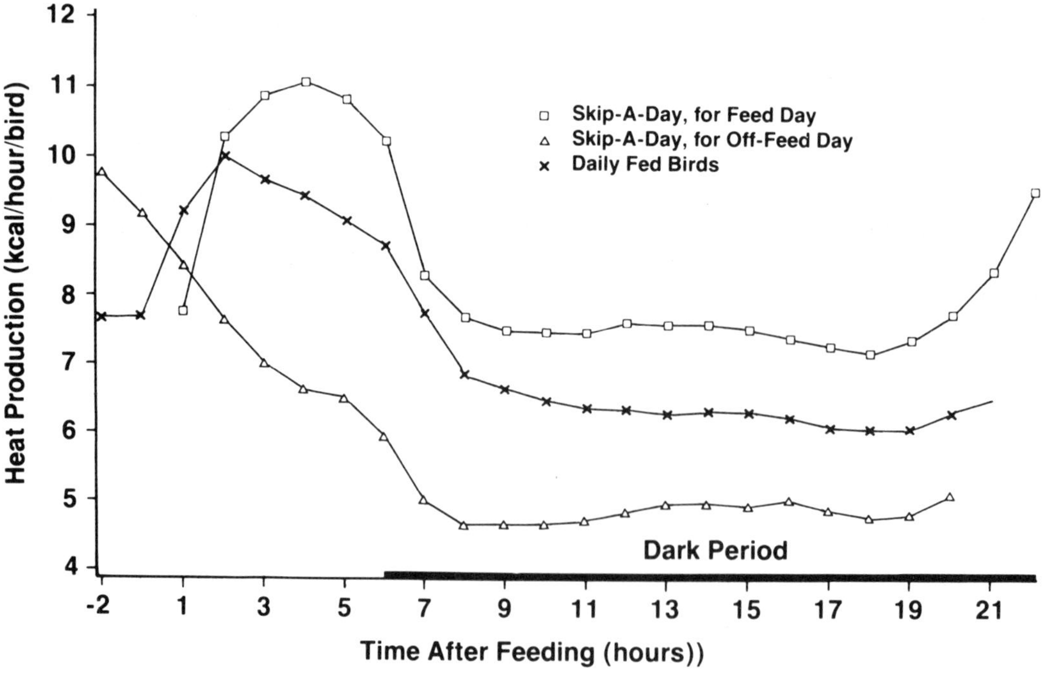

As birds will necessarily be restricted in nutrient intake at some time during rearing, one may question industry practice of utilizing starter diets with high protein/amino acid levels. For example, starter diets with 18 - 20% CP with appropriate levels of methionine and lysine will lead to rapid early growth in most strains of birds. After this initial starter period, one is then faced with the difficult task of tempering growth so as to meet weight guidelines by 6 - 8 weeks of age. There seems no logical reason to enforce rapid early growth, when this must necessarily be followed by severe feed restriction which imposes considerable management problems. A more logical approach is to initiate a much slower initial growth, such that subsequent restriction is less severe, and so less stressful. Our studies have indicated that offering breeder pullets diets of very low protein (13 - 15%) in the starter period will aid in controlling growth (Table 6.4).

TABLE 6.4 Effect of starter diet protein level on body weight of broiler breeder pullets with common grower diet from 21 days (g)

Crude protein in diet 0-21d (%)	Days of age						
	7	14	21	49	77	105	161
20	110	259	477	881	1285	1714	2688
19	110	256	461				
18	109	252	476	875	1280	1703	2648
17	108	250	455				
16	101	236	442				
15	98	196	387	843	1243	1692	2657
14	92	199	382				
13	86	173	337	810	1216	1661	2643

Protein levels as low as 13 - 14% are not usually recommended in commercial practice, because amino acid balance becomes more critical, feathering is often adversely affected, and periodically these diets are more expensive to formulate than diets with less severe protein restriction. However we see no problem with starter diets of 15 - 16% CP under ideal conditions and in fact these can very successfully be used in single-stage starter-grower programs for breeder pullets to 18 - 20 weeks of age. The reason for the success of this simple program, is the flexibility afforded by feed allocation. However, if low-protein starter diets are used, then fatter pullets can be expected at 3 - 5 weeks of age (Table 6.5). Similarly, reduced growth rate caused by lower-protein diets is associated with reduced shank-length (Table 6.6).

TABLE 6.5 Carcass composition of broiler breeder pullets at 21 days		
Crude protein in diet (%)	Crude protein (% DM)	Fat (% DM)
20	50.8	38.9
19	52.3	37.7
18	49.5	41.6
17	50.2	41.3
16	45.7	43.2
15	44.9	45.8
14	44.3	46.8
13	43.1	47.3

TABLE 6.6 Shank length of broiler breeder pullets (cm)						
Crude protein in diet 0-21d	Days of age					
	7	14	21	49	105	161
20	3.7	5.3	6.6	8.7	11.2	11.3
19	3.7	5.3	6.6			
18	3.7	5.2	6.6	8.8	11.3	11.5
17	3.7	5.2	6.6			
16	3.6	5.1	6.5			
15	3.6	4.8	6.1	8.5	11.4	11.4
14	3.5	4.8	6.1			
13	3.5	4.6	5.8	8.4	11.2	11.4

However, both shank-length and carcass composition appear to be normalized by 18 - 20 weeks of age. As previously indicated with Leghorn birds, it is around 12 - 14 weeks of age that such mature characteristics seem to be established. Thus if shank length was adversely affected at say 12 weeks, then this would likely be carried through for the remainder of the breeder cycle. A question often raised concerning lower protein starter diets is the immune competence of the bird. Breeder pullets are subjected to an extensive vaccination program, and so this area is of vital commercial importance. While critical organ weights are reduced at 21d of age with lower protein starter diets, these organ weights subsequently normalize during the growing period, and it seems that immune competence is not affected if diet protein is maintained above 10%. While low-protein diets can therefore be used to temper early growth rate, and so delay the onset and severity of quantitative restriction, continued use of low-protein diets fed ad-lib during rearing has met with only limited success. For example, pullets fed 10% CP diets ad-lib will likely be 0.6 - 0.7 kg heavier at 20 weeks than birds fed a 14% CP diet on a restricted basis. Low-protein starter diets are therefore only designed to temper early growth rate, and some form of restricted feeding will become essential at 2 - 6 weeks depending upon the strain of bird.

With restricted feeding, we have total control over the bird's nutrient intake. Unfortunately, many producers use the primary breeder's feed intake guidelines without considering the energy level of their feeds - this can obviously lead to differences in energy intake, and therefore differences in fat deposition of the pullet. In recent studies we fed pullets quantities of feed according to the breeder's recommendations, although energy level of the feed varied considerably (Table 6.7).

TABLE 6.7 Effect of diet energy level on breeder pullet development and carcass characteristics at first egg

	Low energy diet	Standard energy diet	High energy diet
Age at first egg (days)	185	180	181
Body weight at week 20 (g)	1907a	2059b	2208c
Body weight at first egg (g)	2679a	2872b	3130c
Carcass Fat (g)	395a	501b	624c
Carcass Protein (g)	543a	565a	607b
Carcass Fat-free mass (g)	2284a	2371a	2506b

All pullets matured at about the same time, but had vastly different body composition. Overly fat birds will likely have more reproductive problems, while too lean a bird may not carry sufficient energy reserves to take it through peak egg production. In agreement with other researchers, the data shown in Table 6.7, indicates a relatively constant fat-free mass at time of first egg, and this seems more highly correlated with maturity than either body weight or fat content at this age.

Feeding management of the growing breeder pullet relies on accurate monitoring of body weight. Whatever system of feed restriction is used, and regardless of diet specifications, birds must be fed according to body weight and condition. As previously discussed, there seems no logical reason to promote rapid early growth in the young pullet, because this usually means that subsequent feed restriction is imposed early and/or more severely. Early slow growth rate is therefore recommended, and in fact, there are economic advantages to early slower growth rate. Feeding lower nutrient dense starter diets coupled with regular or with increased degree of restricted feeding means that smaller birds have reduced maintenance costs. However, feed allocation must be adjusted according to body weight, such that the pullet is at the breeder's target weight by 15 - 16 weeks of age (Fig 6.2). This is to ensure that the bird will be of correct weight when it starts egg production. As will be discussed in a subsequent section, the best breeder flocks are invariably those that are slightly on the heavy side as they progress through peak production, the reason for this being a positive energy balance. However if one rears pullets below weight, during the early life-cycle (Fig 6.2) as recommended, catch-up growth must occur prior to sexual maturity. This is essential as it seems as though pullets cannot exhibit catch-up growth at the same time that they are expected to start producing eggs, and start to show increased egg size necessary for production of settable eggs. Underweight birds at sexual maturity most often show a characteristic loss in production immediately after peak egg numbers.

After 14 - 15 weeks of age, feed allowance is becoming quite large, and problems with "choking" (excess feed intake by some birds) starts to occur. If there is adequate feeder space, then birds can be started on alternate restriction programs that entail different sequencing of feed. The idea behind these alternate feeding schedules is to give birds less feed each day, but more feed in total, relative to every-day feeding. Changing to one of these programs also helps in transition from skip-a-day to every-day feeding. Following is an example of various feeding programs where the standard is skip-a-day fed birds consuming 100g at each feeding or 50g/day equivalent (Table 6.8).

Fig. 6.2 Growth curve of broilers breeder pullets

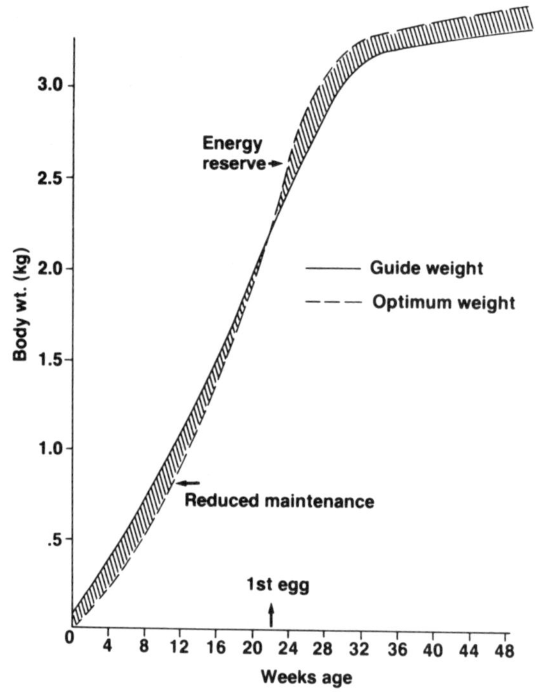

TABLE 6.8 Comparison of feed restriction systems for breeder pullets ≡ 50g/bird/day										
	Grams per bird per feeding each day									
System	1	2	3	4	5	6	7	8	9	10
Skip-a-day	100	-	100	-	100	-	100	-	100	-
2-1	75	75	-	75	75	-	75	75	-	75
5-2	70	70	-	70	70	-	70	70	70	-
6-1	58	58	58	58	58	58	-	58	58	58

Whatever system is used, there needs to be flexibility related to status of the flock as affected by various management decisions. For example, certain vaccinations, and physically moving of birds causes a 1 - 2 day delay in growth rate. These periods of known stress should be counteracted by extra feeding. For example, if skip-a-day fed birds are scheduled to be moved on day 6 as shown in Table 6.8 (non feed day), then birds should be given feed this day regardless of the preplanned schedule.

Coccidiosis is an ever-present problem for those involved with rearing breeder pullets on litter. Because anticoccidials are not usually allowed in adult breeder diets, the bird must develop some immunity during rearing. Such immunity does not develop with anticoccidials commonly used for commercial broilers, and especially the ionophores. This means that if ionophores are used during rearing of breeder pullets, they will most likely prevent clinical coccidiosis, but these birds may develop clinical symptoms as soon as they are transferred to the breeder house. If an anticoccidial is used during rearing, then products such as amprolium are more advantageous. Compounds such as amprolium usually prevent acute clinical symptoms, while at the same time allowing some build-up of immunity. In certain countries, depending upon feed regulations, amprolium can be used throughout the life-cycle of the bird. An alternate approach, and one that requires superior management, is to use non-medicated feed during rearing, and to treat birds as soon as clinical symptoms occur. Since treatment must be immediate, only water-dispensable products, such as amprolium, are recommended. The most recent development in coccidial control is use of attenuated vaccines. These vaccines can be given in the drinking water, or sprayed on the feed when chicks are 2 - 3 days of age. The latest idea is a vaccine within a gel block placed in the chick box at the hatchery, such that chicks arrive at the farm already vaccinated.

6.3 WATER MANAGEMENT

As part of the general management program for feed-restricted birds, it is common to implement some form of water restriction. With feed restriction, birds can consume their feed in 30 mins to 2 hrs depending upon the system and age of bird, and so it is thought that given the opportunity, these birds will consume excessive quantities of water simply out of boredom or to satisfy physical hunger. Certainly birds given free access to water seem to have wetter litter, and there is no doubt that a water restriction program is necessary in order to maintain good litter quality and help prevent build-up of intestinal parasites and maintain foot pad condition. Various water restriction programs are used and there are no universal guidelines that should be adhered to. Pullets develop normally when given as little as ½ hr access to water each day, although longer periods than this are usually recommended. It seems advisable to give birds ½ hr to 1 hr access to water prior to feed delivery, especially with skip-a-day feeding. The reason for this is prevention of a sudden-death type syndrome that occurs with a small proportion of birds that invariably have grossly distended crops full of dry feed. The exact cause of death is unknown although it is possible that the sudden intake of a large volume of dry feed acts as a "sponge" to normal body fluids and so upsets the bird's normal water/electrolyte balance. Giving birds access to water prior to feed delivery often seems to resolve this problem. Table 6.9 gives a general recommendation for water access. These values should be considered as guidelines only, and during periods of heat stress or during disease conditions and around the time of moving etc., time allocations should be increased. With skip-a-day feeding, it is usual to more severely limit water access on off-feed days, based on the assumption that birds tend to drink most water on this day (due to boredom or need to meet physical intake satiety) since they are without feed. However our studies suggest that breeder pullets drink most water on feed days, and seem generally uninterested in water on off-feed days (Table 6.10).

TABLE 6.9 Suggested water access time for growing breeder pullets and cockerels

	Feed-Day		Off-feed Day	
	am	pm	am	pm
skip-a-day feeding	2-3 hr, starting 1 hr prior to feeding	1 hr	1 hr	1 hr
every-day feeding	2 hr, starting 1 hr prior to feeding	1 hr	-	-

TABLE 6.10 Total Water usage of 13-week old skip-a-day or daily fed birds with free or restricted access to water (ml/bird/day)

		Skip-a-day fed birds			
		Water restricted each day	Water restricted on feed days	Free access to water	Daily fed birds, free access to water
Feed day		192	196	273	205
Off-feed day		122	122	37	217
Average		157	161	155	211
	Water:feed	2.38	2.44	2.35	3.20

Bennett, 1989

When birds are daily-fed, there is a fairly consistent pattern of water intake. With skip-a-day feeding and free access to water, pullets surprisingly consume very little water on an off-feed day - for these birds, the largest water intake occurs on the feed day. If this data can be substantiated under field conditions, it suggests that the major emphasis on water-restriction of skip-a-day fed birds should occur on the feed-day rather than the off-feed day. These results are perhaps not too surprising in view of the well established relationship between the intakes of water and feed. It also appears as though daily feeding stimulates overall water usage and increases the water:feed ratio.

6.4 PRE-BREEDER NUTRITION

There has been considerable interest in recent years in the use of the so called pre-breeder diets. Traditionally such diets have been used as a means of introducing a transition phase in terms of calcium metabolism. However, current research interest centers on the potential of diets with relatively high levels of energy and protein/amino acids, as a means of improving early egg production. Thus Cave (1984) suggests high protein pre-lay diets improve early egg production (Table 6.11).

Brake *et al.* (1985) however, indicated no real difference in performance of birds fed high energy and/or high protein diets in the pre-breeder period. In this trial, the high energy (HE) treatment was designed to provide an extra 1145 kcal of ME, while the high protein (HP) treatment provided each bird with an additional 140g CP in the pre-breeder period. These workers showed no effect of pre-breeder diet treatment on performance at any time in the breeder cycle. It is interesting to note that the high protein/energy treatments generally had no effect on body weight (Table 6.12).

TABLE 6.11 Effect of pre-lay diets on breeder performance from 23-56 week of age

Pre-breeder diet (134-181d)	Age 50% production (d)	Egg Prod.	Egg wt. (g)	Hatch of fertile (%)
15.4% CP	192	101a	61.9	83.7
18.1% CP (2780 kcal ME)	193	110b	62.3	82.6

(Cave, 1984)

TABLE 6.12 Effect of pre-breeder nutrition on body weight (kg)

	Pre-breeder treatment			
	HP.HE	HP.LE	LP.HE	LP.LE
18 wk	1.8	1.9	1.8	1.8
24 wk	2.5	2.4	2.4	2.4

HP=High Protein; HE=High Energy; LP=Low Protein; LE=Low Energy

(Brake et al. 1985)

As previously stated, body weight at maturity is critical to optimum breeder performance. Variability in reported response to pre-breeder diets undoubtedly relates to differences in body weight of experimental flocks. The work of Cave (1984) and Brake *et al.* (1985) confirm this hypothesis. Birds detailed in Table 6.11 were underweight relative to most breeder's recommendations and so would be expected to respond to extra nutrients. Birds used by Brake (Table 6.12) were around the breeder's recommended weight and did not respond to pre-breeder nutrition. It can therefore be generalized that pre-breeder diets may be of use where rearing programs have been inadequate and birds are underweight at the time of move to the breeder house. However, when the rearing program has been ideal, and birds are at or above target weight when approaching maturity, then little response to pre-breeder diets can be expected. If birds are under weight at this time, an alternative to introducing a separate pre-breeder diet, is simply to increase allocation of the grower diet.

6.5 ADULT BREEDER PROGRAMS

As with the growing bird, adult breeders must be fed restricted quantities of feed in order to control body weight. After 20 weeks, regardless of rearing program, all birds should be fed on a daily, rather than skip-a-day program. There is little doubt that optimum breeder performance will only be seen following successful rearing management. Data from flocks around the world, housed under various conditions and fed varying types of diet indicate that better performance is invariably achieved when body weight gain is optimum through the late rearing-prebreeder-early breeder transition period. The key nutrient under these conditions is most likely energy, because as with the Leghorn pullet, the broiler breeder is in a somewhat delicate balance regarding energy input and energy expenditure. There is considerable variation in suggested energy requirements for the breeder at this time of early egg production. In fact, reported values vary from 220 - 440 kcal/bird/day.

In attempting to rationalize this obvious range of recommendations, energy requirements were calculated for a commercial strain of broiler breeder (Table 6.13.). In these calculations, values for maintenance energy requirements were extrapolated from our work with breeders. A subsequent factor of 0.82 was used in conversion to ME. An arbitrary 35% activity allowance was included, while growth was assumed to require 5.8 kcal ME/g (50:50, fat:muscle). As shown in Table 6.13, there is concern over the calculated energy requirement in relation to feed allowance, even at the highest feeding level recommended by the breeder. These results suggest that the breeder is in a very precarious situation with regard to energy balance at the critical time of sexual maturity.

TABLE 6.13 Comparison of calculated energy requirement and feed allowance for the breeder pullet. Units are kcal ME equivalents

Weeks of age	Body wt (kg)	Total maintenance (kcal)	Growth (kcal)	Eggs (kcal)	Total daily energy req. (kcal)	Highest feed allowance (kcal)
20	2.07	235	85	-	320	250
21	2.17	245	85	-	330	315
22	2.27	255	105	-	360	330
23	2.39	260	105	-	365	350
24	2.67	290	110	10	410	380
25	2.80	300	90	20	410	420
26	2.91	305	75	40	420	440
27	3.00	310	50	60	420	470
28	3.06	315	30	80	425	480

This problem of energy availability may well be confounded by the nutritionist's overestimation of that portion of diet energy available to the broiler breeder. Most energy levels of diets and/or ingredients, when assayed, are derived using Leghorn type birds. Work at Guelph indicates that broiler breeders are less able to metabolize diet energy, than are Leghorn birds (Table 6.14). Regardless of diet specifications, it would appear that broiler breeders metabolize about 2.5% less energy from feeds than do Leghorns. This relates to some 70 kcal/kg for most breeder diets. Deficiencies of energy around the time of peak egg production will likely reduce egg production at this time, or as often happens in commercial situations, production will decline some 2 - 3 weeks after peak, when a characteristic "dip" in production is seen. It is concluded that optimum breeder performance will occur when the bird is in positive energy balance, and sufficient energy is available for production.

TABLE 6.14 Diet ME determined with Leghorn and broiler breeder pullets

	Diet ME (kcal/kg)		
Diet type	**Leghorn**	**Broiler breeder**	**Δ**
20% CP, 2756 ME	2805	2736	-2.5%
14% CP, 2756 ME	2847	2806	-1.5%
16% CP, 2878 ME	2976	2906	-2.4%
15% CP, 2574 ME	2685	2622	-2.4%

Spratt and Leeson 1987

With energy intakes of 325, 385, or 450 kcal ME/bird/day, Spratt (1987) observed the following partitioning of diet energy intake during a 24-40 week laying period (Table 6.15)

TABLE 6.15 Energy partitioning of broiler breeders (24-40 weeks)

	Daily energy intake (kcal/bird)		
	450	**385**	**325**
Input as:Feed (kcal)	60,000	51,000	32,000
Output as:Body fat (kcal)	21,300	14,700	12,100
Body protein (kcal)	0	1,900	2,400
Eggs (kcal)	11,000	10,000	8,000
% ME into growth	36	32	33
% ME into eggs	18	20	18

It is interesting to observe that even with relatively low energy intakes, birds still produce a reasonable number of eggs and still accrue body mass. In fact, proportional partitioning of "retained" energy into growth or eggs was little affected by energy intake. It is tempting to speculate that energy intake is the controlling factor of egg production of breeders. If this is correct, then it is suggested that the following model applies for the response of the breeder to energy intake (Fig 6.3).

In this scenario, egg production is maximized at a point where some body protein and body fat deposition occurs, ie: as previously suggested, it is essential that the breeder hen continues to gain weight throughout the laying cycle. Fig 6.3 indicates a fine balance between optimum egg output and development of obesity. As will be discussed later, this balance can best be achieved by monitoring feed allocation according to egg production, egg size, body weight and feed clean-up time. If these calculations of energy metabolism are correct, then it is obvious that energy intake and energy balance are critical to breeders that are expected to consistently peak at 82 - 85% egg production. This concept reinforces the statement made earlier regarding optimum body weight and optimum body condition of birds at start of lay. The fact that birds seem to do better when they are on the heavy-side of the breeder's weight guide is likely a factor of such increased body mass acting as a source of additional energy in order to meet the bird's requirements at this time. It is undoubtedly true that any flock that does not gain some weight each week through peak production, will give inferior egg production and hatchability.

Fig. 6.3 Schematic representation of adult breeders response to diet energy intake.

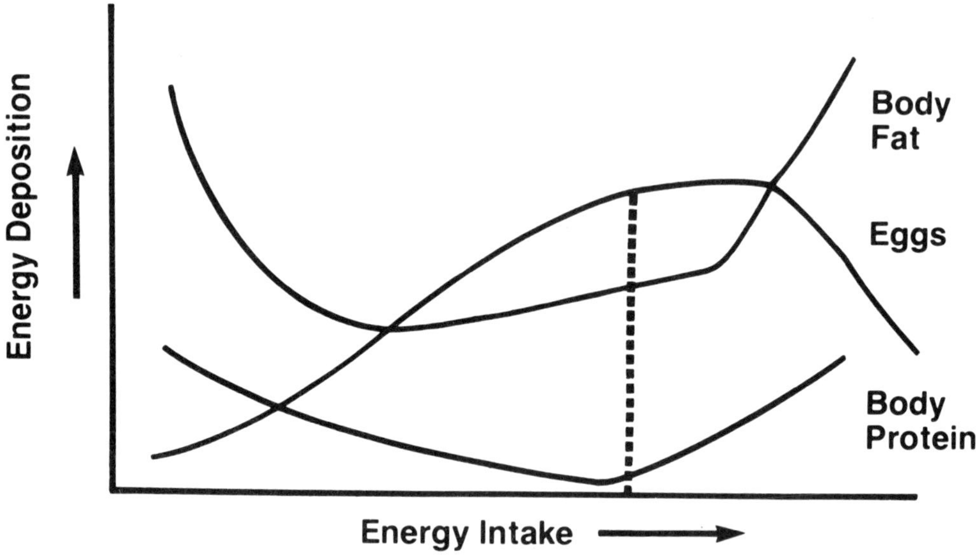

Protein and amino acid needs of the breeder hen have not been clearly established. In general, most breeder flocks will be over-fed, rather than under-fed crude protein, because it is difficult to justify much more than 23 - 25g protein per day. With a feed intake of 165g daily, this means a protein need of only 15%. We have carried out studies involving very low protein diets where levels of methionine + cystine and lysine were kept constant (Fig 6.4). Diets were formulated at 0.82% lysine and 0.59% methionine + cystine in 10, 12, 14 or 16% CP diets. All diets contained the same level of energy and all other nutrients, and quantities fed daily were as suggested by the primary breeder. All roosters were separate-fed a 12% CP male diet. Breeders fed 10% CP performed remarkably well, and although they did not have the highest peak, their better persistency meant no difference in overall egg production (Fig. 6.4).

One of the most surprising results from the study, was better fertility with the lower protein diets. For example, overall fertility to 64 weeks for birds fed 10 vs 16% CP was 95.4 vs 90.6%. The reason for better fertility is thought to be due simply to the fact that these birds gained less weight (Fig. 6.5), because there is a negative correlation between obesity and fertility.

Fig.6.4 Egg Production of Broiler Breeder Hens From 25 to 60 Weeks of Age. (From Lopez and Leeson, 1993)

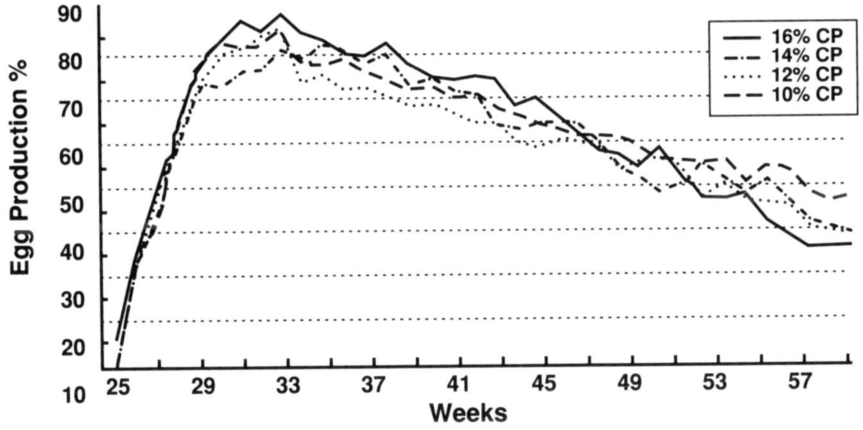

Fig.6.5 Body Weight of Broiler Breeder Hens from 18 to 60 Weeks of Age. (From Lopez and Leeson, 1993)

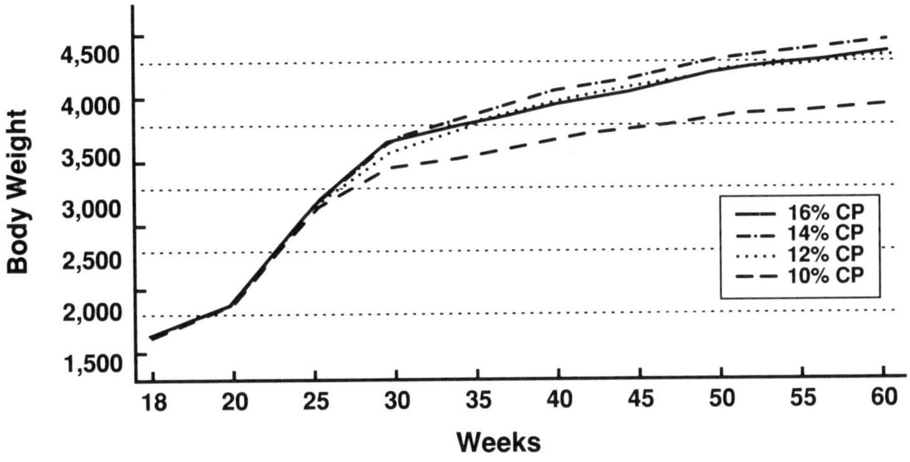

This data suggests that protein/amino acid intake of the breeder hen is related to weight gain, and that excessive weight gain occurring after peak egg production is not merely a factor of energy balance. The results from this study were therefore somewhat unexpected, because birds fed the lowest level of protein produced the most chicks. Although we are not advocating 10% CP diets for breeder hens, this data shows that we could consider lower, rather than higher levels of protein, assuming that adequate amino balance is maintained.

Feed intake during early egg production should be governed by:

a) Egg production (egg mass)

b) Body weight

c) Feed clean-up time

All three of these factors are important in the feeding of breeders. The following discussion sections on each of these factors should not be considered in isolation, but rather all three taken into account prior to increasing or decreasing a feed allocation.

a) Feeding to egg production

Considering the large nutrient output in each egg, it is essential to lead-feed for egg production. It must be remembered that birds are continually producing yolk material, and so a bird at 10% production, should in fact be fed as though it is at 30 - 40% egg production. Failure to lead-feed for egg production, results in the bird utilizing its body fat reserves to supplement the nutrient supply in the feed. If this scenario continues for long enough, all labile energy reserves are utilized, and the bird shows a characteristic loss or stalling in weight that immediately precedes loss in production - assuming the bird has energy reserves to utilize ie. slightly overweight at maturity. An example of a lead-feeding schedule according to egg production is shown in Table 6.16.

TABLE 6.16 Lead-feeding of breeder pullets with a diet providing 16% CP and 2800 kcal ME/kg

	FEED	
Egg production	**(g/b/d)**	**(lbs/100)**
1%	135	30
5%	140	31
10%	145	32
15%	150	33
20%	155	34
25%	160	35
30%	164	36
35%→85%	168	37

Although metric is used throughout this book, we have also included a column in imperial units for Table 6.16 because we realize that producers in many countries still use this measure in describing feed intake of breeders. This schedule means that breeders should be at peak feed allocation by the time they reach 35% egg production. Obviously the figures should be used as guidelines, and if local conditions dictate the use of higher or lower energy density etc., then allocations must be modified accordingly. Also if separate male feeding is not practised, then allowances may need to be increased by 5 - 10% in order to accommodate overconsumption by males.

High and sustained peak egg production can only be achieved with uniform breeder flocks fed to meet their nutritional requirements. With 85 - 88% peaks now possible, it is obvious that we have to carefully plan and execute a feeding program tailored to meet the breeder's nutrient needs. Underfeeding results in very short term peaks, of only 3 - 4 weeks, and these are usually associated with the classical sign of loss or stall-out in body weight for 1 - 2 weeks. On the other hand overfeeding, especially with energy, will result in excessive weight gain, and while peak production may be little affected, there will be precipitous loss in egg production through 34 - 64 weeks of age. The basis of feed allocation at this critical time is obviously to allow genetic potential for increases in both egg numbers and egg size, and also to allow for modest weekly gains in body weight. Managers should consider "challenge feeding" as part of their feed management system at this critical time.

Challenge feeding involves giving the hens extra feed on 2 or 3 days each week, based on need, without changing the base feed quantity scheduled for the flock. For example, a flock may receive 168 g/bird/day at peak, with an additional "challenge" of 7 g/bird/day given three days each week. The challenge feed is, therefore, equivalent to 3 x 7g ÷ 7d = 3 g/bird/day. In reality birds receive the equivalent of 168 g + 3g = 171 g/bird/day. The immediate question is why bother with this more complicated system, rather than just give the flock a base feed allowance of 171 g/bird/day? The advantages of challenge feeding, rather than simply increasing the base allocation are:

- On days of challenge feeding, feeding time will increase, and this helps to improve overall flock uniformity.

- It is easier to make adjustments to nutrient intake based on day-to-day change in needs as may occur with changes in environmental temperature.

- Birds become accustomed to change in feed allocation, which will be important once feed withdrawal is practised after peak.

- Ease of tailoring nutrient needs to individual flocks. For example, a base feed allocation of 165 - 175 g/bird/day may be standardised across all flocks, with individual flock needs at peak being tailored with the quantity and/or frequency of challenge, depending upon actual production, environmental temperature, etc.

The actual quantity and timing of challenge feeds must be flexible if they are to be used efficiently. In practice the challenge should not represent more than 5% of total feed intake, and most often the quantity will be 2 - 4%. On the other hand, the quantity of the challenge should be large enough to meaningfully contribute to the factors listed previously. For this reason there needs to be a balance between the quantity of feed given, and the frequency of this feeding. For example, a daily challenge of 3 g/bird/day given daily will be much less effective than 7 g/bird/day given 3 times each week. In both instances birds are receiving 21 g/week as a challenge, but in the later example the challenge quantity is more meaningful and we are more likely to see a bird response in terms of egg output.

Challenge feeding should start when birds are at 60-70% production, and should be discontinued when egg production falls below 80%. For most flocks, therefore, we can expect to practice challenge feeding from about 29 through 40 weeks of age. The idea of challenge feeding is to more closely tailor feed allocation to breeder hen needs, and so there should be no standardized system. Managers must be given flexibility to alter challenge feeding based on fluctuating needs. In most instances the challenge will be used to lead birds into a sustained peak. Because the concept of challenge feeding is to more closely tailor feed allocation to needs, then it is usual practice to alter the quantity and/or duration of challenge as birds progress through peak egg production. Maximum challenge feeding should coincide with peak egg output, with lesser quantities given prior to, and after actual peak. On this basis we recommend challenge feeding to be reduced (but not discontinued) once birds are 2% below peak egg production. Following are three examples of challenge feeding tailored to three different flock situations. (Tables 6.17 - 6.19).

Table 6.17 Breeders fed a high nutrient dense feed with good ingredient quality control. Expected high-low temperatures of 31°-24°C. Good flock uniformity at 20 weeks of age, and previous flocks show consistent peaks of 85-87%.

Egg production	Base feed		Challenge feed	
35%	162g	(35.6 lb/100)	None	
60%	162g	(35.6 lb/100	5 g/d, 2x/wk	(1.1 lb/100)
80%	162g	(35.6 lb/100)	8 g/d, 2x/wk	(1.8 lb/100)
-2% from peak	162g	(35.6 lb/100)	5 g/d, 2x/wk	(1.1 lb/100)
79%	162g	(35.6 lb/100)	None	-
<79%	Reduce	-	None	-

In this example, because birds are uniform in both weight and maturity and a good quality diet is used, and there is no major temperature stress, the challenge is quite mild. For this flock, a heavier challenge may result in excess weight gain. This type of mild challenge is most frequently used where feed quality is ideal, and there is minimal disease and mycotoxin exposure.

Table 6.18. Breeders fed a high nutrient-dense feed with good ingredient quality control. Expected high-low temperatures 28 - 14°C. Poor to average flock uniformity and previous flocks show variable peaks of 81-87%				
Egg Production	**Base feed**		**Challenge Feed**	
35%	162g	(35.6 lb/100)	None	
60%	162g	(35.6 lb/100)	8 g/d 3 x/wk	(1.8 lb/100)
80%	162g	(35.6 lb/100)	12 g/d 3x/wk	(2.6 lb/100)
-2% from peak	162g	(35.6 lb/100)	8 g/d 3x/wk	(1.8 lb/100)
79%	162g	(35.6 lb/100)	None	-
<79%	Reduce	-	None	-

For this flock there needs to be more challenge feed, because night-time temperature is quite low and there is a problem with maturity related to poorer uniformity. On average this flock may gain a little more weight than example birds in Table 6.17 and this will have to be accommodated with a more vigorous post-peak feed withdrawal program.

Table 6.19 Low nutrient dense feed used, with poor ingredient quality control, and so feed composition may be variable. Expected high-low temperatures, 28-20°C. Average to good flock uniformity at 20 weeks of age, and past flocks show variable peaks at 80-86%.				
Egg production	**Base feed**		**Challenge feed**	
40%	175g	(38.5 lb/100)	None	
65%	175g	(38.5 lb/100)	8 g/d 3x/wk	(1.8 lb/100)
80%	175g	(38.5 lb/100)	14 g/d 3x/wk	(3 lb/100)
-2% from peak	175g	(38.5 lb/100)	8 g/d 3x/wk	(1.8 lb/100)
79%	175g	(38.5 lb/100)	None	-
<79%	Reduce	-	None	-

For this flock, base feed allowance is increased because a low nutrient dense feed is used, and challenge is fairly aggressive again due to concern over feed quality and poor uniformity. In these examples shown in Tables 6.17 - 6.19, it is assumed that managers will continue to maintain breeder body weight through peak, and make necessary adjustments to the challenge if over-or under-weight conditions are seen.

Challenge feeding can also be used post-peak if there are precipitous declines in egg production related to minor disease challenge or management or environmental stress. Under these conditions, challenges of 10 g/bird/day for two consecutive days are recommended. If no immediate response is seen in egg production, then the challenge should be discontinued. If egg production returns to normal, then the challenge should gradually be reduced over the next 2 - 3 days.

Challenge feeding allows tailoring of feed allocation to suit individual flock needs. Managers should be flexible in actual allocations, although maximum challenge feed allocation needs to coincide with peak egg production. Breeder hens will respond to a carefully planned challenge program, with sustained peak production and better post-peak persistency. On the other hand, the challenge should not usually represent more than 5% of total feed intake, because excessive challenge will invariably result in obesity and related loss in post-peak performance. In general, when birds are subjected to such stresses as variable feed quality, mycotoxin challenge and/or fluctuating or extreme environmental temperature, then a high base feed allowance, coupled with aggressive feed challenge, is recommended. On the other hand, lower feed inputs are possible where consistent quality high-energy feeds are used, and where there is good environmental control.

Once birds have peaked in egg production, it is necessary to reduce feed intake. There is often confusion and concern as to how much and how quickly feed should be removed, and this is somewhat surprising, since the same basic rules used pre-peak also apply at this time. This means that birds should be fed according to egg production, body weight and clean-up time. After peak production, feed clean-up time often starts to increase, and this is an indication of birds being overfed. The main problem we are trying to prevent at this time is too large a body weight. If feed is not withdrawn after peak, then because egg production is declining, proportionally more feed will be used for growth. After peak therefore, body weight becomes perhaps the most important parameter used in manipulating feed allocation. It is still important for birds to gain some weight, since loss of weight is indicative of too severe a cut-back in feed.
Feed allocation and withdrawal for breeder hens has to be based upon needs. The hen needs nutrients for four major reasons, namely for growth, egg production, maintaining normal body functions and for daily activity. Each of these needs varies with the age of hen and environmental temperature, and also each need varies with respect to the type of nutrients utilized. Growth, egg production and maintenance all require protein and energy, while activity is only really demanding on energy needs. Actual estimates for these nutrient needs are shown in Table 6.20.

Surprisingly to many people, the maintenance need is by far the largest single factor affecting energy requirements of the breeder. Secondly it is egg production and lastly, growth and activity. In terms of protein needs, egg production and maintenance are the only two meaningful factors. However, as the bird gets older, the actual nutrient needs and the distribution of these needs change (Table 6.20). At 55, rather than 32 weeks of age, therefore, the bird needs less energy and protein for eggs, because egg production has declined (even though egg size has increased) but she needs more of these nutrients for maintenance because over the 23-week period the bird has grown and so needs more feed to maintain herself. At 55 weeks, if all goes well, we will have significantly reduced growth rate, and so both protein and energy needs for growth are greatly reduced. The reduction in nutrient needs for lower egg production and less growth outweighs the needs for more maintenance, and the bottom line is overall reduction in daily need of the hen for both energy (460 vs 510 kcal) and protein (19 vs 21 g).

TABLE 6.20 Protein and energy requirements of breeder hens at 32 and 55 weeks of age

	32 weeks age		55 weeks age	
Nutrient need	**Energy (kcal)**	**Protein (g)**	**Energy (kcal)**	**Protein (g)**
Growth	50	1	5	None
Egg production	80	10	55	8
Maintenance	330	10	370	11
Activity	50	None	30	None
Total	510	21	460	19

We can best achieve reduced nutrient needs, by simply either reducing feed intake, or maintaining feed intake constant but changing the energy and protein levels of the diet. In practice, reducing feed intake after peak production is by far the easiest and most foolproof method of reducing the bird's nutrient intake. Changing to a lower-energy, lower-protein diet means a change of formulation, which itself can be stressful to the bird. On multi-age farms, it is also more hazardous to have multiple diets being delivered to the farm, which can get placed in the wrong feed tank. Changing diet nutrient density is also usually a one-shot deal, whereas in reality the birds' needs are gradually and continually being reduced after peak, a factor that can be accommodated with frequent changes in feed allocation.

The consequences of not reducing nutrient intake of the breeder hen after peak should be fairly obvious. We are not going to force the bird to lay more eggs or become more active by supplying more protein or energy than is required. Oversupply of these nutrients goes directly to increased growth which itself quickly causes increased maintenance requirement. This extra growth will be mainly as fat, but there will also be some muscle (protein) growth. Obesity quickly leads to reduced egg production, diverting even more nutrients into growth (fat). This vicious circle is often responsible for the very sudden drop in egg production seen with flocks that are overfed after peak egg production.

The final questions of course, are how much and how often do we reduce feed allocation after peak production? Regardless of how high a peak production is actually realized, we should not start to reduce feed while birds are $\geq$ 80% production. The main reason for this is that peak egg numbers do not usually coincide with peak nutrient needs for eggs, because egg size is increasing through this period. In most flocks, peak nutrient needs for eggs (production x size) will have been reached by the time birds are down to 79-80% production, at about 39 - 40 weeks of age. At this stage of production we can start to gradually reduce feed intake, and in general the quantity of feed to be removed will depend on peak feed allowance. If birds were peaked on 175 g/bird/day then we likely need to remove more feed than for a flock peaked on 160 g/bird/day. Also, if temperature/seasonal changes are anticipated, then this should be factored into feed allocation - impending warmer weather means we can take more feed away, while if cooler temperatures are anticipated, we may need to take very little feed away (because maintenance needs will naturally increase). Assuming that we have peaked a flock at 175 g/bird/day, and anticipate no major change in environmental temperature, then a feed reduction program, as shown in Table 6.21, is suggested.

TABLE 6.21 Feed allocation program for heavy breeders after peak production

Egg production (%)	Approx. age (wks)	Daily feed intake (g/day)	(lbs/100/day)
80	39	175	38.5
79	40	174	38.3
78	41	174	38.3
77	42	172	37.8
76	43	172	37.8
↓	↓	↓	
74	45	170	37.4
↓	↓	↓	
70	50	165	36.3
↓	↓	↓	
65	55	160	35.2
↓	↓	↓	
60	60	155	34.1
↓	↓	↓	
55	65	150	33.0

With such a low and steady removal in feed, it should be possible to prevent obesity in hens, while at the same time allowing adequate energy and protein for the inevitable slow decline in egg numbers. The reduction in feed intake is necessarily slow and involves small steps because, as shown in Table 6.20, the actual nutrients going into eggs are quite a small proportion of the hen's total needs. Responding to a 5% decline in egg production, therefore, requires very small changes to the feed scale.

Some producers consider a 1 - 2 g/bird/day reduction in feed intake hardly worth bothering about, and either make no adjustment, or few much larger reductions. Sudden large reductions $\geq$ 5 g/bird/day can often be very stressful and result in sudden drops in egg production. Making no adjustments and continuing near peak allocation to 64 weeks, will be uneconomical in terms of birds becoming overweight with associated loss of egg production. In the example shown in Table 6.21, a bird fed according to this suggested schedule will eat about 30.8 kg to 65 weeks. Feeding 175 g through to 65 weeks, with no feed withdrawal, will result in an extra 2.3 kg feed intake. This quantity of extra feed will likely result in an additional 0.3 - 0.4 kg body weight gain, most of which will be fat.

b) Feeding to body weight

It is essential that producers continue to monitor body weight throughout the adult life of the breeder. All too often, the monitoring of body weight stops when birds enter the breeder house and so birds are fed solely according to egg production. We have already emphasized the importance of body weight and body reserves of breeders through peak production, and this means continual monitoring of body weight. It is essential that birds continually gain some weight through peak production. Loss of weight or stall-out in weight usually implies that birds are not getting enough nutrients, and that loss in egg production will occur within 7-10 d. In this context, monitoring of body weight will give an earlier indication of impending problems. From 20 - 32 we^ks of age, pullets should ideally be weighed weekly.

Feeding to body weight assumes that birds are at ideal weight around 22 weeks of age (≃ 2.2 kg). If birds are over- or underweight, then adjustments to intake must be made. If birds are underweight, then they should obviously be given more feed than the standard recommended allowance in an attempt to stimulate growth. Overweight birds should also be given more feed (Table 6.22).

TABLE 6.22 Example of energy allowance for breeders (kcal ME/day)

Age wks	Underweight	Ideal Weight	Overweight
18	230	240	250
20	270	250	280
22	310	295	325
24	345	345	380
26	430	430	470

This apparent dichotomy of ideas is based on the fact that heavier birds have a larger maintenance requirement and so need more feed to meet their overall energy (nutrient) requirement. This is a difficult concept for farm managers to accept, since they are afraid of overweight birds getting even heavier. There is obviously a fine-line between overfeeding and feeding to requirement for this overweight bird, but as previously discussed, the ideal 20 week-old pullet is slightly overweight in comparison to most primary breeder guidelines. Under ideal conditions pullets will not lose weight after 20 weeks of age, rather they show continued small increments of weight gain each week and hopefully are around 3.5 kg at the end of their first laying cycle. Environmental temperature will also have a major effect on feed requirements, because as shown in Table 6.20, the largest portion of the feed is used for maintenance needs. These maintenance needs are greatly affected by environmental temperature (Table 6.23).

TABLE 6.23 Examples of feed peak needs of breeders at various environmental temperatures (g)

Feed need	18°C	24°C	34°C
Growth	10	10	10
Maintenance	140	125	110 (130)
Eggs	30	30	30
TOTAL	**180**	**165**	**150 (170)**

As temperature increases, so feed need is reduced. In this example two values are shown for maintenance feed need at 34°C. The value in brackets (130g) represents feed need when the bird is under stress, and panting etc., where it needs energy to drive the cooling mechanisms in the body. In this situation, total feed need becomes 170g, which is actually greater than suggested at 24°C. It is often difficult to get breeders to eat more feed under heat-stress conditions, yet this increased energy intake is critical if egg production is to be maintained.

c) Feeding according to feed clean-up time

Feed clean-up time should be used as an indication of adequacy of feed allocation. Major changes in clean-up time are an indication of over-or under-feeding, and as such, are an early warning of subsequent changes in body weight and egg production. As a routine management procedure, the time taken to clean-up most of the feed allocation should be recorded each day, to the nearest 30 minutes. If clean-up time varies by more than 60 minutes on a daily basis, then bird weight should be measured immediately. However, major changes in feed allocation should not be made solely on the basis of feed clean-up time, rather these values should be used as a guide to investigate feed needs through more precise monitoring parameters. Feed clean-up time with newer high-yield strain of breeder hens is often greater (+ 1 hr) compared to the more conventional strains and merely reflects a less aggressive feeding behaviour.

Clean-up time for feed can vary considerably from flock to flock for no apparent reason. For example, one flock may take 4 hr to clean-up feed, whereas a sister flock of the same age etc. can take 2 hr. For this reason absolute time taken to clean-up feed cannot be used as a management guide - the only useful parameter is change in clean-up time. Sudden changes in clean-up time often precede changes in body weight by 2 - 3 d, and changes in egg production by 10 -12 d.

6.6 MALE FEEDING PROGRAMS

Condition and performance of the breeder male are critical for optimum chick yield per breeder hen. Most of our losses in hatchability are due to infertility. If a breeder hen produces an egg, then infertility is usually due to simple absence of sperm in the oviduct, and this itself is directly related to mating frequency and/or mating success. In many situations, therefore, loss of hatchability is caused by incorrect body condition of hens and/or roosters, such that mating activity is reduced. For hens this is usually due to overfeeding and obesity, and in roosters is caused by both over-, and under- feeding. Just as we take great care to meet the hen's nutrient requirements with continual adjustments to diet or feed allocation, so we have to carefully monitor the males condition and environment and to feed them accordingly. In many respects it is easy to predict the male's nutrient requirements, because we do not have the complication of egg production as occurs with the hen. The feeding program therefore has to meet just two basic needs namely, growth and maintenance of body functions. The major criteria of our male feeding programs, therefore, are monitoring body weight and body condition and controlling frame size and uniformity.

a) Day-old to 19 weeks

Males can be grown with the hens or grown separately, but in both situations they will almost exclusively be fed starter and grower diets designed for the female birds. This poses no major problem, because there are no large differences in nutrient requirements of the sexes up to maturity, and we can manage with a "compromise" situation. Where males and females are grown together, the onset of restriction programs and general feed allocation systems are usually dictated by progress in hen weight and condition. Male growth and condition cannot be controlled as well under these conditions, and this has to be an accepted consequence of this management decision.

When males are grown separately we have the best opportunity to dictate and control their development. As with commercial broilers, the male breeders will respond more to high-protein starter diets or to more prolonged feeding of these diets. The opposite situation also applies, in that male breeder chicks will be more adversely affected by low-protein or low-amino acid starter diets. For example, under ideal conditions, a well-balanced 15% CP diet can be used as a starter for female chicks and this results in slower early growth rate with the added advantage of delay in introducing restricted feeding. Male chicks can also be grown on such diets, although it is not usually recommended, because there will be poorer early feathering and perhaps more uneven growth rate. These problems resolve themselves over time, but as a general rule it is better to start male breeder chicks on a 17 - 18% CP diet. The male breeder chick is also more sensitive to the effects of low-protein diets that contain anticoccidials, such as monensin. Again, poor feathering will result if starter diet protein is much less than 18% CP. Poor early feathering has no long-lasting effect on subsequent breeder performance, although the chicks obviously look different and they may suffer more from early cold stress.

Male chicks will usually be started on a controlled feeding program at around 3 weeks of age. The starting time for a skip-a-day program is based upon body weight. As with the female chicks, it is important not to have birds overweight early, because this is costly on feed, and subsequent attempts at bringing birds back "on to standard" are usually associated with loss of uniformity. Starting at 3 weeks of age, groups of 10 chicks can be weighed together to give an idea of body weight, and in turn the introduction of controlled feeding. Starting at 4 weeks of age, sample birds should be weighed individually, just as occurs with the hens, and mean weight and uniformity plotted to give a visual image of flock progress. Ideally, feed allocation will be increased on a weekly basis, although this should be dictated by the weekly body weight measurements. Changes in environmental temperature or unprogrammed changes in diet energy (due to ingredient variability, etc.) will affect nutrient needs, feed intake and/or growth rate. Usually the body weight and uniformity of the birds represents their true feed needs at that time and so there needs to be flexibility in feed allocation to account for these variables.

Skip-a-day feeding is usually the preferred system up to time of transfer to the breeder house. However in some situations, choking can occur with males (and females) after 14 - 16 weeks of age, and this is caused by rapid and excessive feed intake on feeding days. The choking problem causes 0.5% mortality per day in extreme situations, but can sometimes be resolved by ensuring that water is available for at least 2 hours prior to feeding. Where this technique fails to correct the problem, it may be necessary to change to a 5:2 or even a 6:1 feeding program (see Table 6.8). These programs provide the same amount of feed on a weekly basis, but this is given as smaller quantities, more often. There seems to be less gorging when birds eat smaller quantities of feed more often. A potential problem with changing to 5:2 or 6:1 feeding programs is loss of uniformity, because daily feeding time will be very short. If males are grown separately from females, then this uniformity problem can sometimes be resolved by changing to a lower nutrient density diet, and giving proportionally more on feed days so as to maintain normal nutrient intake (however under these conditions, daily feed intake will still be less than with skip-a-day feeding). Whatever feeding system is used, it is essential to provide adequate feeder space such that all birds can eat at one time.

b) 20 - 34 weeks

The period during early maturity is probably the most critical in the adult life of the breeder male. Up to about 30 weeks of age the breeder male is still expected to grow quite fast. For example a weight gain of around 1.4 kg is expected between 10 and 20 weeks of age, and this is only slightly reduced to around 1.2 kg weight gain between 20 and 30 weeks. It is, therefore, very important to maintain this growth potential through to 30 weeks, and so continued monitoring of body weight is critical.

The major complication of feeding the breeder male at this time relates to the separate male-female feeding systems now commonly used. Grills on the female feeders are usually around 43 mm in width. Unfortunately 19 - 21 week old male breeders, when first moved to the breeder facilities, will have head width slightly less than this. The males will, therefore, eat from the female feeders while they still have smaller head size. Different males will grow at different rates, and their head width will reach > 43 mm, on average around 26 - 28 weeks. The larger birds usually have larger heads, and so there is a self-limiting system that evolves with exclusion over time of males from the female lines. However we are faced with the problem of trying to estimate the males feed and nutrient intake. One answer to this problem has been the use of so called "nose-bars" which are plastic rods inserted through the nostrils of the bird. This "nose-bar" effectively excludes the male from the female line almost immediately, and so males will only take feed from their own feed line. The effectiveness of nose-bars has been reported as quite variable, and like many situations with broiler breeders, there undoubtedly needs to be a desire by the flock supervisors to make the system work. Another potential solution to the problem of male access to the hen feeders, is to delay placement of the males in the breeder house until 22 - 23 weeks, when the males head width will naturally be wider. This management decision should not affect fertility, because eggs are rarely saved for hatching until 27 - 28 weeks of age, and by this time there will be normal male activity in the breeder house. If males are held in the growing facilities until 22 - 23 weeks, it is important to still light-stimulate them according to the hen lighting schedule. This will ensure that roosters are as mature as the hens when introduced at this later date.

Leaving males un-dubbed also helps in earlier exclusion of males from the female line. Sometimes this causes problems of roosters getting their combs caught in mechanical equipment, and here just dubbing the back 20% of the comb seems beneficial, without really affecting the "size" of the comb. Consideration of comb size raises another important consideration of feeder design. Much emphasis has been placed on grill width (≃ 43 mm) although too often grills provide too much height, such that roosters will force their way into the hen feeders. If roosters are not dubbed, then grill height should be no more than 70 mm, and ideally closer to 65 mm. If roosters are dubbed, then grill height should be no more than 60 mm.

The other major variable affecting breeder male feed intake, is environmental temperature. Because maintenance plays such a major role in nutrient needs, then environmental temperature can greatly influence the amount of energy needed to maintain body temperature. Birds will need more energy in cooler environments, and less energy under warmer conditions. Unfortunately it is difficult to differentiate energy from the other nutrients in a diet, and so meeting fluctuating energy needs can only practically be accommodated by varying overall feed intake.

Table 6.24 gives examples of feed intake for breeder males, with emphasis on the critical period up to 36 weeks of age. Because in most cases males will have some access to the female feeders, we have emphasized this system in Table 6.24, and shown suggested intakes under various environmental conditions. Table 6.24 also shows suggested feed intake for males excluded from female feeders, using techniques such as nose-bars. Under comparable environmental conditions, these birds should be given more feed, because this allocation is their only source of feed.

When roosters have access to hen feeders, we have a major feeding management decision to make at around 28 - 30 weeks of age. At this time almost all roosters will be unable to get into the hen feeder, and so they are suddenly faced with a potential major reduction in feed intake. At this time the roosters can start to lose weight and/or start to become very aggressive. One management decision, as shown in Table 6.24, is to increase the rooster feed at this time, and then more gradually wean them off of this extra allowance over the next few weeks. Roosters that previously had access to the hen feeder line are given more feed, especially from 30 - 36 weeks, compared to those birds with nose-bars etc. By 40 weeks of age, all roosters should be fed about the same amount of feed, regardless of whether or not they previously had access to the hen feeders.

The feed intake numbers shown in Table 6.24 correspond to diets of around 2900 kcal ME/kg. If diet energy level is much different to 2900 kcal, then daily feed intake should be adjusted so as to maintain these energy intake values. Corresponding daily energy intake values can be calculated for birds kept at < 15^0C or > 35^0C. The data shown in Table 6.24 are only guidelines and again it must be stressed that any pre-set feeding schedule must be flexible based on actual growth achieved by the breeder males.

c) 35 - 64 weeks

After 30 - 35 weeks of age we hope to see a slow-down in the males growth, and so there can be corresponding reduction in feed intake. Because maintenance requirement assumes almost the entire nutrient needs of this slower growing older bird, the mature males feeding needs are greatly affected by environmental temperature. After 30 weeks of age the temperature effects on feed intake as shown in Table 6.24 are accentuated relative to that occurring prior to this time. As at other times in the production cycle, it is critical to weigh sample birds in order to determine actual growth and development.

TABLE 6.24 Examples of feeding schedules for male breeders consuming a diet of around 2900 kcal ME/kg (grams/bird/day)

Weeks age	Assuming males have access to hen feeders until approximately 28 weeks age				Assuming males totally excluded from hen feeders	
	>35°C	20-28°C	(kcal ME/day)	<15°C	20-28°C	(kcal ME/day)
20	108	110	319	120	115	334
22	110	115	334	125	118	342
24	112	118	342	130	120	348
26	120	125	363	135	130	377
28	124	130	377	140	135	392
30	130	135	392	150	135	392
32	135	140	406	155	130	377
34	130	135	392	152	130	377
36	125	130	377	148	128	371
40	125	128	370	145	128	371
50	120	126	365	140	126	365
60	120	126	365	140	126	365

The usual problem encountered in this period is roosters becoming overweight, and especially becoming overly fat. The simple reason for this is excessive nutrient intake in relation to needs for maintenance. Extra growth will only occur if extra nutrients are available. The most important nutrients at this time are again energy and protein. After 35 weeks of age, the rooster needs only the equivalent of around 10% crude protein, albeit well-balanced in important amino acids. Energy needs are shown in Table 6.24, although sample weighing of birds will quickly tell if the allocation is correct. If roosters become excessively overweight/obese there should be an attempt at reducing their nutrient intake. If roosters are 200 - 400g overweight, then body weight control can be achieved by reducing daily feed allowance by 5g/bird/day each week until desired weight and condition are achieved. If roosters are > 500g overweight, it may be essential to use a low-nutrient dense feed (see next section) as well as reducing allocation over time. The reason for using a low-nutrient dense feed is to maintain weight uniformity because proportionally more feed can be given daily (albeit at reducing quantities weekly). Any manager facing these problems should seriously evaluate the feeding management strategy of birds in the critical 19 - 30 week period.

d) Separate male feeds

Male and female breeders will usually be fed the same diets up to maturity. In the breeder facilities there is the choice of using the breeder hen diet for all birds, or a separate diet specifically formulated for males. Such male diets will usually be much lower in crude protein, amino acids and calcium compared to the breeder hen diet. The advantage of a separate male diet is that it more closely meets the male's nutrient requirements and allows for a slightly more generous feeding allowance. The protein and amino acid needs of the mature male are very low, being in the range of 10% CP. Such low protein diets are often difficult and expensive to formulate, but body weight control, and subsequent fertility, will usually be improved. A practical compromise formulation is around 12% crude protein or to use a 14 - 15% pullet grower diet. When low protein diets are used it must be remembered that protein quality is still very important. For these low protein diets, methionine should be maintained at 2% of the protein, and lysine at around 5% of protein. Using a lower energy level, such as 2600 kcal/kg, together with the lower protein, means that we can give males more feed, which will prolong feeding time and help maintain body weight uniformity. The calcium present in the hen breeder diet is also excessively high for the male. Because it is not producing eggshells, the male needs only 0.8 - 0.9% calcium in the diet. This extra calcium intake may pose additional stress on the kidney, although under most farm conditions, the roosters can handle this extra calcium load. However when combined with other stressors to the kidney, such as high protein, or high mineral intakes, or mycotoxins such as ochratoxin, there can be problems with the general metabolism of the bird's kidney.

There are distinct advantages to using a separate male diet and so why is it not used universally? The major disadvantage is the practical problem of handling relatively small quantities of feed, often in bags. The potential advantages for a separate male diet can be quickly lost if male feed is stored on-farm for extended periods. If male feed is stored for more than 6 weeks, especially in hot humid weather, then any advantages may be counterbalanced by potential loss of nutrients and/or mold growth.

6.7 NUTRITION AND HATCHABILITY

Successful hatching of an egg depends upon a fertile egg having adequate nutrients and environmental conditions, such that the embryo can develop into a viable chick. From a nutritional point-of-view, hatchability can be influenced by fertility of both male and female breeders, the nutrients deposited in the egg for the embryo, and certain physical egg characteristics that can affect gas and water exchange during incubation. Traditionally, vitamin status of breeders is often considered the major nutritional factor influencing hatchability, although we now know that imbalance or excess of a number of nutrients can affect embryo viability. In the following discussion it is assumed that incubation conditions are ideal, and also that eggs are stored and transported under ideal environmental and sanitary conditions.

Fertility

There is surprisingly little information available on the effect of nutrition on fertility, and especially for the hen. With hens it is assumed that if a bird is capable of producing eggs, and if viable sperm are available, fertility will occur. Nutritional effects on female fertility are, therefore, assumed to be quite minor in relation to nutritional effects on egg formation *per se*. While this is true for nutrients such as vitamins and minerals, it may not be true for nutrients affecting general body size and body composition, such as diet protein and diet energy. In a recent study Lopez and Leeson (1995) indicated that protein level of the diet of breeder hens had a significant effect on fertility (Table 6.25).

TABLE 6.25 Diet protein and female fertility to 64 weeks of age

Diet protein (%)	Fertility (%)
16	91.6[b]
14	93.3[a]
12	95.1[a]
10	95.4[a]

In these diets, methionine and lysine levels were kept constant, as was energy level, and only diet crude protein was varied. All roosters were fed a separate male diet at 12% CP, and so the data shown in Table 6.25 is a true female effect. Lopez and Leeson (1995) conclude that this apparent crude protein effect was simply due to body weight, because hens on the lower protein diets were smaller throughout the experiment. Birds fed 10% CP were some 500g smaller than birds fed 16% CP at 64 weeks, even though feed and energy intake were similar for all treatment groups. Limiting excess body weight after peak production is, therefore, important in maintaining greater mating activity of these smaller more active birds. In this respect, overfeeding both protein and energy is expected to reduce fertility, simply by making birds obese, and so less willing to mate with the roosters.

This same concept also applies to roosters, where overfeeding of protein and/or energy is likely to result in reduced fertility. Overfeeding of male Leghorn breeders results in a dramatic decline in total sperm production with associated increase in production of dead spermatozoa. The introduction of separate male feeding systems has also resulted in better fertility, simply because of better control over feed intake of the rooster. However even with separate male feeding, it seems advantageous to use low protein diets (McDaniel, 1986, Fig. 6.6).

Fig. 6.6 Diet protein level and percentage of roosters producing semen (from McDaniel, 1986).

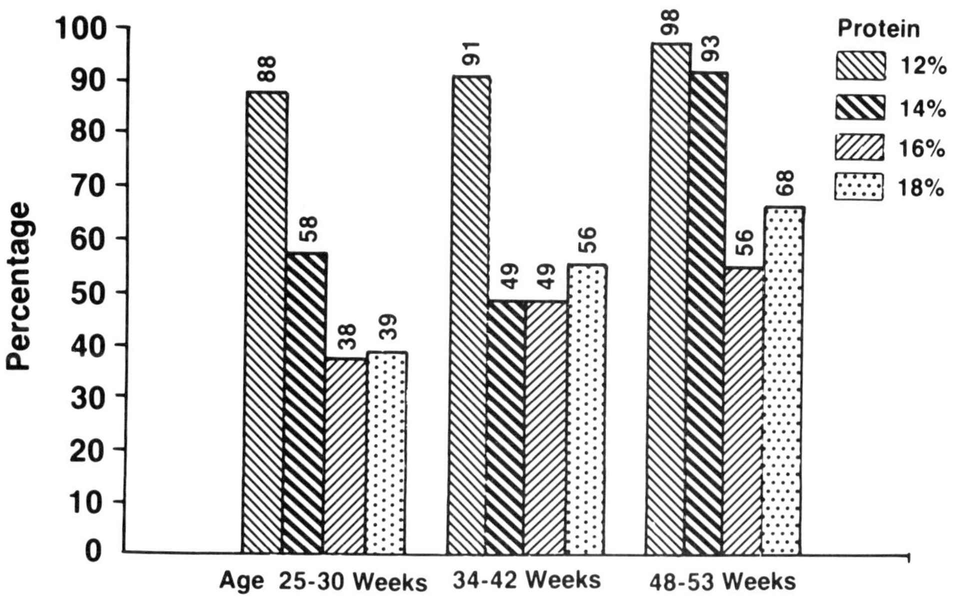

Hatchability

Nutritional effects on hatchability of fertile eggs are not easily quantified, apart from the effect of gross deficiencies of vitamins and some other nutrients. Table 6.26 provides a summary of common embryo deficiency symptoms for selected vitamins and minerals. It should be emphasized that classical deficiency symptoms of individual vitamins are rarely seen. More often, multiple vitamin deficiencies occur when vitamin premixes are inadvertently omitted from the diet, or more commonly, deficiencies are induced by some other nutrient or toxin. These latter effects are obviously difficult to diagnose, since diet analysis reveals a correct vitamin level, even though a deficiency of that vitamin is evident.

TABLE 6.26. Common embryo deficiency symptoms for vitamins and minerals

Nutrient	Deficiency symptoms
Vitamin A	Early embryo mortality (48 hours) with failure to develop circulatory system.
Vitamin D_3	Depending on reserves in dams, stunted chicks and soft bones. Usually associated with shell defects and hence changes in porosity of the shell.
Vitamin E	Usually see early embryo mortality at 1-3d. Encephalomalacia may be seen in the embryo and exudative diathesis is common.
Riboflavin	Excessive embryo mortality 9-14 or 17-21d. Embryos show edema and/or clubbed down. Chicks may show a curling of the toes.
Pantothenic acid	Subcutaneous hemorrhages in unhatched embryos.
Biotin	Reduced hatch without reduced egg production. Peak in embryo mortality during first week and last 3 days of incubation. May see skeletal deformities and crooked beaks.
Vitamin B_{12}	Embryo mortality around 8-14 days, with possibly edema, curled toes and shortening of the beak.
Thiamin	There are two stages of embryo mortality - one very early and the other at 19-21d. Many dead chicks appear on the trays although there are few, if any, deformed chicks. Mortality can be high for 10-14 days for those chickens that do hatch. Injecting the chicks with thiamin results in an almost instantaneous recovery. Certain types of disinfectants, anticoccidials and poor quality fish meal have been implemented in thiamin deficiencies. There is also recent evidence to suggest that thiamin requirements are increased in the presence of some Fusarium molds.
Calcium and phosphorus	As maternal deficiency progresses, embryo mortality shifts from later to earlier stages of incubation. Shortened and thickened legs are seen with shortened lower mandible, bulging forehead, edema of neck and protruding abdomen. Shell quality is usually affected.
Zinc	Numerous skeletal deficiencies, and feather down may appear to be "tufted".
Manganese	Late embryo mortality (18 to 21 days). Embryos show shortened wings and legs with abnormal head and beak shape. Edema is common and feather down is usually abnormal.

In situations of complex vitamin deficiency, caused for example by accidentally failing to add the vitamin premix, then riboflavin deficiency is often the first to appear, and this has the most dramatic effect on breeders, with hatchability reaching very low levels in 3 - 4 weeks (Table 6.27).

TABLE 6.27 Hatchability of eggs produced by caged breeders fed corn-soybean diets devoid of supplemental vitamins (% fertile eggs)

	Vitamin omitted from control diet							
Week on diet	**None (control)**	**Biotin**	**B_{12}**	**E**	**Folacin**	**Niacin**	**Panto-thenate**	**Riboflavin**
1	95	86	97	97	97	96	94	95
3	97	83	95	84	89	87	81	55
5	98	63*	84	67	30*	61*	74*	19*
7	92	54*	61*	62*	19*	69	26*	1*
13	88	52	27*	95	38*	50	54	0*
15**	90	96	21*	75	70	38*	56	0*
17	95	90	50*	58*	85	61	40*	57*
19	97	99	99	92	99	98	97	96

* **Significantly different from control ($P < .05$).**
** **Vitamins reintroduced.**

In this study, hens were fed corn-soy diets where the premix was formulated without individual vitamins as detailed. For some vitamins, therefore, corn and soybean meal will provide some base level of vitamins and this may be the reason for differential results to the diets. As already indicated, the response to riboflavin is most severe, with hatchability down to zero in seven weeks. After 15 weeks of deficient diets, we reintroduced a regular fortified diet, and as shown in Table 6.27, for all treatments hatchability returned to normal within 4 weeks. Hatchability problems related to vitamin deficiencies therefore seem reversible once adequate diets are fed, and there seems to be no long lasting effect. A practical problem with on-farm nutritional deficiencies is that hatchability declines are not seen until three weeks after deficient diets are consumed. For this reason, weekly checks on embryo survival will give a much quicker indication of potential problems. This type of data is shown in Table 6.28, which relates to the same study detailed in Table 6.27.

TABLE 6.28 Incidence and age of dead germs from eggs produced by breeder diets devoid of supplemental vitamins (% fertile eggs)

Week on diet	Vitamin omitted from control diet																	
	None (control)			Biotin			B_{12}			E			Pantothenate			Riboflavin		
	E	M	L**	E	M	L	E	M	L	E	M	L	E	M	L	E	M	L
1	5	0	0	11	0	2	3	0	0	0	1	0	6	0	0	4	0	1
3	2	0	0	11	0	5	3	0	1	6	0	6	18*	0	0	10	7*	24*
5	0	0	0	23*	1	5	10*	1	1	7	15*	6	19*	0	4	34*	31*	5
7	2	0	2	15*	6*	5	4	21*	1	10	14*	4	16*	36*	6	70*	24*	1
13	9	0	2	33	1	3	0	39*	18	2	2	0	28	3	3	47*	36*	0
15[1]	0	0	10	0	0	4	0	0	79*	0	0	25	12*	3	4	28*	16*	39
17	1	3	0	7	2	0	0	50*	0	0	3	35	39*	4	0	7	5	16*
19	2	0	1	0	0	1	1	0	0	1	0	7	1	0	0	1	0	1

*Significantly different from control within each age group ($P < 0.05$)
**E, 1-7 days; M, 8-14 days; L, 15-21 days of incubation
[1]Vitamins reintroduced

The pattern of embryo mortality in the treatment groups should be compared with that occurring in the control-fed birds. Within 3 weeks in the case of riboflavin, and within 5 weeks for most other vitamins, there is dramatic increase in incidence of dead embryos which coincides with the loss in hatchability shown in Table 6.27. The most characteristic feature of this study is embryo mortality occurring in the mid, 8 - 14 day period of incubation. For the control birds there is virtually no mortality in this period. Likewise for the treatment birds in the first week of the study, or in the last week of the study 4 weeks after reintroduction of the control diet, there is again no 8 - 14 d embryo mortality. This data suggests that 8-14d embryo mortality can be used as a diagnostic tool for establishing hatchability problems related to inadequate breeder vitamin nutrition.

Another interesting aspect to this study was the differential pattern of embryo mortality that occurred over time with certain vitamins. Extreme differences in pattern of embryo mortality are seen with niacin and riboflavin deficiency. These details are shown in Table 6.28, but can be more clearly seen when plotted over time as shown in Figs. 6.7 and 6.8. During the initial stages of riboflavin deficiency (Fig. 6.8) when embryo mortality is relatively low, birds are able to transmit sufficient riboflavin to the egg for the embryo to develop for 15 - 21 days. As deficiency is prolonged, age of mortality decreases, such that after 14 wk of treatment, 100% mortality of 1 - 7 day embryos occurs. When riboflavin was added to the diet, there was a rapid decline in early embryonic mortality, although prior to resumption of normal hatchability, there were resurgent peaks in both mid- and late-term embryonic deaths (Fig. 6.8). Such peaks are compatible with a proportion of the adult population failing to totally respond to added dietary vitamins, and transferring to the egg an increased, but still insufficient, quantity of vitamin. With niacin deficiency, the overall pattern of embryo mortality was similar to that recorded with riboflavin, although there was a reversal in the age period of the various peaks (Fig. 6.7). In the initial stages of these deficiencies, early dead germs are noticed (late dead in the case of riboflavin), while with duration of treatment, this peak is superseded by consecutive peaks of mid and finally late dead embryos. When niacin was added to the diet resurgent peaks were again noted in embryo deaths characteristically occurring earlier, i.e., 1 - 7, 8 - 14 day peaks. Overall hatchability did not decline to 0% with niacin as occurred with riboflavin, suggesting a greater degree of variability in individual bird requirements for this vitamin. It is speculated that the progressive peaks shown with niacin correspond to bird groups having decreasing requirements for niacin. On this basis, it must be assumed that birds with a high niacin requirement transfer sufficient vitamin for only 1 - 7 day embryo development (initial peak) when fed deficient diets. Conversely, birds with the smallest requirement are not affected until 12 wk of treatment, and at this time are still able to transfer sufficient vitamin to the egg to enable 15 - 21 day embryo growth. It might be expected that hatchability would have declined to 0% with a more prolonged period of deficiency, when even those birds of low niacin requirement would have been depleted of the vitamin. If this overall hypothesis is correct with respect to niacin, the embryo mortality pattern as shown for this vitamin (Fig. 6.7) may be useful in identifying those nutrients whose requirement by groups or individuals, is variable. These results indicate striking differences in the pattern of embryo mortality age for various vitamins. Observations on malformations and malpositioned embryos in this study indicate no clear trend, inferring limitations of these parameters as diagnostic indicators of practical-type vitamin deficiencies in breeder diets.

Fig. 6.7 Pattern of embryo mortality in birds fed diets devoid of synthetic niacin.

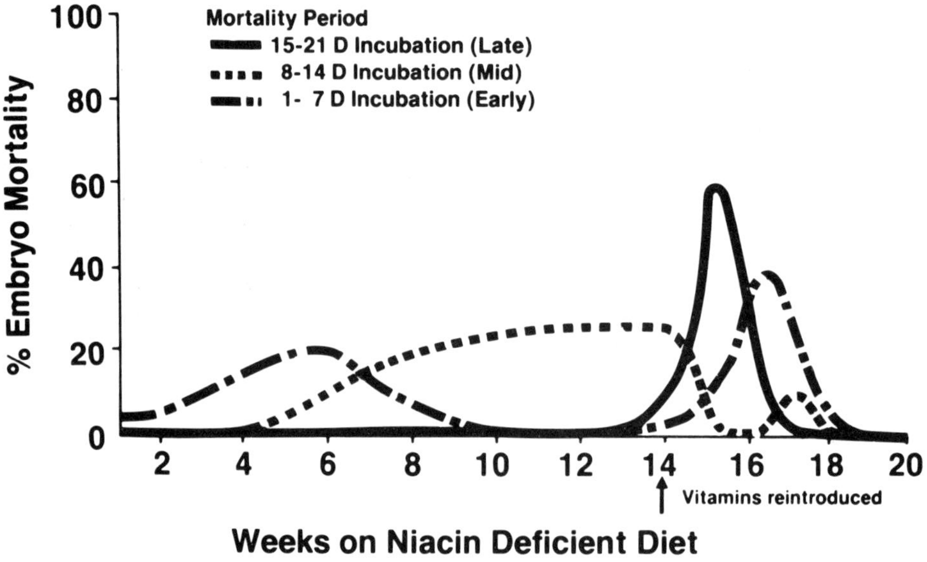

Fig. 6.8 Pattern of embryo mortality in birds fed diets devoid of synthetic riboflavin.

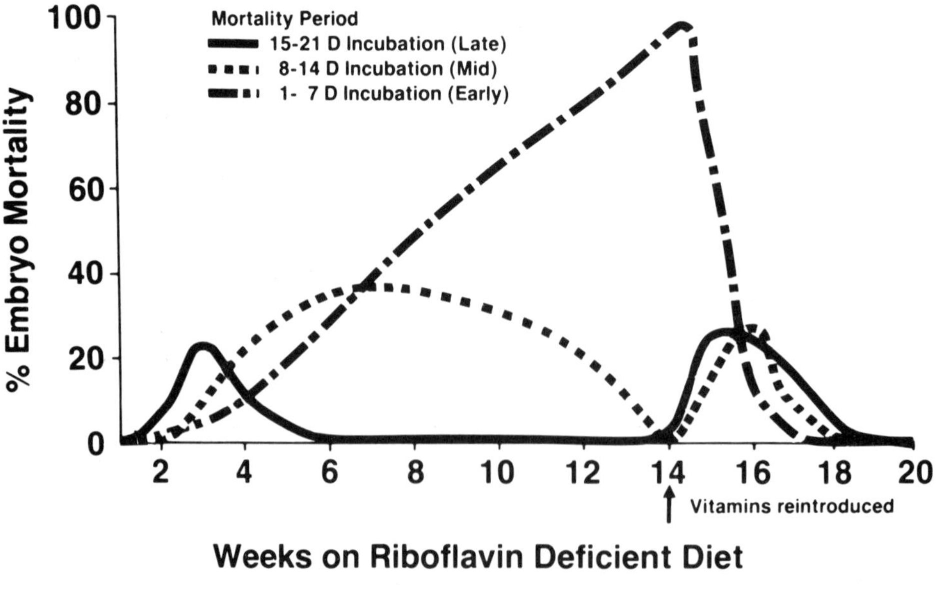

Vitamin deficiencies of course should not occur under commercial conditions, because all requirement needs should be met as synthetic sources in the premix. In fact breeder diets often contain the highest levels of supplemental vitamins of any class of poultry, and this is sometimes questioned as being too costly. In feeding breeders we not only want to prevent signs of deficiency as detailed previously, but also to ensure optimum production and hatchability. The superior performance of breeders that we routinely see today, with peaks of 85 - 88%, will only be achieved by feeding relatively high levels of vitamins as part of a balanced nutritional program.

One reason for higher vitamin fortifications relative to standards, such as NRC (1994), is the loss in potency of vitamins that can occur between feed manufacture and consumption by the bird. Different vitamins are susceptible to various stresses to varying degrees, but as a generalization it can be stated that the major causes of loss of vitamin potency are storage time, storage temperature, and storage humidity of the premix before mixing, and of the feed after mixing. Another major loss of vitamins occurs if they are premixed with minerals and stored for any length of time prior to incorporation in feed. Also conditions within the premix and feed can cause loss of potency. For example, some vitamins are acidic whereas others break-down under acidic conditions. Finally, to really cause problems to vitamin stability, we sometimes pellet feed, and here the temperature and humidity can cause vitamin breakdown.

An obvious question is the cost effectiveness of adding relatively high levels of vitamins so as to ensure good egg production and optimum hatchability. A major producer of pure vitamins, recently conducted a survey of breeder operations in the USA and, as anticipated, found a range of vitamin levels being used. This industry data covered 62 broiler companies and represents about 90% of the industry. Table 6.29 shows the highest and lowest vitamin levels being used, compared to NRC (1994). There is about a 100% difference between the top 25% and bottom 25% of reported levels for vitamin fortification. Table 6.29 also shows the cost of obtaining these individual vitamin levels, and the total cost per tonne of mixed feed.

Obviously the bottom line is how well do breeders perform based on high vs low-industry values. We do not have this data, but we can see that the higher vitamin levels cost about 10¢ more per breeder, which in most markets is equivalent to 0.5 chicks per breeder. Considering that a number of the low-industry vitamin levels (Table 6.29) are even less than NRC (1994) we can be sure that this will affect hatchability, and from our experience, this will be much more than 0.5 chicks per breeder. Marginal vitamin levels can easily result in loss of 2 - 5 chicks per breeder, which is 4 - 10 times the cost of the extra vitamins in the feed. With these extra levels of vitamins in the feed, there should be no need to use supplemental vitamins in the water, except in situations of environmental or disease stress when feed intake is not optimal, or there is evidence of enteritis.

If vitamin levels are to be adjusted according to local conditions, then which ones are most likely to influence feed cost? Table 6.30 is a repeat of the industry high-level breeder vitamin recommendations, but in this situation they are listed according to contribution to cost of the premix. This data shows vitamin E, biotin and vitamin A to be the most expensive, and together make up over 50% of the cost of the vitamin premix. These three vitamins should, therefore, receive most attention when additions to, or deletions from premixes are contemplated based upon knowledge of specific needs within a feeding program.

TABLE 6.29 Breeder vitamin levels and costs

	Industry High (Top 25%)[1]		Industry Low (Bottom 25%)[1]		NRC 1994	
Vitamin	**Level**	**$/t feed**	**Level**	**$/t feed**	**Level**	**$/t feed**
Vit. A (IU/kg)	12,800	0.68	8,100	0.43	3,000	0.16
Vit. D_3 (IU/kg)	3,500	0.08	2,100	0.05	300	0.01
Vit. E (IU/kg)	36	1.08	14.3	0.43	10	0.30
Vit. K_3 (mg/kg)	3.1	0.13	0.74	0.03	1.0	0.03
Thiamine (mg/kg)	3.2	0.11	1.0	0.04	0.7	0.03
Riboflavin (mg/kg)	9.9	0.53	5.6	0.29	3.6	0.29
Pantothenic acid (mg/kg)	17.3	0.38	9.3	0.20	7.0	0.20
Niacin (mg/kg)	43	0.27	23	0.14	10	0.14
Pyridoxine (mg/kg)	6.0	0.31	1.4	0.07	4.5	0.07
Folic acid (mg/kg)	1.3	0.13	0.63	0.06	0.35	0.06
Biotin (μg/kg)	220	0.77	88.0	0.31	100	0.31
Vit. B_{12} (μg/kg)	17.5	0.09	10	0.05	8	0.05
Total		**$4.56/t**		**$2.10/t**		**$1.65/t**
Cost/breeder hen 20-64 weeks		19.1¢		8.8¢		6.9¢

[1]**BASF Technical Bulletin #KC9305**

Vitamins are expensive, with the premix representing about 2 - 3% of the cost of breeder feed. However the effects of deficient or marginal levels in breeder feed can be far greater than the 10¢ per breeder potentially saved by feeding at the low vs high end of current industry standards.

While vitamin deficiencies should only occur under unusual circumstances, there is an indication that lower hatchability, related to vitamin metabolism, may occur when high protein diets are fed. Although there has been no definitive research to implicate high-protein intakes with induced vitamin deficiencies, there is substantial circumstantial evidence from a number of studies showing that with high crude protein diets, there is often reduced hatchability and embryo mortality which suggest problems with vitamin adequacy of the egg. It is possible that deamination/transamination involved in "excretion" of excess nitrogen could impose a very high metabolic demand for B-vitamins. Work out of the UK and our work with both Leghorn and broiler breeders shows a consistently reduced hatchability for birds fed high vs low protein diets.

TABLE 6.30 Vitamin recommendations ranked by $ cost/tonne feed		
Vitamin	**$/tonne feed**	**% contribution to premix cost**
Vitamin E	1.08	23.7
Biotin	0.77	16.9
Vitamin A	0.68	14.9
Riboflavin	0.53	11.5
Pantothenic acid	0.38	8.3
Pyridoxine	0.31	6.8
Niacin	0.27	5.9
Vitamin K_3	0.13	2.9
Folic acid	0.13	2.9
Thiamine	0.11	2.4
Vitamin B_{12}	0.09	2.0
Vitamin D_3	0.08	1.8
Total	**$4.56**	**100.0**

In relation to industry practice, "high protein" levels as discussed here are really not too high in relation to normal feeding practice. In our work "high-protein" refers to 25 g protein per bird per day. In order to achieve a "high" protein intake of 25 g/b/d, a flock of breeders consuming 160 g/b/d would achieve this intake with a diet protein content of only 15.7%. A comparable protein intake by Leghorns consuming 100 g/b/d would be achieved with a diet level of 25% CP. Obviously the breeder hen requires more protein than does the Leghorn for maintenance, although with a 1 kg body weight difference this should amount to only 3 g/b/d. If these calculations are correct, then we are overfeeding the breeder by some 5 g CP/d.

6.8 DWARF BREEDERS

Dwarf broiler breeders have been available commercially for a number of years, yet they only seem popular in certain countries such as France. For the most part, the reluctance to use dwarf females related to difficulty of management when birds were housed on the floor - the dwarf is ideally suited to a cage environment, yet most producers at this time are reluctant to install these expensive systems. With floor rearing, past experience showed that the breeder males literally dominated the small females, such that mating was infrequent, and so fertility was unacceptable. However, with the introduction of separate-male feeding, it should now be possible to more closely control the body size of the breeder males, so preventing a major discrepancy in size of the two sexes. A more recent management problem with the dwarf female, is the popularity of aggressive high-yield roosters.

The main attribute of the dwarf is a smaller body size ($\simeq$ 30%) that requires less feed for maintenance (20% less feed). Offspring from the dwarf are smaller at day-old, and this persists up to 42 d of age. The dwarf is ideally suited therefore to locations where feed prices are very high, and broiler offspring are grown-out to 49 - 56 d of age. Characteristics of the dwarf relative to a conventional breeder are shown in Table 6.31.

TABLE 6.31 Characteristics of dwarf vs regular size broiler breeders

	Dwarf	Regular Breeder
Rearing Period (0-20 wks)		
Mature body wt	1.7 kg	2.2 kg
Feed intake	7.2 kg	9.0 kg
Stocking density		
0 - 6 wk	10/m^2	8/m^2
6 - 20 wk	7/m^2	5/m^2
Breeder Period (40 wks)		
Age 1st egg	160 d	160 d
Stocking density	5-7/m^2	3-5/m^2
Feed intake	33 kg	40 kg
Egg Production	168	165
Chicks	140	135
Broiler offspring		
42 d wt ♂	2150 g	2350 g
50 d wt ♂	2650 g	2900 g
42 d wt ♀	1820 g	1950 g

From a nutritional management point of view, the dwarf breeder is much easier to handle than the conventional sized bird. The dwarf is more timid and docile, and there seems to be less stress associated with restricted feeding. Around peak production, the maximum feed allowance will likely be about 122 g/b/d, and this almost represents ad-lib intake for this bird, and so clean-up time is quite prolonged. Because feed lasts so long in the feeders, it is obvious that separate male feeding is essential with the dwarf.

It is often stated that dwarf strains are slower to mature than non-dwarf strains, although care must be taken in interpretation of this data. As previously discussed for breeders, mature body size and composition are the factors likely responsible for sexual maturity. However with open-sided buildings, and good rearing management, there may be little difference in actual age at first egg for dwarf and regular-sized breeders reared under the same general conditions. Dwarfs are claimed to have better persistency, since many flocks in Europe are taken on to 48 - 50 weeks of lay. However the converse situation is seen in many commercial trials in N. America, where egg production falls off very quickly after peak. One interesting feature of birds carrying the dwarf gene, is their apparent greater tolerance to heat stress. This effect may in part be due to a smaller body size, the fact that the bird has a higher plasma sodium content, and/or the fact that dwarfs are hypothyrotic. The future of the dwarf breeder hen lies in either good management of separate male feeding, or a general move by the industry to cages and artificial insemination, or cages with natural mating.

6.9 CAGED BREEDERS

The dwarf bird seems an ideal candidate for cage management, although there are serious problems seen when regular size breeders are caged for any length of time. Commercial trials with regular breeders, involving artificial insemination, have generally proven unsuccessful due to lack of uniformity and foot pad lesions. Both of these problems seem to relate to feeding management, and the propensity of the regular sized breeder to become overweight. Few mechanical systems are able to accurately dispense feed to each cage, and so over/undernutrition becomes a problem. Our experience with field trials indicate that most often, in a cage containing three breeder hens, while "average weight" may be ideal, there will often be three distinct sized birds likely related to aggression and dominance behavior within the group.

Feed allocation is also difficult to regulate as mortality progresses, since for most systems it means physically moving birds so as to maintain a constant number per cage. Foot pad lesions often develop after 35 weeks of age, especially with overweight birds. Until a simplified and accurate feed allocation system is developed for cage systems, it is doubtful that they can be made to operate economically under commercial conditions. We have experienced similar problems with a new colony cage system involving twenty breeder hens and two roosters per cage. Again foot pad lesions become problematic after peak egg production.

CHAPTER 6

Selected References

Attia, Y.A., W.H. Burke, K.A. Yamani and L.S. Jensen 1995. *Daily energy allotments and performance of broiler breeders.* Poult. Sci. 74:247-260, 261-270.

Bartov, I. 1994. *Attempts to achieve low-weight broiler breeder hens by severe growth depression during various periods up to 6 weeks of age and food allocation below the recommendations thereafter.* Br. Poult. Sci. 35:573-584.

Bennett, C.D. and S. Leeson 1989. *Water usage of broiler breeders.* Poult. Sci. 68:617-621.

Bennett, C.D., S. Leeson and H.S. Bayley 1990. *Heat production of skip-a-day and daily fed broiler breeder pullets.* Can. J. Anim. Sci. 70:667-671.

Bowmaker, J.E. and R.M. Gous 1989. *Quantification of reproductive changes and nutrient requirements of broiler breeder pullets at sexual maturity.* Br. Poult. Sci. 30:663-675.

Brake, J., J.D. Garlich and E.D. Peebles 1985. *Effect of protein and energy intake by broiler breeders during the prebreeder transition period on subsequent reproductive performance.* Poult. Sci. 64:2335-2340.

Cave, N.A.G. 1984. *Effect of a high-protein diet fed prior to the onset of lay on performance of broiler breeder pullets.* Poult. Sci. 63:1823-1827.

Fancher, B.I. 1993. *Developing feeding programs for broiler breeder nutrition.* Poult. Digest, p. 18.

Fontana, E.A., W.D. Weaver and H.P. VanKrey 1990. *Effects of various feeding regimens on reproduction in broiler breeder males.* Poult. Sci. 69:209-216.

Harms, R.H. 1992. *A determination of the order of limitation of amino acids in a broiler breeder diet.* J. Appl. Poult. Res. 1:410-414.

Harms, R.H. and G.B. Russell 1995. *A re-evaluation of the protein and lysine requirement for broiler breeder hens. Poult. Sci.* 74:581-585.

Hocking, P.M. 1993. *Optimum size of feeder grids in relation to the welfare of broiler breeder females fed on a separate sex basis.* Br. Poult. Sci. 34:849.

Hocking, P.M. 1994. *Effects of body weight at photostimulation and subsequent food intake on ovarian function at first egg in broiler breeder females.* Br. Poult. Sci. 35:819.

Hocking, P.M., D. Waddington, M.A. Walker and A.B. Gilbert 1989. *Control of the development of the ovarian hierarchy in broiler breeder pullets by food restriction during rearing.* Br. Poult. Sci. 30:161-174.

Leeson, S., B.S. Reinhart and J.D. Summers 1979. *Response of White Leghorn and Rhode Island Red breeder hens to dietary deficiencies of synthetic vitamins. 1. Egg production, hatchability and chick growth.* Can. J. Anim. Sci. 59:561-567.

Lopez, G. and S. Leeson 1994. *Nutrition and broiler breeder performance. A review with emphasis on response to diet protein.* J. Appl. Poult. Res. 3:303-312.

Lopez, G. and S. Leeson 1995. *Response of broiler breeders to low-protein diets. 1. Adult breeder performance.* Poultry Sci. 74:685-695.

Lopez, G. and S. Leeson 1995. *Response of broiler breeders to low-protein diets. 2. Offspring performance.* Poultry Sci. 74:696-701.

Reis, L.H. 1995. *Extra dietary calcium supplement and broiler breeders.* Appl. Poult. Res. 4:276-282.

Samara, M.H. 1996. *Interaction of feeding time and temperature and their relationship to performance of the broiler breeder hen.* Poult. Sci. 75:34-41.

Spratt, R.S. and S. Leeson 1987. *Broiler breeder performance in response to diet protein and energy.* Poult. Sci. 66:683-693.

CHAPTER 7. FEEDING PROGRAMS FOR TURKEYS

The turkey continues to show improved market performance in terms of weight-for-age and feed efficiency. It was only about 20 years ago that tom turkey standards were 1 lb/week of age - today this is fast approaching 1 kg/week of age. Some of this improvement in body weight has been achieved through increased selection pressure for weight on female lines, and as will be discussed later, this has presented some unique problems in managing the turkey breeder hen.

As a greater proportion of turkey carcasses are used for further processing, then there has been a definite trend back to production of large-weight birds. However management of such birds is not without its problems, and as already described for broiler chickens, there is some potential for modification of the feeding program as a means towards alleviating such factors as skeletal abnormalities and undesired quantities of carcass fat. Of all the classes of birds discussed in this book, the turkey is perhaps changing the most rapidly. For this reason, the reader should be adaptable in selecting a diet program, and be prepared to continuously monitor its economic effectiveness.

7.1 MARKET TURKEY FEEDING PROGRAMS

Feeding programs for Large White turkeys must be adaptable in terms of perceived market goals. If the main criteria is maximum weight gain, then the ideal protein sequencing is one starting at 28% CP with gradual step-down to around 16% CP at market age. As with most classes of poultry, if protein/amino acid intake is adequate, then the bird is most responsive to energy. With a conventional step-down protein series, optimum weight of males is achieved with a series of diets ranging from 2900 - 3300 kcal ME/kg. Birds do respond well to low-energy diets, or more correctly, high-energy (3100 - 3300 kcal ME/kg) diets are less effective when the protein-amino acid ratio is not adjusted. As expected poorer performance with some high-energy programs is, therefore, a consequence of protein deficiency. Dietary specifications for market turkeys are shown in Table 7.1, and examples of some of these diets are shown in Table 7.2. A conventional program for toms and hens is subsequently shown in Table 7.3.

TABLE 7.1 Dietary specifications for commercial turkeys

	Starter		Grower		Developer	Finisher
	1	2	3	4	5	6
Approximate protein level (%)	28	26	23	21.5	18	16
Amino acids (% of diet)						
Arginine (%)	1.60	1.55	1.40	1.25	1.02	0.95
Lysine (%)	1.70	1.60	1.50	1.30	1.15	1.00
Methionine (%)	0.62	0.55	0.50	0.47	0.42	0.34
Methionine + cystine (%)	1.00	0.90	0.80	0.76	0.67	0.58
Tryptophan (%)	0.28	0.26	0.22	0.21	0.18	0.16
Histidine (%)	0.57	0.55	0.48	0.45	0.37	0.32
Leucine (%)	1.95	1.90	1.60	1.45	1.25	1.15
Isoleucine (%)	1.13	1.05	0.90	0.85	0.72	0.65
Phenylalanine (%)	1.03	0.95	0.82	0.79	0.67	0.60
Phenylalanine + tyrosine (%)	1.80	1.70	1.50	1.40	1.20	1.00
Threonine (%)	1.00	0.95	0.85	0.78	0.67	0.61
Valine (%)	1.20	1.15	1.00	0.90	0.75	0.65
Metabolizable energy (kcal/kg)	2900	3000	3050	3100	3200	3300
Calcium (%)	1.4	1.3	1.2	1.3	1.0	0.90
Available phosphorus (%)	0.7	0.6	0.5	0.6	0.5	0.40
Sodium (%)	0.18	0.18	0.17	0.17	0.17	0.17

	Starter (1–2)	Grower (3–4)	Developer/Finisher (5–6)
Vitamins (per kg of diet)			
Vitamin A (I.U.)	9500	8500	7000
Vitamin D_3 (I.U.)	2700	2400	2200
Choline equivalents (mg)	1900	1700	1500
Riboflavin (mg)	6.0	5.0	5.0
Pantothenic acid (mg)	17.0	15.0	15.0
Vitamin B_{12} (mg)	0.014	0.014	0.012
Folic acid (mg)	1.0	0.75	0.5
Biotin (mg)	0.25	0.20	0.15
Niacin (mg)	80.0	70.0	60.0
Vitamin K (mg)	2.0	2.0	1.5
Vitamin E (I.U.)	40.0	30.0	20.0
Thiamin (mg)	3.0	3.0	2.5
Pyridoxine (mg)	6.0	5.0	3.0
Trace minerals (per kg of diet)			
Manganese (mg)	80.0	80.0	80.0
Iron (mg)	110.0	110.0	110.0
Copper (mg)	10.0	10.0	10.0
Zinc (mg)	80.0	80.0	80.0
Selenium (mg)	0.3	0.25	0.015
Iodine (mg)	0.45	0.45	0.45

TABLE 7.2 Examples of turkey diets (kg)

	Starter		Grower		Developer	Finisher
	1	2	3	4	5	6
Corn	450	500	577	617	679	741
Soybean meal (48%)	385	336	265	222	180	146
Meat meal (50%)	50	50	100	100	50	30
Fish meal (60%)	50	50	---	---	---	---
Fat	21	30	29	29	55	46
Ground limestone	10	8	6	7	10	9
Calcium phosphate	20	12.5	10	12	13	14
Salt	2.0	2.0	2.0	2.0	2	2
Vitamin:Mineral premix[1]	10	10	10	10	10	10
Methionine	1.3	1.0	0.7	0.7	---	0.8
Calculated analysis						
Crude protein (%)	27.9	26.0	23.2	21.5	17.0	15.0
Digestible protein (%)	25.5	23.7	21.2	19.6	15.5	13.7
Crude fat (%)	4.4	5.5	5.6	5.8	8.5	6.4
Crude fiber (%)	3.5	3.4	3.4	3.4	3.4	3.4
Metabolizable energy (kcal/mg)	2893	3000	3066	3090	3250	3275
Calcium (%)	1.58	1.34	1.20	1.27	1.07	0.90
Av. Phosphorus (%)	0.85	0.69	0.63	0.66	0.50	0.46
Sodium (%)	0.18	0.18	0.18	0.18	0.17	0.16
Methionine (%)	0.61	0.56	0.50	0.47	0.78	0.69
Methionine+cystine (%)	1.05	0.96	0.85	0.79	0.64	0.56
Lysine (%)	1.77	1.62	1.41	1.28	0.89	0.75
Tryptophan (%)	0.37	0.34	0.30	0.27	0.22	0.19
Threonine (%)	1.18	1.10	0.99	0.92	0.87	0.64

[1]Use additional choline if premix does not contain this vitamin

TABLE 7.3. Turkey diet specifications according to age

Diet	Comments	CP	ME	Meth	M&C	Lys	Ca	Av.P
Starter 1	0-2 wks	28	2850	.70	1.1	1.70	1.4	.8
Starter 2	2-4 wks	26	2950	.62	1.0	1.60	1.3	.7
Grower 1	4-8 wks	23	3050	.60	.9	1.45	1.2	.6
Grower 2	8-12 wks	20	3150	.55	.82	1.30	1.1	.5
Developer 1	12-15 wks	18	3225	.50	.79	1.00	.9	.45
Finisher 1 ♀	12-16 wks	16	3300	.45	.65	.9	.9	.45
Finisher 2 ♂	15-18 wks	16	3350	.40	.62	.8	.8	.40
Finisher 3 ♂	15-20 wks+	16	3250	.38	.60	.75	.9	.40

a) Starter diets and early poult viability

Early poult viability continues to be a concern in the industry, and mortality rates of 1.5% in the first 7 d are still very common. A contributor to this problem is undoubtedly the severe processing conditions imposed on the poult, where beaktrimming, detoeing, vaccination and antibiotic injection can all be performed at the hatchery. These combined stresses must influence the bird's general metabolism, and in particular energy status has come under close scrutiny.

Because there are virtually no carbohydrates in the egg, then the poult is hatched with very low available energy reserves, and glycogen is most likely synthesized by gluconeogenesis from protein. Attempts at improving glycogen/energy reserves in the poult have generally had little beneficial effect. In the past, poults have been given glucose solutions prior to transport from the hatchery. Recent data suggests that while this may have very short term benefits (2 - 3 hrs maximum) the glucose uptake likely suppresses key enzymes, so suppressing glycogen synthesis, and so can be detrimental to subsequent longer-term health status. Injecting alanine, a non-essential amino acid, has been shown to elevate blood glucose levels without the concomitant reduction in glycogen reserves. However the long term benefits with alanine injection are difficult to quantitate, even in terms of 7 d mortality. Propionate is also a precursor of glucose, and has been fed to young poults. However propionate is also an anorexic agent, and so this is counterproductive to ensuring long-term benefits for the poult.

There has also been concern about the vitamin E status of the young poult. Vitamin E levels in the liver and serum of poults reach alarmingly low levels 2 - 3 weeks after hatch. For example, while poults may show 80 μg Vit E/g liver at hatch, the normal levels at 21 d are closer to 0.5 μg/g. Sell and co-workers at Iowa State have investigated this problem, and while significant treatment differences are sometimes seen, normal blood and liver values are still in the order of magnitude as described previously. For example, feeding the medium chain fatty acids as found in coconut oil, rather than tallows or even sucrose, does seem to change vitamin E status, yet after 21 d, liver levels are still less than 1 μg/g tissue. It therefore seems very difficult to stop this "natural" decline in vitamin E status, and obviously the poults immune status is being questioned relative to these changes. Because vitamin E plays a number of roles in the body, it is possible that fat levels and fat oxidation may influence general health status of the young poult. However, adding antioxidants has not been beneficial. Likewise adding bile salts to the diet does little to improve Vitamin E status of the poult, and so absorption *per se* is not thought to be a limiting factor.

b) *Heavy turkey programs*

As genetic potential of the bird continues to increase, so there is the possibility of increasing market weight and/or market age. Most of our modern strains of turkey are now capable of significant daily gains through to 20 - 22 kg live weight. At this end of the spectrum, nutritional programs aimed at sustaining skeletal integrity and manipulating the balance of carcass fat: protein become more critical. As a generalization, the young turkey is most responsive to amino acids, while economic growth of the larger bird is more related to energy intake. Over the last few years, there has been a major changes in the type of turkeys available for production, although today this difference is less evident as all breeders strive to aim for larger birds with older market ages. This later maturing type of turkey, which has been very common in Europe for many years, is now becoming the standard "type" worldwide, and so today there is less emphasis on "strain-specific" feeding programs. However, even with this common goal, there are still some differences in growth characteristics of certain strains, as recently shown by Lilburn and co-workers (Table 7.4).

TABLE 7.4 Growth characteristics of two commercial strains of Large White Turkey

	Strain 1	Strain 2
Age at reduction in linear growth (d)	99	89
B. Wt at time of reduced growth (kg)	10.5	9.3
Age at 90% of asymptote growth (d)	136	122
B. Wt at asymptote (kg)	28.5	25.3

Adapted from Lilburn and Emmerson (1993)

Even though both strains can conceivably be grown to 20 kg liveweight, they do show differences in apparent maturity of growth. Strain 2 would be expected to deposit excess fat at later ages, and this fact should be accommodated in the late grow-out feeding schedule.

In diets composed essentially of corn and soybean meal, methionine (and/or TSAA) is likely to be the most limiting amino acid. Requirements for methionine will obviously vary with energy level of the diet, although it is possible to make general recommendations of around 2.5, 2 and 1.5 mg methionine per kcal ME for starter, grower/developer and finisher diets respectively. With later maturing turkey strains that are now used almost exclusively, higher levels of lysine seem beneficial. Lysine levels are therefore around 5, 4 and 2.5 mg/kcal ME for starter, grower and finisher diets respectively. Most nutritionists consider the turkey to be very responsive to lysine levels, although as a percentage of crude protein the levels used in practice are little different than for other meat birds. There is some evidence to suggest that toms and hens are not too responsive to higher levels of lysine assuming balanced protein is being used, although there is some indication of strain differences in this respect. Hurwitz and co-workers have developed a model for determination of amino acid requirements of the growing turkey. In corn-soy or milo-soy diets, methionine is obviously the first limiting amino acid, although after this, lysine does not seem to be second limiting for 1 - 4 week old poults. At these ages, a low-protein diet is required to obtain a response to lysine. A significant response to lysine was shown in protein adequate diets after 16 weeks of age. The model of Hurwitz indicates TSAA requirement for 1 and 5 week old poults at 3.4 and 3.1 mg/kcal ME respectively. For lysine, corresponding values are 4.8, 2.7 and 2.2 mg/kcal for 1, 12 and 16 week old birds.

Traditionally, the concept of optimum energy:protein has been considered for most classes of turkeys. Recent evidence suggests that this concept is no longer applicable, or at least not always economically viable. Sell and co-workers in an extensive series of studies, working with growing tom turkeys from 9 - 20 weeks of age, concluded that increasing CP or ME improved weight gain and feed:gain, but that the CP effect was independent of ME. It is perhaps pertinent that the energy response in this work in fact relates to added fat. Increasing the energy concentration of the diet reduced the quantity of protein consumed per kg of body weight gain, although it had no effect on ME consumed per kg of gain. Increasing the protein content of the diet reduced protein efficiency in relation to gain, although efficiency for ME was improved. Interestingly, these changes in diet specification had little effect on carcass composition. These workers conclude that optimum CP:ME as a constraint in formulation may be inappropriate, and that it may be better to consider independent effects for both protein and energy.

In meeting the nutrient requirements of the bird, changes in diet specification with age are obviously a compromise in attempting to accommodate reduced requirements of older birds. With a 16 - 20 week growing period, the potential for diet change is much greater, although they must obviously be balanced against practical considerations of feed manufacture and on-farm handling. Salmon *et al.* (1982) suggest that frequency of diet change has little effect on rate of growth as long as protein intake is adequate. With adjustment each 3 weeks, optimum growth to 12 weeks was recorded with diet protein levels of 29.5, 27.4, 25.3 and 24.2% respectively. However, with a single diet to 12 weeks, comparable growth was obtained only with at least 27.4% CP. Comparable values from 12 to 24 weeks at 3 week intervals, were 19.8, 17.9, 15.8 and 15.0%. Again, when these diets were fed singly to 24 weeks, then comparable growth to the above system was attained only with 19.8% crude protein. To some extent, these values argue against any form of compensatory growth as will be discussed later.

If diet changes occur each 2 weeks up to 10 weeks of age, then optimum levels are 29.3, 29.0, 27.4, 25.8 and 24.8% respectively. From 10 - 22 weeks, corresponding optimum values are 22.5, 21.8, 15.6, 13.9 and 13.9% CP respectively. The fewer diets that are used therefore, the greater the need to over-feed protein to meet requirements during the bird's "youngest" time period that it consumes the feed. It is interesting to note that such "overfeeding" of protein has no effect on carcass composition, and one must therefore suspect a relatively efficient utilization of protein for energy purposes.

The use of fats in diets for turkeys has always been a controversial topic, and certainly one that has received considerable attention in recent years. In many instances, research protocols fail to differentiate between the effects of fat and energy. Considering the dominant role that energy plays in controlling growth, it is perhaps not too surprising that turkeys respond to supplemental dietary fat. At fixed energy levels, there is often improvement in feed:gain with added fat, and this effect increases with increase in age of the bird. From 0-20 weeks, F:G is improved by about 1.5% for each 1% added fat. From 12 - 20 weeks, a corresponding value of 3.5% is seen. It is often noted that if fat is removed from the diet of older birds, then any improvements to that time are often lost. These data suggest little return in use of fat for young birds, and that economic response is maximized after 8 weeks of age. The age response is likely a reflection of digestibility of saturated fatty acids. Earlier work of Sell and co-workers indicates 17 - 20% improvement in digestibility of fat by 9 vs 3 week old toms. Similarly, we have recently shown around 65% digestion of tallow with 10 d old poults vs 89% for corn oil. By 4 weeks of age, corresponding values of 78 vs 95% were recorded. If fats are used with young poults, then unsaturated sources seem beneficial. Another beneficial effect of adding fat to the diet is stimulation of feed intake. With isoenergetic diets containing added fat, there are reported increases in feed intake of 1.7, 1.8 and 2.2% during the 8 - 12, 12 - 16 and 16 - 20 week periods.

Owings and Sell (1982) tested diets with 0 - 8% added fat, to give diets from 2800 - 3100 kcal ME/kg for 0 - 21 d old poults. A 1.7% improvement in weight was recorded for each 1% additional fat, with this effect independent of energy:protein. Consequently, comparable responses were recorded for adjusted and non-adjusted protein levels. These data confirm the previous suggestion that the birds' responses to protein and energy are independent. The same workers conclude that 0.6-1.5% improvements in 140 d weight per 1% added fat are independent of the ratio of energy to CP or amino acids, and that there is little advantage to pelleting of the diet under these conditions. Hurwitz *et al.* (1983) indicate that with very young poults, weight gain is linearly related to diet energy up to 2750 kcal/kg and that after this level the response is still positive, but less pronounced. With higher energy contents, growth improved by 1.66% per 100 kcal increase in energy content of the diet. As the bird gets older, then the break-point in response to fat moves from 2750 to 2900 kcal/kg.

The turkey's response to energy is to some extent influenced by environmental temperature. Maximum weight gain for market weight birds is achieved at 10 - 18°C. At 27°C and 35°C, gain is reduced by about 6% and 12% respectively. Feed:gain is improved by 1.2% for each + 1°C up to 27°C. It is generally recognized that amino acid levels in the diet should increase as temperature increases, because feed intake will likely decline. However, Hurwitz *et al.* (1980) give evidence for a non-linear relationship between amino acid requirements and environmental temperature when expressed in terms of mg/cal ME. These workers suggest a maximum need at 26 - 28°C, followed by a decline caused by an increase in the bird's energy requirement, presumably related to heat dissipation. In noting the bird's response to energy at 10°C or 27°C, Hurwitz and co-workers suggest that the energy contribution of fat inclusive of the "extracaloric" response is not temperature dependent. However, in studies involving isoenergetic addition of fat, there is evidence to suggest that the growth response to fat *per se* is obtained only at higher environmental temperatures. This finding confirms previous suggestions that the extraenergetic effect of fat is most pronounced in hot weather. When the energy level of the diet is increased by use of carbohydrate (glucose), then greater responses are seen at low rather than high temperatures. Hurwitz concludes that in diet formulation, no differentiation is made in the energy values of feedstuffs to be fed in hot or cold environments. Their results suggest that while this may be valid for ingredients of high-fat content, values assigned to carbohydrate-rich ingredients (cereals?) are dependent upon the environment in which they will be used.

c) Compensatory growth

The type of program previously shown in Table 7.3, involves conventional diets where growth rate is maximized at all ages. As previously described for broiler chickens (Chapter 5), there is potential for compensatory gain where a slower initial growth rate is accepted (Table 7.5).

TABLE 7.5 Compensatory growth program

Diet	Toms	Hens	CP	ME	Meth	M+C	Lys	Ca	Av. P
	(wks)		(%)	(kcal)	(%)	(%)	(%)	(%)	(%)
Starter 1	0-3	0-3	25	2850	.55	.90	1.35	1.4	.7
Starter 2	3-6	3-6	23	3000	.50	.80	1.2	1.3	.65
Grower 1	6-10	6-9	21.5	3100	.47	.76	1.1	1.2	.6
Grower 2	10-13	9-12	20	3175	.45	.72	1.0	1.0	.5
Developer	13-15	12-14	18	3250	.40	.64	.9	.9	.45
Finisher	15+	14+	16	3275	.35	.56	.8	.8	.40

The diets used in such a program can be those used in conventional programs, albeit used at different ages. However care must be taken that low-protein starter diets *per se* are used, rather than conventional grower diets, since the latter do not usually contain adequate levels of calcium and available phosphorus needed for early skeletal development. With lower-protein starter diets, there is a suggestion that amino acid levels other than methionine and lysine should be more closely studied. With starter protein levels as low as 22% CP, threonine and valine levels may be equally as important as lysine, and interaction among branched-chain amino acids may be problematic. For example, high levels of leucine seem to cause growth depression in low-protein starter diets, and this effect is only partly alleviated by additions of valine. Such data suggest caution in the use of high-leucine ingredients such as corn gluten or blood meal in compensatory growth-type feeding programs. Confirming such data, Jackson and Potter (1984) indicate that low protein starter diets containing in excess of 1.9% leucine will adversely affect growth, likely via an induced deficiency of valine. These workers conclude that with a 22% CP starter diet for poults, assuming adequate methionine fortification, lysine is the most limiting amino acid - however valine adequacy may decline as leucine or isoleucine levels increase.

More recent studies on compensatory gain have generally failed to show any distinct advantage in terms of overall feed usage or cost/kg gain, and in fact complete growth compensation to a specific age is sometimes not realized. For example, feeding low-protein starter diets for just the first 3 weeks, followed by normal diets, has been shown detrimental to 18 week body weight. It seems as though the later maturing strains common today are not ideal candidates for compensatory growth. Their genetically inherent slower initial growth in effect parallels the concept of compensatory growth, and it appears that these birds are unable to fully recover from a period of early undernutrition. For market ages of 16 - 20 weeks compensatory growth therefore has limited application. The concept will likely be re-visited as market weights increase further.

d) Diet self-selection

As with most other classes of poultry, the turkey seems able to balance its own nutrient intake when offered a selection of different diets or a range of individual ingredients. However, results from studies with turkeys are encouraging in that leaner carcasses are produced with this type of system. Following are selected data from our studies conducted some 10 years ago. In this study, turkeys were offered a series of conventional diets, or simultaneous access to two diets that were essentially corn or soybean meal balanced with respect to amino acids (as % CP) and minerals and vitamins (Table 7.6).

TABLE 7.6 Body weight of turkeys allowed diet self selection expressed as percentage of weight achieved by control fed birds

	Weeks of age				
	4	8	12	16	19♀ 21♂
Male	89	92	102	106	103
Female	89	100	95	99	99

Turkeys eating from a self-selection system were as heavy as conventionally fed birds, with no difference in feed efficiency. However, these test birds were leaner (Table 7.7) which was a consequence of greater protein and/or lower energy intake.

TABLE 7.7 Eviscerated carcass composition of male and female turkeys reared with diet self-selection

	% DM	Fat (% DM)	Protein (% DM)
Male			
Control	33.7	26.0a	58.6b
Self-selection	32.7	25.3a	62.8c
Female			
Control	35.6	35.2b	53.3a
Self-selection	34.7	31.2a	56.2ab

Although such a feeding system presents obvious technical difficulties at the farm level, it does offer a very simple one-stage diet program that warrants further attention in view of the experimental results seen with improved lean meat yield.

e) Broiler Turkeys

There has been little work conducted recently on the nutrient requirements of turkey broilers, and in large part, this is a reflection of the economic importance of this bird. In general, it seems as though nutrient requirements are quite similar to the Large White strain, with perhaps earlier introduction of a higher energy finisher diet 2-3 weeks prior to expected market age of around 10 - 13 weeks. Moran indicated optimum growth with Small White turkeys using a 32-30-28-26% CP diet series at 2920 kcal ME/kg, with diet change occurring each 2 wks. Hulan and Proudfoot (1981) tested the turkey broiler response to a range of protein levels in the grower and finisher stages with diet changes from starter to grower at 14 or 21 d and grower to finisher at 56 or 70 d. Optimum biological and economic response were obtained with birds reared on a 24% CP, 3100 kcal ME/kg grower, changed from starter at 21 d of age. Earlier moves to either grower or finisher resulted in smaller weight birds. These workers indicated that for each 1% increase in protein level of the finisher diet, there was an improvement in body weight of 0.15 kg for males and 0.07 kg for females. They indicated that the finisher diet should contain at least 16% CP.

Salmon (1986) indicated that decreasing the ratio of energy to protein in the diet, by increasing diet protein, resulted in increased growth rate and improved feed utilization. This data is consistent with the hypothesis previously outlined for Large White birds, suggesting that effects of protein and energy are independent and that less emphasis should be placed on energy:protein during formulation. An example of a feeding program for turkey broilers is shown in Table 7.8.

TABLE 7.8 Feeding program for turkey broilers (% or kcal/kg ME)

Diet	Comments	CP	ME	Meth	M+C	Lys	Ca	AvP
Starter	0-2 wks	29	2850	.62	1.0	1.8	1.4	.8
Grower 1	2-4 wks	26.5	3000	.55	.9	1.5	1.3	.7
Grower 2	4-8 wks	23	3200	.50	.8	1.35	1.2	.6
Developer	8-2 wks prior to marketing	19	3250	.45	.75	1.2	1.1	.5
Finisher	Last 2 wks	17	3300	.32	.55	1.0	.9	.45

In practice, females of the heavy strain are often used for the broiler market. In this situation, if there are grading restraints regarding fat cover, then it may be necessary to use even higher levels of energy for the finisher diets. Compensatory gain programs do not seem to be economical for the turkey broiler, and methionine and lysine levels of starter and grower are often the major factors influencing the ability to finish the bird at desired weight/age.

f) *Flushing syndrome and turkey "knockdown"*

Two reoccurring problems in the industry are so-called flushing syndrome, which appears as a diarrhea, and turkey knockdown which disables older turkeys. Both problems may have a nutritional component, although it is obvious that other, as yet unknown, factors are also involved.

As its name implies, flushing syndrome is characterised by wet, runny excreta that is seen in commercial flocks from 6 - 14 weeks of age, although most commonly is centred in the 8 - 12 week period. Because this timing coincides with removal of anticoccidials from the diet, there has been speculation about associated changes in intestinal microflora. The associated problem is development of wet litter and potential for increase in leg disorders and breast blemishes. Wetter excreta can be caused by high levels of minerals, and especially salt and also by excess protein which both relate to increased water consumption. However the flushing syndrome is associated with a "sticky" type of excreta, whereas extra salt and protein usually cause watery and urate-dense excreta respectively.

Over the past few years, we have been testing various diet situations that influence cecal activity and consistency of cecal and regular excreta. Surprisingly, diet fiber level and source had little effect on cecal and excreta appearance. The occurrence of flushing at 8-12 weeks coincides with increase in fat content of the diet, and so various levels and sources of fat, and also fats with various degrees of rancidity were tested, again without any consistent effect. The only diet nutrient that consistently affected the degree and consistency of cecal excreta was copper. Adding 500 g copper sulphate/tonne feed resulted in greater cecal evacuation and the cecal excreta was of much greater viscosity. Cecal excreta contained as much as 14,000 ppm copper. Certainly not all turkeys are fed additional copper sulphate, although it does appear to contribute to abnormal excreta consistency. Other attempts at diet modification to treat or prevent flushing syndrome have generally met with little success. There are some reports of benefit to adding 2 kg betain/1000 litres of drinking water.

Turkey knockdown also occurs at around 10-14 weeks of age where affected birds are unable to stand or walk. The condition resembles ionophore toxicity, but this has largely been ruled out as a single causative factor. Knockdown is most severe when there is any on-off lighting program and turkeys are seen to "gorge" feed when lights are turned on after a prolonged period of darkness. This latter situation is especially relevant when turkeys have limited access to water. Because of the implications of large feed intake over a short period of time, associated high intakes of ionophores have been suspected. However under controlled studies when turkeys were encouraged to gorge on feed containing even 140 ppm monensin, no knockdown was observed.

g) Carcass composition

With an increased proportion of large turkey carcasses being cut-up or further-processed, there is a continued demand for information on carcass yield and composition of turkeys. Due to dramatic changes in weight-for-age of the turkey, information regarding carcass composition for specific ages of bird becomes virtually redundant overnight. For this reason, we have presented the following carcass data based on expected changes based on age of turkey hens and toms (Table 7.9). In order to use this information, the coefficients for a particular parameter are multiplied by "age in days" and "(age in days)^2."

eg: % total viscera for a 100 d tom is calculated as

$$\begin{aligned} &17.87 - (0.158 \times 100) + (.00054 \times 100^2) \\ &= 17.87 - 15.8 + 5.4 \\ &= \underline{7.47\%} \end{aligned}$$

Similarly , Table 7.10 gives expected changes in cut-up yields for toms and hens.

TABLE 7.9 Relationships between turkey age (days) and body weight and organ proportions. For each parameter the first line represents toms, the second line hens

	Constant	±	Age	±	Age^2	Coefficient significance	
						Age	Age^2
As % of body weight:							
Total viscera ♂	17.870	-	.158	+	.000544	**	**
♀ etc.	18.696	-	.180	+	.000772	**	**
Liver	2.792	-	.0204	+	.0000673	**	**
	3.023	-	.0247	+	.0000824	**	**
Heart	.798	-	.00653	+	.0000267	**	**
	.766	-	.00681	+	.0000284	**	**
Alimentary tract	14.168	-	.129	+	.000400	**	**
	14.924	-	.152	+	.000548	**	**
Gizzard+proventriculus	5.089	-	.0284	+	.0000258	**	NS
	5.455	-	.0458	+	.000134	**	**
Alimentary tract (cm/100 g body weight)	48.064	-	.899	+	.00387	**	**
	49.135	-	.906	+	.00396	**	**
Composition data							
Viscera fat, % DM	21.363	-	.193	+	.00214	**	**
	18.719	+	.0529	-	.00161	NS	**
Carcass fat, % DM	26.040	-	.310	+	.00248	**	**
	25.399	-	.188	+	.00226	**	**
Viscera CP, % DM	62.719	+	.0561	-	.00111	**	NS
	64.659	-	.116	-	.000828	*	**
Carcass CP, % DM	62.270	+	.189	-	.00164	**	**
	63.811	+	.0962	-	.00156	NS	**
Viscera DM, %	25.295	+	.0392	+	.000298	**	NS
	23.187	+	.123	+	.000357	**	*
Carcass DM, %	23.416	+	.124	-	.000313	**	**
	22.272	+	.164	-	.000250	**	NS
Carcass fat, g	62.990	-	5.309	+	.0929	**	**
	-55.970	+	.749	+	.0596	NS	**
Carcass CP, g	-103.10	+	8.738	+	.0357	**	**
	-122.46	+	10.704	-	.00863	**	NS
Viscera fat, g	3.516	-	.421	+	.0173	NS	**
	-26.375	+	.737	+	.0182	NS	**
Viscera CP, g	-32.15	+	3.478	-	.00524	**	**
	-21.23	+	2.806	-	.00733	**	**
Total body fat, g	66.506	-	5.729	+	.110	**	**
	-82.344	+	1.486	+	.0778	NS	**
Total body CP, g	-135.269	+	12.216	+	.0305	**	**
	-143.692	+	13.510	-	.0160	**	NS
Total body fat, %	5.708	-	.0565	+	.000698	**	**
	4.704	+	.00588	+	.000724	NS	**
Total body CP, %	15.438	+	.102	-	.000545	**	**
	15.250	+	.0996	-	.00577	**	**
Viscera fat as % body fat	41.123	-	.323	+	.00122	**	**
	41.383	-	.290	+	.00127	**	**

* P<.05, ** P<0.01, NS, Not significant

TABLE 7.10 Relationships between turkey age (days) and body weight and organ proportions. For each parameter the first line represents toms, the second line hens

	Constant	±	Age	±	Age²	Coefficient significance	
						Age	Age²
% Neck ♂	9.126	-	.0267	+	.000121	**	*
♀ etc.	9.141	-	.0199	+	.0000192	*	NS
% Drumsticks	14.498	+	.00183	-	.000115	*	NS
	14.158	-	.0108	-	.0000366	NS	NS
% Thighs	16.013	-	.0194	+	.000104	NS	NS
	16.471	-	.0287	+	.000183	**	**
% Wings	14.852	+	.0458	-	.000493	**	**
	16.067	+	.00711	-	.000285	**	NS
% Back	18.398	-	.113	+	.000468	**	**
	19.203	-	.131	+	.000652	**	**
% Breast	26.653	+	.1233	-	.000167	**	NS
	26.525	+	.172	-	.000502	**	**
% Yield	58.260	+	.352	-	.00123	**	**
	55.932	+	.426	-	.00170	**	**

*** P<.05, ** P<0.01, NS, Not significant**

Nutrition and feeding management can play a significant role in attempting to meet the processors demand for leaner turkey carcasses/meat. While previous discussion has detailed the importance of energy and protein, rather than energy:protein in terms of conventional growth parameters, one must be aware of the importance of this balance in finisher diets as it influences carcass fat deposition. There seems little doubt that the turkey responds in a classical manner in this respect, in that wider energy:protein will lead to increased carcass fat and *vice versa*. In reducing the ratio of energy:protein, one obviously has the option of increasing protein/amino acids in relation to energy, or reducing energy while maintaining normal protein levels. Diet self-selection for turkeys as previously discussed, also offers some interesting possibilities for manipulating carcass composition.

7.2 TURKEY BREEDER

a) Hens

Over the last few years, there has been an increase in selection pressure for growth in the breeder hen. To some extent, reproductive characteristics have been "neglected", meaning that managers are faced with new problems in hen management. To indicate the type of change that has occurred in breeders, one only has to review research data that is 10 years old and compare this with current achievements. Review articles written 10 years ago, most often, showed no benefit to reduced nutrient intake of growing turkey hens while today, for some strains, it is the only way to realize optimum reproductive performance. Diet specifications for turkey breeders are shown in Table 7.11, while Table 7.12 gives examples of some diets.

TABLE 7.11 Dietary specifications for breeder turkeys

	Holding diet	Breeder diets			Tom	Force-molt	
		1	2	3		1	2
Approximate protein level (%)	10.5	16.0	15.0	14.0	13.0	17.0	12.0
Amino acids (% of diet)							
Arginine	0.47	0.95	0.85	0.70	0.62	0.95	0.55
Lysine	0.45	0.85	0.77	0.60	0.58	0.85	0.52
Methionine	0.17	0.34	0.30	0.23	0.21	0.38	0.21
Methionine+cystine	0.31	0.58	0.53	0.44	0.38	0.62	0.37
Tryptophan	0.08	0.15	0.14	0.13	0.11	0.15	0.10
Histidine	0.12	0.30	0.28	0.25	0.20	0.31	0.15
Leucine	0.55	1.10	0.90	0.82	0.70	1.05	0.65
Isoleucine	0.35	0.65	0.60	0.55	0.42	0.65	0.40
Phenylalanine	0.25	0.60	0.50	0.42	0.35	0.55	0.30
Phenylalanine+tyrosine	0.50	1.00	0.90	0.80	0.70	1.00	0.60
Threonine	0.40	0.60	0.57	0.55	0.50	0.60	0.45
Valine	0.40	0.65	0.61	0.58	0.52	0.58	0.45
Metabolizable energy (kcal/kg)	2850	2800-3100	2800-3000	2800-3000	2750	2750	2800
Calcium (%)	0.90	2.5	2.7	2.9	0.9	0.9	0.9
Available phosphorus (%)	0.40	0.45	0.40	0.35	0.40	0.35	0.35
Sodium (%)	0.17	0.17	0.17	0.17	0.17	0.17	0.17
Vitamins (per kg of diet)							
Vitamin A (I.U.)			8000			6000	
Vitamin D_3 (I.U.)			3500			2500	
Choline equivalents (mg)			1100			800	
Riboflavin (mg)			6.0			4.5	
Pantothenic acid (mg)			17.0			13.0	
Vitamin B_{12} (mg)			0.014			0.01	
Folic acid (mg)			1.0			0.75	
Biotin (mg)			0.3			0.1	
Niacin (mg)			70.0			60.0	
Vitamin K (mg)			3.0			1.5	
Vitamin E (I.U.)			40.0			20.0	
Thiamin (mg)			3.0			2.5	
Pyridoxine (mg)			4.0			3.0	
Trace minerals (per kg of diet)							
Manganese (mg)			80.0			80.0	
Iron (mg)			80.0			80.0	
Copper (mg)			10.0			10.0	
Zinc (mg)			80.0			80.0	
Selenium (mg)			0.30			0.30	
Iodine (mg)			0.45			0.45	

TABLE 7.12 Examples of turkey breeder diets (kg)

	Holding diet			Tom	Force-molt	
		1	2		1	2
Corn	695	711	752	580	538	558
Wheat shorts	250	---	---	300	236	210
Soybean meal (48%)	---	200	152	58	172	180
Fat	10	10	10	16	10	10
Ground limestone	14	55	62	16	14	14
Calcium phosphate	16	11	11	16	16	16
Salt	2	2	2	2	2	2
Vitamin:Mineral premix[1]	10	10	10	10	10	10
Methionine	---	0.2	---	---	1	---
Lysine HCl	2	---	---	1.2	---	1.5
Calculated analysis						
Crude protein (%)	10.4	16.0	14.1	13.0	17.0	12.0
Digestible protein (%)	9.0	14.6	12.8	11.3	15.1	10.4
Crude fat (%)	5.0	4.0	4.2	5.4	4.4	4.5
Crude fiber (%)	4.4	3.4	3.3	4.7	4.5	5.1
Metabolizable energy (kcal/kg)	2860	2933	2947	2762	2748	2810
Calcium (%)	0.9	2.7	3.0	1.0	0.93	0.92
Av. Phosphorus (%)	0.44	0.45	0.44	0.45	0.46	0.44
Sodium (%)	0.15	0.17	0.17	0.15	0.15	0.15
Methionine (%)	0.19	0.30	0.26	0.22	0.38	0.21
Methionine+cystine (%)	0.31	0.53	0.45	0.38	0.62	0.37
Lysine (%)	0.45	0.82	0.67	0.57	0.80	0.56
Tryptophan (%)	0.12	0.21	0.18	0.16	0.22	0.15
Threonine (%)	0.40	0.68	0.60	0.50	0.68	0.46

[1]**Use additional choline if premix does not contain this vitamin**

As with any female bird, body weight and condition at maturity seem to be the key to successful reproductive performance. For most strains of Large White hens, it is no longer possible to use commercial-type meat bird rearing programs, because birds become overweight within the first 4 - 6 weeks of growth. The type of program as shown in Table 7.13 has proven successful in the rearing of most strains of turkey hens. If reduced early growth rate is required, then it is tempting to start birds on lower-nutrient dense grower diets. This concept has been used successfully with broiler breeder pullets. However, for the young poult we are faced with large changes in specification of many nutrients in grower compared to starter diets. This is most critical with levels of calcium and phosphorus, and so a common consequence of starting poults on grower diets, is the occurrence of rickets. If slower early growth is required in breeder hens, then it is necessary to formulate specific low protein/energy diets.

TABLE 7.13 Feeding program for Large White turkey breeders

	Diet	Age (Wks)	CP (%)	ME (kcal)	Meth (%)	M+C (%)	Lys (%)	Ca (%)	AvP. (%)
Large White Breeder ♀	Starter	0-2	28	2750	.55	.9	1.4	1.4	.7
	Grower 1	2-4	24	2800	.50	.83	1.2	1.3	.6
	Grower 2	4-8	21	2800	.42	.7	1.1	1.1	.55
	Developer 1	8-12	18	2850	.38	.65	1.0	1.0	.45
	Developer 2	12-16	16	2900	.35	.6	.9	.9	.40
	Holding	16-27	14	2950	.30	.5	.8	.9	.38
	Prebreeder	27-31	15	3000	.30	.45	.7	.9	.4
	Breeder 1	31-40	16	3000	.35	.55	.9	3.0	.45
	Breeder 2	40->	15	2900	.30	.5	.8	2.5	.40
Large White Breeder ♂	Starter	0-2	28	2750	.55	.9	1.6	1.4	.7
	Grower 1	2-4	24	2800	.50	.75	1.4	1.3	.6
	Grower 2	4-8	20	2800	.40	.65	1.0	1.1	.55
	Developer 1	8-12	17	2800	.32	.55	.8	1.0	.45
	Developer 2	12-16	14	2900	.30	.45	.7	.9	.40
	Holding-Breeder	16->	10	2950	.21	.40	.5	.9	.38

The key to successful rearing of the breeder hen is in monitoring of body weight, and scheduling diets according to weight-for-age, similar to the concept previously described for Leghorn pullets (Chapter 3). Depending upon individual flock circumstances, managers should be flexible in diet selection within a program (Table 7.13). For example, if hens do not gain weight for two weeks, then it may be necessary to feed a higher protein diet until desired weight-for-age is achieved. Due to its large body size, the breeder hen has a very large energy maintenance requirement. It is for this reason that the bird is greatly influenced by changing environmental conditions. For example, a small Leghorn bird is expected to be slightly smaller when grown in hot vs cool environments. However, the same environment for the turkey breeder hen can mean the difference between the need to restrict feed vs the need to stimulate growth. Managers should be acutely aware of this effect and be prepared for flexibility in feeding programs in summer vs winter conditions.

If growth rate of the breeder hen continues to increase, then some form of physical feed restriction may become necessary in order to realize optimum weight and condition at maturity. In recent studies (Miles and Leeson, 1990), we have shown that daily restriction is the most practical way of controlling growth (Table 7.14). In this study, hens were fed a conventional series of diets *ad-libitum,* or by skip-1 or skip-2 days per week, or by physical daily restriction to 85% of ad-lib intake. While all systems of feed restriction resulted in lower body weight, the greatest degree of control was achieved with daily physical restriction. Daily feed restriction also resulted in a significant reduction in carcass fat content of these hens at 29 weeks of age. When hens from each treatment group were monitored through a laying cycle, most settable eggs were obtained from the group restricted to 95% of ad-lib intake during rearing. It was interesting to note that these birds did not exhibit the characteristic decline in feed intake immediately following the move to the breeder facilities and light stimulation. If hens are to start producing eggs at around 32 - 33 weeks of age, then it seems necessary to induce a partial molt at 24 - 25 weeks of age, and to light stimulate all birds around 30 weeks of age.

TABLE 7.14 Effect of feed restriction on body weight of turkey breeder hens (kg)

Treatment	Weeks of age				
	4	8	16	24	29
1. Ad-lib	1.06a	3.9a	7.4a	9.6a	10.7a
2. Skip-1 day/wk	1.10a	3.1b	6.8bc	9.0b	10.2b
3. Skip-2 day/wk	1.09a	3.0b	6.7bcd	8.8b	9.9b
4. [1]95% ad-lib	0.96b	3.1b	6.9b	8.8b	10.0b
5. [2]90% ad-lib	1.07a	3.1b	6.7cd	8.4c	9.3c
6. [3]85% ad-lib	1.09a	3.1b	6.5d	8.2c	9.1c

[1]95% starting at 2 weeks age; [2]95% to 8 weeks, then 90%; [3]95% to 8 weeks, 90% to 12 weeks, then 85% of ad-lib. Skip-1-day at 2 weeks; skip-2-days at 8 weeks
a-d means followed by different letters are significantly different

Such a molt is best achieved with a sudden reduction in daylength from 14 - 16 hours down to 6 - 8hours for a 6 - 8 week period. During this molt, hens should ideally lose all primary wing feathers, although in practice, the 10th primary is often carried over. During this period of molt, hens will invariably eat some of the dropped feathers, especially if no straw is used as litter. This behaviour seems normal, and is not indicative of a sulfur-amino acid deficiency. Obviously, birds should be fed insoluble grit at this time.

The use of a pre-breeder diet is open to some debate. The type of high nutrient dense pre-breeder shown in Table 7.13. is often used on the assumption that it will advantageously pre-condition the bird immediately prior to lay. This may be true if birds are underweight at this age, due to poor rearing management. However, for birds of ideal weight and condition, there seems no advantage to using pre-breeder diets. With other classes of stock, pre-breeder diets are often used in an attempt to stimulate medullary bone development as a prerequisite to shell calcification. However in relation to shell output, the turkey consumes significant quantities of calcium from breeder diets, and so there is likely less emphasis on medullary bone calcium. The turkey hen does however exhibit an unusual pattern of feed intake in relation to egg production (Fig 7.1.).

As the hen increases her egg production up to peak, she consumes a diminishing quantity of feed. In extreme cases, she may reduce feed intake from a peak of 300 g/d down to close to 200 g/d. At the same time, she is obviously increasing egg mass output, and so she will be in negative energy balance in terms of feed input:egg output. Such imbalance is obviously accommodated for by loss in body weight (Fig 7.1). The ability to lose up to 1 kg of body weight seems critical for optimum egg output, and this emphasizes the reasoning behind the importance of body weight and condition of the hen at 30 weeks of age. Unless nutritionists are aware of this inherent problem in breeder hens, and of the importance of tailoring diets according to mature weight, then variable responses to diets can be recorded. It is suggested that such lack of understanding is largely responsible for the vast range of nutrient specifications currently used for turkey hens.

Fig. 7.1 Feed intake pattern of adult breeder turkey hens.

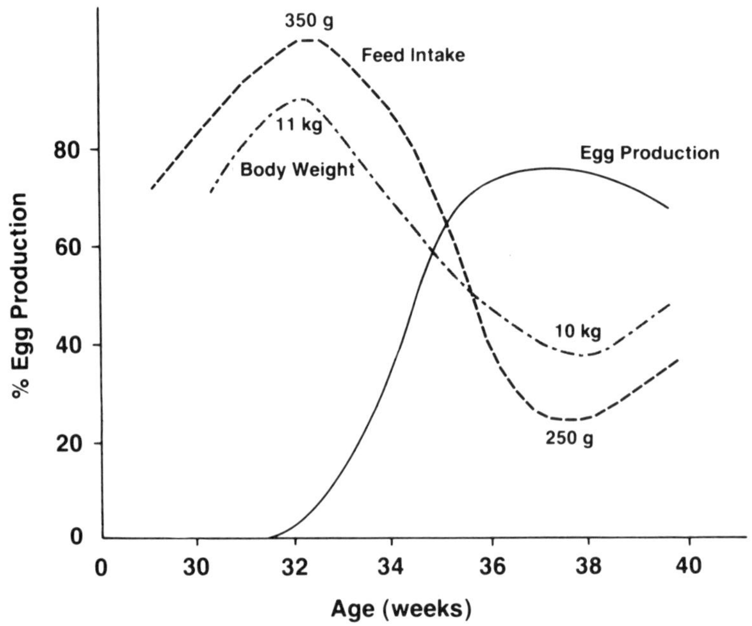

The positive response of turkey hens to added fat in the diet agrees with the general assumption of energy insufficiency with this type of bird. Harms and Wilson (1983) indicate improvements in egg production and number of viable poults per hen, resulting from the addition of up to 8% fat in the diet. These improvements in performance can be related to increases in intake of energy from these high-fat, high-energy diets (Table 7.15, Harms and Wilson 1983).

TABLE 7.15 Response of turkey breeder hens to dietary fat

Diet fat (%)	Diet ME (kcal/kg)	Eggs in 126d	Energy intake (kcal ME/d)
0	2900	54a	577b
4	3100	59b	602a
8	3300	63c	607a

a-c means followed by different letters are significantly different

Harms, *et al.* (1984) also indicate a positive effect on reproductive performance in response to adding fat in the grower diet. It is likely that this response is manifested via increases in energy (fat) reserves as the hen initiates egg production. There are very few reports in the literature of research conducted with the broiler-type turkey hen. However, data that is available, suggests that this smaller bird responds in a similar way in terms of energy balance. Thus Robel (1985) indicates that Small-White hens respond favourably to added fat in the diet, recording 49, 58 and 65 eggs/16 wks for birds fed 0, 5 and 10% fat respectively. Robel (1985) also recorded greater responses to added fat in warm summer months, confirming the general hypothesis of energy insufficiency of these breeder hens. High nutrient-dense pre-breeder diets are essential at this time, especially where birds have failed to achieve optimal mature weight under hot weather conditions. After peak egg production, there is usually normalization of feed intake and resumption of weight gain. At this time, breeder diets of reduced protein content can be considered.

In addition to egg production, the feeding program must also accommodate optimum hatchability and poult quality. So called "first-AI" syndrome is common with the first few hatches from young flocks, where hatchability is sub-optimal, and poult quality quite variable. Some tenuous relationships between this condition and nutrition of the breeder have been developed, although the condition invariably persists regardless of feeding program. The condition seems much worse with underweight birds, where presumably nutrient supply is limited. Under such conditions, vitamin supply is often questioned, although response to extra vitamins in the diet or water seems quite variable. There is some logical basis for extra vitamin supplementation at this time, because the content of some vitamins in the egg such as biotin seem suboptimal for the first few eggs produced by the turkey hen (Fig 7.2, Robel, 1983).

Fig. 7.2 Dry egg biotin content (Robel, 1983).

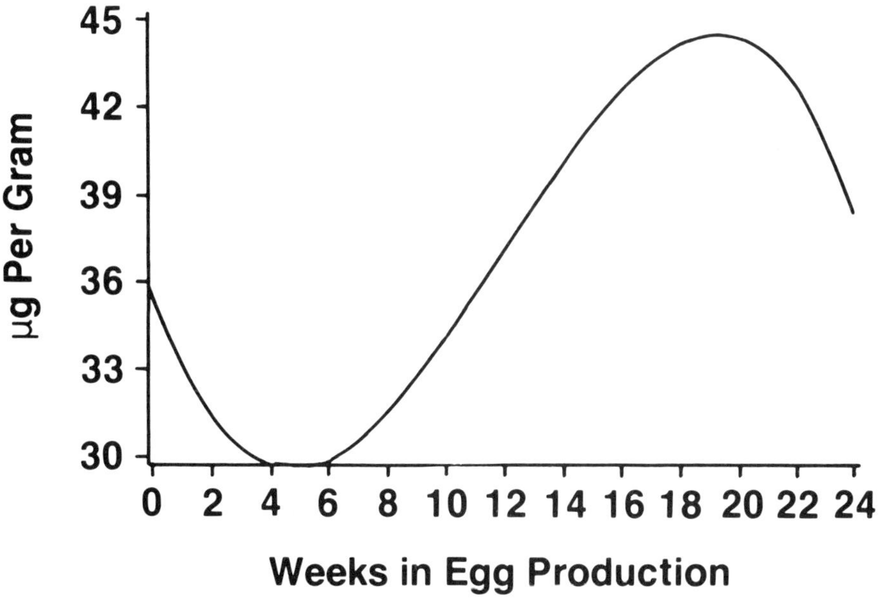

Such changes in egg vitamin content may however be independent of nutritional status of the bird. Robel (1983) concludes that vitamin levels in the diet may be inadequate to sustain the original nutrient levels in eggs of turkey hens over the production season, and that such inadequate levels may be causative in seasonal declines in hatchability.

A considerable number of turkey hens are being force-molted and used for a second cycle. An example of a molting program is given in Table 4.39. Guidelines to be followed during such a program are similar to those involved in rearing young hens, since body weight and condition at the start of the second cycle are again important. As shown in Fig 7.1, the hen regains weight after peak production, such that at the end of the second cycle she may be 0.75 - 1 kg heavier than her 30 week body weight. Ideally, the molting program will ensure that this weight is lost, so that the second cycle starts with a body weight once more at around 10.5 kg. In practice, hens are often heavier than this weight.

b) Toms

Considering the genetic potential of the male breeder turkey, it is obvious that some form of restricted feeding is essential. Such restriction results in smaller toms that are easier to handle in stud pens as well as birds of superior reproductive capacity. Restricted feeding programs usually involve diets of relatively low nutrient content (Table 7.13).

Seasonal decline in semen quality and quantity experienced with breeder toms can often be largely prevented through restricted feeding during rearing. Such males are easier to handle, and up to a 50% reduction in feed intake can be achieved. Hulet and Brody (1986) also observed that toms were easier to handle following restriction to achieve 80 or 60% of the body weight of control-fed birds. These reductions in body weight were achieved with approximately 30 and 50% levels of feed restriction. These authors recorded no loss in reproductive performance.

Studies have been conducted with young growing tom breeder candidates in which protein levels have varied significantly. As expected, the lower the protein content, the smaller the bird, yet effects are not pronounced until diets of less than 12 - 13% CP are used after 10 weeks of age. Most turkey multipliers carry out quite severe selection at 16 - 17 weeks of age, at which time only 50 - 60% of toms may be considered as potential breeder candidates. Such selection is based on body weight, leg condition and general conformation. A criticism of using too severe a feed restriction program prior to this time is that there can be no selection against those birds exhibiting adverse characteristics associated with fast growth - the assumption being that their offspring may also show such characteristics. It is obviously more difficult to control growth after 17 weeks of age if high nutrient dense diets are used prior to this time in an attempt to simulate commercial growing conditions. Lower-protein diets used during rearing of toms also delays the onset of semen production, and this should be accommodated in placement time of toms and hens (Fig 7.3, Cecil, 1984).

A problem sometimes encountered with breeder toms, is production of so-called yellow semen. Such semen is of inferior quality, and an incidence of up to 15% has been recorded in commercial flocks. There is some evidence linking a higher incidence with those toms fed low-protein, low-energy diets although the relationship is far from clear. When a high incidence of yellow semen occurs other environmental factors should also be investigated.

Fig. 7.3 Effect of diet crude protein on proportion of males producing semen.

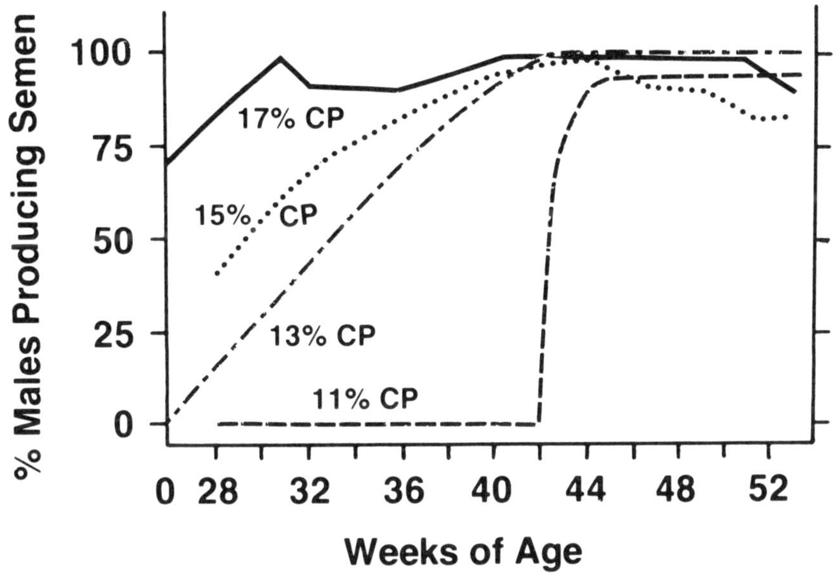

In those situations where physical daily feed restriction is not possible, then slowing growth through use of low nutrient dense diets after selection must be considered. It seems that 8-10% CP diets, of adequate amino acid balance, are suitable, although such specifications are often difficult to achieve in many geographical locations. Such low protein diets are most easily achieved with corn, and it is likely that this is the basis for recommending high energy levels (3000 kcal ME/kg) for such diets. While low protein-high energy diets may temper growth of toms in warm climates, these birds often overconsume energy in more temperate environments. Under these later conditions, lower energy (2800 kcal ME/kg) levels are recommended.

In the stud pens, physical feed restriction is again an ideal management practice where most Large White strains will consume about 400 - 450 g each day. However, body weight monitoring must continue throughout this period, since it seems advantageous that the breeder tom gain weight, albeit very little (150 g/wk after 40 wks age). This can best be achieved by very small, but gradual increases in feed allowance each week.

c) Predicting nutrient specifications

There has been relatively little research on turkey breeder nutrition in the last few years, and in part this is due to the high cost of such research. An alternative system for defining nutrient requirements is to make predictions based on simple assumptions of need according to known inputs. For both breeder hens and toms body weight and growth are by far the largest input, and for the hen there is need to account for nutrients needed for egg production.

In the following calculations, maintenance nutrient requirements were taken from data presented by Moran *et al.* (1983) and for the hen estimates of egg size, egg production and egg composition are used to calculate corresponding production requirements. Predictions for major nutrient needs of hens are shown in Table 7.16 and for toms in Table 7.17. Hens were assumed to have a body weight of 11kg and be at 60% production with a 100g egg. Tom needs were calculated based on body weight of 20 kg.

TABLE 7.16 Predicted daily nutrient requirements of turkey breeder hens

Nutrient	Prediction	Diet specifications @ 250g daily feed intake	Typical diet
ME	790 kcal/d	3070 kcal/kg	2970 kcal/kg
CP	24.9g	12.0%	17.0%
Arginine	1.9g	0.91%	0.95%
Isoleucine	1.5g	0.73%	0.75%
Lysine	1.3g	0.63%	0.85%
Methionine	0.5g	0.24%	0.42%
Cystine	1.3g	0.61%	0.24%
Meth+Cyst	1.8g	0.85%	0.66%
Threonine	1.2g	0.59%	0.55%
Tryptophan	0.3g	0.13%	0.14%

TABLE 7.17 Predicted daily nutrient requirements of turkey breeder toms

Nutrient	Prediction	Diet specifications @ 450g daily feed intake	Typical diet
ME	1034 kcal/d	2300 kcal/kg	2750 kcal/kg
CP	35.3g	9.5%	12.0%
Arginine	2.8g	0.74%	0.61%
Isoleucine	2.2g	0.59%	0.53%
Lysine	1.6g	0.43%	0.60%
Methionine	0.5g	0.13%	0.20%
Cystine	2.2g	0.59%	0.18%
Meth+Cyst	2.7g	0.72%	0.38%
Threonine	1.7g	0.44%	0.42%
Tryptophan	0.3g	0.07%	0.11%

Data in Tables 7.16 and 7.17 shows two major discrepancies between predicted levels and those provided by a typical commercial diet. For both hens and toms we seem to be overestimating needs for crude protein and greatly underestimating requirement for the amino acid cystine. The situation with cystine is of greatest concern, and the model prediction estimates are high because of the extensive need for maintenance related to feather status. If these values are correct, then the only practical way of meeting such a requirement is to include feather meal in the diet and/or to use more synthetic methionine.

CHAPTER 7

Selected References

A) MARKET TURKEYS

Chen, F. 1994. *Dietary biotin and turkey breeder performance.* Poult. Sci. 73:682-686.

Donaldson, W.E. 1994. *Administration of propionate to day-old turkeys.* Poult. Sci. 73:1249-1253.

Donaldson, W.E. 1995. *Carbohydrate, hatchery stressors affect poult survival.* Feedstuffs: p.16.

Grimes, J.L. 1994. *The effect of protein level fed during the prebreeder period on performance of large white turkey breeder hens after an induced molt.* Poult. Sci. 73:37-44.

Hocking, P. 1992. *Effects of photostimulation at 18, 24, and 30 weeks of age on the productivity of female turkeys fed ad lib. or restricted until point of lay.* Br. Poult. Sci. 33:253.

Hocking, P., D. Waddington and M. Walker 1992. *Changes in ovarian function of female turkeys photostimulated at 18, 24 or 30 weeks of age and fed ad lib. or restricted until point of lay.* Br. Poult. Sci. 33:639.

Hulan, H.W. and F.G. Proudfoot 1981. *Bioeconomic effects of feeding turkey broilers grower-finisher diet combinations differing in protein level.* Poult. Sci. 60:358-364.

Hurwitz, S., Y. Frisch, A. Bar, U. Eisner, I. Bengal and M. Pines 1983. *The amino acid requirements of growing turkeys. I. Model construction and parameter estimation.* Poult. Sci. 62:2398-2042.

Jackson, S. and L.M. Potter 1984. *Influence of basic and branched chain amino acid interactions on the lysine and valine requirements of young turkeys.* Poult. Sci. 63:2391-2398.

Kagan, A. 1981. *Supplemental fats for growing turkeys: A review.* World's Poult. Sci. J. 37:203-210.

Leeson, S. and J.D. Summers 1978. *Dietary self-selection by turkeys.* Poult. Sci. 57:1579-1585.

Leeson, S. and J.D. Summers 1980. *Production and carcass characteristics of the Large White turkey.* Poult. Sci. 59:1237-1245.

Lilburn, M.S. and D. Emmerson 1993. *The influence of differences in dietary amino acids during the early growing period on growth and development of Nicholas and British United Turkey toms.* Poult. Sci. 72:1722-1730.

Mamputu, M. 1992. *Performance of turkeys subjected to day and night feeding programs during heat stress.* J. Appl. Poult. Res. 1:296-299.

Moran, E.T., Jr. 1995. *Performance of turkeys at 110 vs. 115% of NRC (1994) protein recommendation.* J. Appl. Poult. Res. 4:No. 2, 138-147.

Oju, E.M., P.E. Waibel and S.L. Noll 1988. *Early protein undernutrition and subsequent realimentation in turkeys. 1. Effect of performance and body composition.* Poult. Sci. 67:1750-1759.

Owen, J.A., P.W. Waldroup, C.J. Mabray and P.J. Slagter 1981. *Response of growing turkeys to dietary energy level.* Poult. Sci. 60:418-424.

Potter, L.M. and J.P. McCarthy 1985. *Varying fat and protein in diets of growing Large White turkeys.* Poult. Sci. 64:1941-1949.

Renema, R.A. et al 1994. *The use of feed restriction for improving reproductive traits in male-line large white turkey hens. 1. Growth and carcass characteristics.* Poult. Sci. 73:1724-1738.

Rivas, F.M. and J.D. Firman 1994. *The influence of energy and protein on turkeys during the finisher period.* J. Appl. Poult. Res. 3:327-335.

Salmon, R.E. 1986. *Effect of nutrient density and energy to protein ratio on performance and carcass quality of small white turkeys.* Br. Poult. Sci. 27:629-638.

Sell, J.L., R.J. Hasiak and W.J. Owings 1985. *Independent effects of dietary metabolisable energy and protein concentrations on performance and carcass characteristics of tom turkeys.* Poult. Sci. 64:1527-1535.

Sell, J.L. 1993. *Influence of metabolizable energy feeding sequency and dietary protein on performance and selected carcass traits of tom turkeys.* Poult. Sci. 72:521-534.

Sell, J.L. and M.J. Jeffrey 1994. *Influence of amino acid supplementation of low-protein diets and metabolizable energy feeding sequence on performance and carcass composition of toms.* Poult. Sci. 73:1867-1880.

Soto-Salanova, M. and J.L. Sell 1995. *Influence of supplemental dietary fat on changes in vitamin E concentration in livers of poults.* Poult. Sci. 74:201-204.

Waldroup, P.W. 1993. *Effects of amino acid restriction during starter and grower periods on subsequent performance and incidence of leg disorders in male large white turkeys.* Poult. Sci. 72:816-828.

Watkins, K.L. 1993. *Effects of feed restriction and subsequent gorging with limited access to water on male turkeys fed graded levels of monensin.* Poult. Sci. 72:677-683.

B) BREEDER TURKEYS

Cecil, H.C. 1984. *Effect of dietary protein and light restriction on body weight and semen production of breeder male turkeys.* Poult. Sci. 63:1175-1183.

Ferket, P.R. and E.T. Moran, Jr. 1986. *Effect of plane of nutrition from starting to and through the breeder period on reproductive performance of hen turkeys.* Poult. Sci. 65:1581-1590.

Harms, R.H., R.E. Buresh and H.R. Wilson 1984. *The influence of the grower diet and fat in the layer diet on performance of turkey hens.* Poult. Sci. 63:1634-1637.

Leeson, S., L.J. Caston and B. Rogers 1989. *Restricted water access time as a means of growth control in turkey tom breeder candidates.* Poult. Sci. 68:1236-1238.

Owings, W.J. and J.L. Sell 1980. *Effect of restricted feeding from 6 to 20 weeks of age on reproductive performance of turkeys.* Poult. Sci. 59:77-81.

Robel, E.J. 1983. *The effect of age of breeder hen on the levels of vitamins and minerals in turkey eggs.* Poult. Sci. 62:1751-1756.

Robel, E.J. 1984. *Weight loss as related to body composition in aging white turkey hens.* Poult. Sci. 63:1281-1283.

Robel, E.J. 1985. *Added fat in breeder hen turkey diets.* Nutr. Rep. Int. 31:1281-1289.

CHAPTER 8. FEEDING PROGRAMS FOR WATERFOWL

As consumers look for new meat products, and producers look for more hardy species requiring less capital input, there is an increasing interest in the production of various waterfowl. Ducks, and to a lesser extent geese, are the main species favoured on a large scale, although there is interest in some of the more exotic species.

8.1 DUCKS

There has been considerable interest in the growth potential of meat-type ducks, such that their overall performance is often better than that of chickens. Carcass quality continues to be a major limitation, because the duck characteristically deposits considerable quantities of subcutaneous fat. Tables 8.1 and 8.2 outline dietary specifications for commercial meat ducks (Pekin type strains) and associated breeder stock. Examples of some diets for meat ducks and breeders are given in Tables 8.3 and 8.4.

In formulating diets for meat ducks, care must be taken in adjusting the balance of protein:energy to try and minimize carcass fat deposition. The duck seems to respond in a similar way to protein:energy as previously described for the broiler chicken and turkey, such that higher protein diets in relation to energy generally result in less carcass fat. The duck seems to be able to digest fiber slightly better than does the chicken, and as such, metabolizable energy values for ducks may be 5-6% greater than corresponding values for chickens - such differences should be considered in setting energy specifications of diets.

TABLE 8.1 Dietary specifications for commercial meat ducks

	Starter		Grower-finisher	
	1	2	3	4
Approximate protein level (%)	22.0	20.0	18.0	16.0
Amino acids (% of diet)				
Arginine	1.20	1.04	0.94	0.85
Lysine	1.10	0.96	0.86	0.78
Methionine	0.48	0.43	0.39	0.35
Methionine+cystine	0.83	0.72	0.66	0.60
Tryptophan	0.22	0.18	0.16	0.15
Histidine	0.44	0.37	0.33	0.29
Leucine	1.56	1.28	1.16	1.04
Isoleucine	0.84	0.69	0.63	0.56
Phenylalanine	0.78	0.64	0.58	0.52
Phenylalanine+tyrosine	1.52	1.24	1.12	1.01
Threonine	0.76	0.62	0.56	0.50
Valine	0.93	0.77	0.69	0.62
Metabolizable energy (kcal/kg)	2850	2900	3100	3125
Calcium (%)	0.8	0.83	0.76	0.75
Available phosphorus (%)	0.4	0.42	0.38	0.35
Sodium (%)	0.18	0.18	0.18	0.18
Vitamins (per kg of diet)				
Vitamin A (I.U.)	8000			
Vitamin D_3 (I.U.)	2500			
Choline equivalents (mg)	800			
Riboflavin (mg)	4.0			
Pantothenic acid (mg)	12.0			
Vitamin B_{12} (mg)	.010			
Folic acid (mg)	0.5			
Biotin (mg)	0.2			
Niacin (mg)	60.0			
Vitamin K (mg)	1.5			
Vitamin E (I.U.)	20.0			
Thiamin (mg)	2.0			
Pyridoxine (mg)	3.0			
Trace minerals (per kg of diet				
Manganese (mg)	60			
Iron (mg)	80			
Copper (mg)	8			
Zinc (mg)	60			
Selenium (mg)	0.2			
Iodine (mg)	0.4			

TABLE 8.2 Dietary specifications for heavy duck breeders

	Holding diets		Breeder diets	
	1	2	3	4
Approximate protein level (%)	11.0	13.0	16.0	18.0
Amino acids (% of diet)				
Arginine	0.59	0.65	0.67	0.75
Lysine	0.44	0.50	0.80	0.95
Methionine	0.20	0.23	0.34	0.37
Methionine+cystine	0.42	0.45	0.58	0.64
Tryptophan	0.11	0.13	0.13	0.15
Histidine	0.13	0.15	0.16	0.17
Leucine	0.71	0.79	0.82	0.91
Isoleucine	0.49	0.55	0.57	0.63
Phenylalanine	0.37	0.41	0.43	0.47
Phenylalanine+tyrosine	0.65	0.71	0.75	0.83
Threonine	0.49	0.55	0.57	0.63
Valine	0.54	0.60	0.64	0.70
Metabolizable energy (kcal/kg)	2500	2600	2700	2850
Calcium (%)	0.95	1.10	2.8	2.9
Available phosphorus (%)	0.38	0.42	0.38	0.42
Sodium (%)	0.15	0.15	0.16	0.16
Vitamins (per kg of diet)				
Vitamin A (I.U.)	5000		10,000	
Vitamin D_3 (I.U.)	1250		2500	
Choline equivalents (mg)	400		800	
Riboflavin (mg)	2.7		5.5	
Pantothenic acid (mg)	2.5		5.0	
Vitamin B_{12} (mg)	0.07		0.014	
Folic acid (mg)	0.2		0.5	
Biotin (mg)	0.12		0.25	
Niacin (mg)	25		50	
Vitamin K (mg)	1.0		2.0	
Vitamin E (I.U.)	12.5		25.0	
Thiamin (mg)	1.0		2.0	
Pyridoxine (mg)	1.5		3.0	
Trace minerals (per kg of diet)				
Manganese (mg)	30		60	
Iron (mg)	46		80	
Copper (mg)	4		8	
Zinc (mg)	30		60	
Selenium (mg)	.15		0.3	
Iodine (mg)	0.2		0.4	

TABLE 8.3 Duck starter and grower diets (kg)

	Starter		Grower-Finisher			
Corn	412	---	522	---	---	317
Wheat	---	464	---	584	670	315
Barley	150	150	150	150	100	100
Wheat shorts	50	50	50	50	100	100
Dehydrated alfalfa	20	20	---	---	---	---
Soybean meal (48%)	332	280	240	178	95	133
Ground limestone	11	11	12.5	12.5	12.5	12.5
Calcium phosphate	12.5	12.5	13	13	10	10
Salt	2.0	2.0	2.0	2.0	2.5	2.5
Vitamin:mineral premix[1]	10	10	10	10	10	10
Methionine	1.0	1.25	0.5	0.75	0.88	0.75
<u>Calculated analyses</u>						
Crude protein (%)	22.7	22.7	18.9	19.0	16.2	16.1
Digestible protein (%)	20.4	20.3	17.0	17.0	14.4	14.4
Crude fat (%)	2.5	1.6	2.8	1.6	1.7	2.4
Crude fiber (%)	3.8	4.0	3.8	3.7	3.6	3.4
Metabolizable energy (kcal/kg)	2771	2730	2868	2814	2873	2906
Calcium (%)	0.80	0.80	0.82	0.83	0.75	0.79
Available phosphorus (%)	0.39	0.41	0.39	0.40	0.33	0.33
Sodium (%)	0.16	0.18	0.17	0.19	0.18	0.17
Methionine (%)	0.46	0.46	0.36	0.36	0.33	0.33
Methionine+cystine (%)	0.81	0.82	0.64	0.66	0.58	0.59
Lysine (%)	1.29	1.27	0.98	0.96	0.74	0.75
Tryptophan (%)	0.32	0.34	0.26	0.28	0.24	0.23

[1]Use additional choline chloride if premix does not contain this vitamin

TABLE 8.4 Duck holding diet and duck breeder diets (kg)

	Holding	Breeder	
		1	2
Corn	451	518	---
Wheat	---	---	580
Barley	110	100	100
Wheat middlings	210	78	78
Wheat bran	120	---	---
Dehydrated alfalfa	20	20	20
Soybean meal (48%)	50	195	133
Ground limestone	17.5	66.5	66.5
Calcium phosphate (20% CP)	8	10	10
Salt	2.5	2.5	2.5
Vitamin:mineral premix[1]	10	10	10
Methionine	1.00	0.5	0.75
Calculated analyses			
Crude protein (%)	12.7	17.0	17.2
Digestible protein (%)	11.0	15.2	15.2
Crude fat (%)	3.8	2.8	1.6
Crude fiber (%)	5.3	3.7	3.9
Metabolizable energy, (kcal/kg)	2544	2772	2719
Calcium (%)	0.92	2.9	2.9
Available phosphorus (%)	0.42	0.40	0.43
Sodium (%)	0.19	0.18	0.19
Methionine (%)	0.21	0.34	0.33
Methionine+cystine (%)	0.40	0.60	0.60
Lysine (%)	0.54	0.87	0.86
Tryptophan (%)	0.20	0.24	0.25
Threonine (%)	0.50	0.70	0.60

[1]**Use additional choline chloride if premix does not contain this vitamin**

Methionine and lysine are likely to be the most limiting amino acids in diets for ducks, and the normal base level of 2 and 5% of crude protein respectively seem applicable to the duck. Growth characteristics of Pekin ducks are shown in Table 8.5.

TABLE 8.5 Growth rate, feed efficiency and feed consumption of White Pekin ducklings

	Average weight (g)		Feed intake: body weight gain	
Weeks	♂	♀	♂	♀
1	300	290	1.2	1.2
2	1000	800	1.6	1.6
3	1500	1300	1.7	1.8
4	2100	1900	1.8	2.0
5	2700	2500	2.0	2.2
6	3400	3100	2.1	2.3
7	3700	3400	2.3	2.5

In developing feeding programs for ducks, carcass composition must be taken into account, especially for late-grower and finisher diets. Table 8.6 outlines the yield and commercial portions of Pekin ducks, while Table 8.7 details the fat and protein development in the carcass to 49 d of age. As shown in Table 8.7, fat represents in excess of 60% of the carcass dry matter at 49 d of age. At 49 d of age, abdominal fat represents only some 2% of body weight, which is comparable to that found in chickens - this data confirms that the major problem with fat in the body of the duck is subcutaneous fat depots, and so this is impossible to remove during processing for the whole-bird market.

TABLE 8.6 Yield and commercial cuts of Pekin ducks

	35 d		42 d		49 d	
	♂	♀	♂	♀	♂	♀
Eviscerated carcass (g)	1850	1750	2300	2100	2600	2300
Carcass yield (%)	70.4	72.1	73.2	74.5	76.9	77.7
% Thighs	14.0	12.4	13.7	13.0	12.2	11.2
% Drumsticks	13.7	13.5	12.5	12.2	10.6	10.3
% Wings	12.1	12.4	12.3	11.8	11.5	11.6
% Breast	17.0	18.9	20.5	21.7	25.7	26.5

TABLE 8.7 Carcass composition of male pekin ducks

	28 days	35 days	42 days	49 days
MALE				
Offal fat, % DM	38.9	42.0	55.5	54.1
Carcass fat, % DM	56.9	61.7	64.0	64.1
Offal CP, % DM	47.7	46.5	37.6	36.2
Carcass CP, % DM	31.3	32.1	30.5	28.8
Offal DM, %	35.4	36.0	41.3	40.6
Carcass DM, %	43.5	44.5	47.4	47.9
Carcass fat, g	321.4	480.3	632.1	785.0
Carcass CP, g	175.3	100.1	66.4	72.8
Offal fat, g	66.5	90.5	140.6	160.0
Offal CP, g	79.1	98.2	94.5	106.5
Total body fat, g	388	570.9	772.8	945.0
Total body CP, g	254	343.3	394.7	459.0
Total body fat, %	19.2	24.4	28.7	35.0
Total body CP, %	14.8	14.8	14.6	17.0
Carcass fat as % body fat	84.1	84.2	81.8	83.0
Carcass CP as % body CP	67.8	71.3	76.0	76.7

If fast growth rate is desired, then there seems to be a distinct advantage to feeding good quality pellets. Ducks do not perform adequately on mash diets, a factor likely related to their inability to efficiently pick-up feed, and to do so without causing major feed wastage. The trend towards higher energy diets over the last few years, through use of dietary fat and higher levels of corn, leads to more problems in creating quality pelleted feed. Recent work with "wet" mash diets suggests improvements of liveweight and feed efficiency of around 5%. However these advantages are somewhat offset by the "dirty" condition of the ducks, especially around the vent area.

High energy diets are often blamed for the high levels of fat seen in the carcass. However the duck seems to eat to its energy requirement over quite a wide range of diet energy levels, and so it is not so obvious that high diet energy levels will lead to increased energy intake. In most instances, such high energy diets are not adjusted for crude protein content, and it is the balance of protein to energy that is most often the culprit that leads to increases in carcass fat observed with high energy diets. There is reason to believe that the net energy of fat is increased when considerable portions of fat are being deposited in the carcass, and this situation does detract from the use of high energy diets. Due to the duck's apparent superior utilization of crude fiber, and the duck's ability to adjust feed intake to diet energy concentration, there seem to be advantages to using diets of medium-low energy concentration. In addition to tempering diet energy concentration as a means of controlling carcass fat content, there have also been reports of restricted feeding regimes, especially in the late-grower and finishing periods. Feed restriction *per se* does not seem to be as useful for controlling carcass fat as does attention to protein:energy in the diet, although a combination of the two systems may be feasible. As with any restricted feeding program, growth rate will be reduced, therefore producers must realize higher monetary returns from such leaner carcasses.

There seems to be a real advantage to feed restriction in growing breeder candidates. Most breeding stock is selected from within commercial flocks at normal market age, and so there is a great challenge to "hold" birds up to time of sexual maturity. Low-nutrient dense holding diets (Table 8.2) fed on a restricted basis according to desired body weight seems to be the only practical method of both delaying sexual maturity and controlling mature body size. Without such control, egg production is often very poor, and fertility of males may be virtually non-existent. Under such conditions, breeder candidates should be fed according to body weight as was previously described for broiler breeder stock. Heavy breeder strains can also successfully be molted as discussed previously for chickens and turkeys. Table 4.39 gives a general outline of a molting program. As with other species the initial requirement is for loss of 25 - 30% of body weight, and this is achieved by feed withdrawal and reduction in daylength. The body reserves of the breeder, and her ovary and oviduct are then re-established through gradual return to ad-lib intake of a breeder diet over a 5-6 week period.

Discussion on the nutrition of ducks, invariably includes their need for water. Water intake values are shown in Table 2.40. Ducks do have relatively high water requirements, and this is likely associated with the increased rate of passage of digesta. Reducing access time to water, as a means of controlling litter moisture, most often results in reduced feed intake and reduced growth rate, although two 4 hour periods of water access seems to be a compromise situation. Contrary to popular belief, there is no need to provide water such that ducks can either swim or immerse their heads, and so bell-type or even nipple drinkers are acceptable.

While most discussion to date has centered on the Pekin-type strains of meat bird, there is growing interest in the production of Muscovy ducks. This genetically distinct strain is most easily identified by the large sexual dimorphism in body weight (Table 8.8) with the male being at least 50% heavier at a slightly later marketing age. Muscovy ducklings seem to have a slightly lower protein requirement of around 21 and 19% CP respectively in starter diets for male and females. Requirements seem to decline to 14 - 17% CP in finisher diets for both sexes, although females are usually marketed some 2 - 3 weeks earlier than males in order to limit carcass fat deposition. The Muscovy may be an ideal candidate for programs involving compensatory growth.

TABLE 8.8 Performance of muscovy ducks

Age weeks	Body wt (g)		Feed intake:body wt gain	
	♂	♀	♂	♀
1	100	100	1.10	1.10
2	250	240	1.20	1.20
3	700	600	1.40	1.42
4	1400	1200	1.60	1.75
5	2000	1400	1.80	1.90
6	2500	1800	1.95	2.05
7	2900	2100	2.05	2.15
8	3400	2350	2.25	2.40
9	3700	2500	2.40	2.60
10	4000	2700	2.50	2.72
11	4250	---	2.65	---
12	4500	---	2.80	---

The most recent development in commercial duck production has been the acceptance of the Mule duck, which is a cross between the Muscovy male and a conventional female strain such as the Pekin. The main advantage of the sterile hybrid is that the males and females are of comparable weight, so removing the main obstacle associated with marketing of smaller Muscovy females. Typical growth rates of such hybrids are shown in Table 8.9.

TABLE 8.9 Growth and feed efficiency of hybrid mule ducks

Age (wks)	Body weight (kg)		Feed efficiency	
	♂	♀	♂	♀
3	0.8	0.7	1.3	1.5
4	1.4	1.1	1.4	1.6
5	1.9	1.6	1.5	1.7
6	2.5	2.1	1.6	1.8
7	2.9	2.4	1.7	1.9
8	3.5	3.0	1.8	2.0
9	3.7	3.3	1.9	2.1
10	3.9	3.5	2.0	2.2

All ducks are very susceptible to mycotoxins, and in particular aflatoxin. Levels as low as 30 - 40 ppb have been shown to impair protein utilization, while levels of 60 - 80 ppb can cause a dramatic loss in growth rate. With low-protein diets the symptoms are greatly accentuated, and onset occurs more quickly. Heavy metal toxicity has also been studied extensively with ducks, although much of this research seems directed at application of natural contamination in wild species. Most species of ducks seem very susceptible to such heavy metals as cadmium, lead and arsenic, although toxic levels should not normally be encountered in uncontaminated commercial feeds. There are few research reports detailing the nutrient requirements of egg-type duck strains such as the Khaki Campbell and Indian Runner. In fact, a review of the literature suggests that strains such as the Khaki Campbell can be offered diets comparable to brown-egg strain chickens and that feeding management is also similar.

8.2 GEESE

While geese can exhibit the most rapid growth rate of all domesticated poultry species, a major limitation of expanded commercial production is undoubtedly carcass quality. As for most waterfowl species, the goose has a propensity to deposit fat in the body, and it is evident that a large proportion of this very rapid growth appears as skin, feathers and body fat. When economically feasible, such concerns must be accommodated in the development of feeding programs. Diet specifications for commercial meat-type geese and breeders are shown in Tables 8.10 and 8.11 respectively, while corresponding diet examples are shown in Tables 8.12 and 8.13 respectively.

TABLE 8.10 Dietary specifications for commercial meat geese

	Starter diet		Grower-finisher diets	
	1	2	3	4
Approximate protein level (%)	22.0	17.0	19.0	15.0
Amino acids (% of diet)				
Arginine	1.15	1.03	0.93	0.80
Lysine	1.06	1.00	0.90	0.73
Methionine	0.48	0.47	0.43	0.32
Methionine+cystine	0.83	0.74	0.68	0.57
Tryptophan	0.21	0.18	0.16	0.14
Histidine	0.42	0.37	0.33	0.29
Leucine	1.49	1.22	1.10	1.03
Isoleucine	0.80	0.65	0.59	0.55
Phenylalanine	0.75	0.63	0.57	0.52
Phenylalanine+tyrosine	1.45	1.21	1.06	1.01
Threonine	0.73	0.70	0.64	0.50
Valine	0.89	0.74	0.67	0.63
Metabolizable energy (kcal/kg)	2760	2850	2900	2950
Calcium (%)	0.80	0.75	0.71	0.71
Available phosphorus (%)	0.40	0.42	0.38	0.38
Sodium (%)	0.18	0.18	0.18	0.18
Vitamins (per kg of diet)				
Vitamin A (I.U.)	8000			
Vitamin D_3 (I.U.)	2000			
Choline equivalents (mg)	800			
Riboflavin (mg)	4.0			
Pantothenic acid (mg)	12.0			
Vitamin B_{12} (mg)	.010			
Folic acid (mg)	0.5			
Biotin (mg)	0.2			
Niacin (mg)	60.0			
Vitamin K (mg)	1.5			
Vitamin E (IU)	15.0			
Thiamin (mg)	2.0			
Pyridoxine (mg)	3.0			
Trace minerals (per kg of diet)				
Manganese (mg)	60			
Iron (mg)	80			
Copper (mg)	8			
Zinc (mg)	60			
Selenium (mg)	0.2			
Iodine (mg)	0.4			

TABLE 8.11 Dietary specifications for geese holding and breeder diets

	Holding diets		Breeder diets	
	1	2	3	4
Approximate protein level (%)	12.0	14.0	14.0	16.0
Amino acids (% of diet)				
Arginine	0.54	0.56	0.63	0.69
Lysine	0.50	0.56	0.59	0.65
Methionine	0.23	0.25	0.32	0.34
Methionine+cystine	0.43	0.47	0.54	0.60
Tryptophan	0.11	0.13	0.12	0.14
Histidine	0.66	0.72	0.76	0.84
Leucine	0.46	0.50	0.52	0.58
Isoleucine	0.34	0.38	0.39	0.43
Phenylalanine	0.40	0.43	0.49	0.50
Phenylalanine+tyrosine	0.77	0.80	0.82	0.90
Threonine	0.46	0.50	0.52	0.58
Valine	0.50	0.56	0.59	0.65
Metabolizable energy (kcal/kg)	2350	2600	2650	2850
Calcium (%)	0.80	0.90	1.90	2.10
Available phosphorus (%)	0.32	0.35	0.35	0.38
Sodium (%)	0.18	0.18	0.18	0.18
Vitamins (per kg of diet)				
Vitamin A (I.U.)	5000		10,000	
Vitamin D_3 (I.U.)	1000		2000	
Choline equivalents (mg)	400		800	
Riboflavin (mg)	2.6		5.0	
Pantothenic acid (mg)	5.0		10.0	
Vitamin B_{12} (mg)	.007		.015	
Folic acid (mg)	0.25		0.5	
Biotin (mg)	.10		.20	
Niacin (mg)	30		60	
Vitamin K (mg)	1.0		2.0	
Vitamin E (IU)	10.0		20.0	
Thiamin (mg)	1.0		2.0	
Pyridoxine (mg)	1.5		3.0	
Trace minerals (per kg of diet)				
Manganese (mg)	30		60	
Iron (mg)	50		80	
Copper (mg)	4		8	
Zinc (mg)	40		60	
Selenium (mg)	0.15		0.3	
Iodine (mg)	0.20		0.42	

TABLE 8.12. Examples of goose starter and grower-finisher diets (kg)

	Starter		Grower-Finisher			
Corn	434	218	548	548	621	---
Wheat	---	242	---	100	---	695
Barley	100	100	103	50	100	100
Wheat shorts	50	50	50	50	50	50
Wheat middlings	50	50	50	50	50	50
Dehydrated alfalfa	20	20	20	20	---	---
Soybean meal (48%)	310	284	215	220	142	70
Ground limestone	11	11	9.5	9.5	12	12
Calcium phosphate (20%P)	13	13	12.5	12.5	12	11
Salt (iodized)	2	2	2	2	2	2
Vitamin:mineral premix[1]	10	10	10	10	10	10
Methionine	.5	.63	.5	.88	.5	.88
Calculated anlayses						
Crude protein (%)	21.8	21.9	17.9	18.1	15.1	15.2
Digestible protein (%)	19.6	19.6	16.1	16.3	13.2	14.1
Crude fat (%)	2.6	2.1	2.9	2.7	3.1	1.7
Crude fiber (%)	3.8	3.9	3.3	3.5	3.3	3.5
Metabolizable energy (kcal/kg)	2771	2755	2890	2943	2940	2880
Calcium (%)	0.80	0.81	0.70	0.71	0.76	0.75
Avail. phosphorus (%)	0.40	0.44	0.37	0.39	0.35	0.35
Sodium (%)	0.18	0.17	0.18	0.18	0.18	0.18
Methionine (%)	0.40	0.40	0.35	0.39	0.31	0.32
Methionine+cystine (%)	0.85	0.85	0.72	0.80	0.65	0.66
Lysine (%)	1.22	1.21	0.89	0.91	0.68	0.68
Tryptophan (%)	0.31	0.32	0.24	0.25	0.23	0.23
Threonine (%)	0.90	0.86	0.74	0.75	0.51	0.51

[1]**Use additional choline chloride if premix does not contain this vitamin**

TABLE 8.13 Examples of goose holding and breeder diets (kg)

	Holding	Breeder	
Corn	547	295	567
Wheat	---	287	---
Barley	100	100	100
Wheat shorts	50	50	50
Wheat middlings	---	50	50
Wheat bran	200	---	---
Dehydrated alfalfa	20	20	20
Meat meal (50%)	---	---	20
Fish meal (60%)	---	---	20
Soybean meal (48%)	45	130	113
Ground limestone	18	46	45
Calcium phosphate (20%P)	7	7	-
Salt	3	3	3
Vitamin:mineral premix[1]	10	10	10
Methionine	0.6	1.0	0.63
Calculated anlayses			
Crude protein (%)	12.4	15.7	15.7
Digestible protein (%)	10.8	14.0	14.0
Crude fat (%)	3.6	2.3	3.1
Crude fiber (%)	5.2	3.8	3.6
Metabolizable energy (kcal/kg)	2650	2763	2814
Calcium (%)	0.9	2.05	2.10
Avail. phosphorus (%)	0.32	0.35	0.35
Sodium (%)	0.18	0.19	0.19
Methionine (%)	0.26	0.35	0.35
Methionine+cystine (%)	0.51	0.57	0.58
Lysine (%)	0.54	0.76	0.78
Tryptophan (%)	0.20	0.23	0.20
Threonine (%)	0.48	0.59	0.64

[1]Use additional choline chloride if premix does not contain this vitamin

As most production systems for geese involve some time spent on pasture, then the actual choice of feeding program must accommodate the availability and quality of pasture. Regardless of rearing system, early growth rate is best optimized through use of pelleted starter diets for 3 - 4 weeks during which time goslings are usually confined. Subsequent grower and finisher diets can be fed as sole dietary sources, and it is under such conditions that optimum growth rate is often achieved. Differences in performance of Chinese x Embden geese related to rearing system is shown in Table 8.14.

TABLE 8.14 Growth rate and feed consumption of White Chinese x Embden geese (mixed sex)

	Confinement reared			Range reared		
Age weeks	Average weight (kg)	Cumulative feed consumption (kg)	Feed intake: body weight gain	Average weight (kg)	Cumulative feed consumption (kg)	Feed intake: body weight gain
3	1.68	2.65	1.55	1.59	2.64	1.66
6	4.20	8.40	2.00	3.80	6.08	1.60
9	5.74	17.16	2.99	4.98	9.61	1.93
12	6.71	23.89	3.56	5.80	16.23	2.75
14	7.10	28.61	4.03	5.95	18.68	3.14

Growth rate and feed intake data for sexed Pilgrim geese reared under confinement are shown in Table 8.15. While confinement rearing, involving the sole use of complete diets, may result in the most rapid growth, less extensive systems may be more economical. Alternate systems often involve a period on pasture and/or supplemental grain feeding. Because geese have voracious appetites, they are able to consume large quantities of forage, and in eating such large quantities their nutrient intake from this poorer quality feed is maintained at near normal levels. Nutrient intake under such conditions will obviously depend upon both quantity and quality of pasture, and in the latter context pasture management as it relates to ruminant animals can be applied. Geese seem to accept a range of quality roughages, including clovers, mixed grasses, cereals and corn-silage.

TABLE 8.15 Growth rate and feed consumption of Pilgrim geese

Age (weeks)	Average weight (kg)		Cumulative feed (kg)		Feed intake: body weight gain	
	♂	♀	♂	♀	♂	♀
2	0.70	0.70	0.88	0.88	1.26	1.26
4	2.20	2.10	3.20	3.40	1.45	1.61
6	3.60	3.00	6.70	7.00	1.86	2.33
8	4.70	3.90	11.70	11.60	2.48	2.97
10	5.00	4.50	15.10	15.00	3.02	3.33
12	6.10	5.10	20.00	18.50	3.27	3.63
14	6.45	5.40	25.00	23.00	3.87	4.26

Depending upon the time of year, it is often difficult to get adequate finish on "younger' birds on pasture. Full-feeding of finisher diets for the last 10 d is often necessary to provide adequate feathering with a minimum of pin-feathers. When weaning geese from, or to pasture systems, the allocation of complete feed should not change abruptly, rather the transition should occur over a 2 - 3 d period. For geese on pasture, or for birds supplemented with whole grains, insoluble grit should be offered at about 1 kg/100 geese/wk.

The energy level of all diets should be considered in relation to the propensity for carcass fat deposition. It appears as though growth rate, and hence carcass fat content, is more sensitive to diet energy concentration than to inputs of protein and/or amino acids (Table 8.16).

TABLE 8.16 Effect of diet protein and amino acid supplementation on performance of Embden geese

Diet crude protein (%)	Methionine (%)	Lysine (%)	Body wt (kg)				Carcass fat (% DM)	
			21d		63d			
			♂	♀	♂	♀	♂	♀
22	.36	1.25	1.76	1.68	4.8	4.3	49	54
20	.34	1.10	1.80	1.63	5.0	4.4	50	53
18	.31	0.96	1.75	1.64	4.9	4.4	49	50
16	.29	0.81	1.60	1.55	4.7	4.4	51	53

In this study, Embden geese fed single-diet programs varying in protein content from 22 - 16% exhibited comparable body weights and carcass composition at 9 weeks of age. Even when all diets were supplemented with additional methionine and lysine, there was no response in these performance characteristics. Failure to show a response to amino acid supplementation beyond 3 wk of age suggests that with a 16% protein, corn-soybean meal diet the lysine requirement for maximizing subsequent gain is no higher than 0.8%. In this study, calculation of the ratio of protein consumed to protein appearing as edible carcass protein gave values ranging from 5 to 9% for the various diets, which is markedly lower than the values of around 15 and 21% reported for the chicken broiler and turkey, respectively. While feed wastage was relatively high in this study (Table 8.16), it would be expected to account for only a relatively small portion of the difference in protein utilization. Thus, while geese may be very fast growing animals, they appear to be extremely inefficient in converting dietary protein to edible carcass protein. However, it is obvious that in the present study, as well as many other reports, the level of dietary protein fed was in excess of that required for optimum gain.

Geese seem to derive about the same amount of energy as do chickens from most feedstuffs. Their ability to perform adequately on high-fiber diets is therefore a factor of increased feed intake, rather than improved digestibility. Geese are quite sensitive to a number of mycotoxins as previously described for ducks, and are also greatly influenced by anti-nutrients such as the trypsin inhibitor found in raw soybeans. This increased susceptibility to anti-nutrients is manifested as decreased feed efficiency rather than any direct effect on feed intake. In situations of reduced performance, such as occurs with raw soybeans, poor feathering is often an early indicator to potential problems. As with most species of waterfowl, stage and degree of feathering can have economic significance in terms of yield of feathers and/or incidence of pin-feathers. Most data suggest that geese are able to perform most economically on low-energy, high-fiber diets. This scenario usually implies access to pasture and/or whole grain as well as complete feeds. The capacity of the goose to consume large quantities of dry matter enables it to meet its nutrient requirement from a diet very high in fiber. The goose might provide an interesting alternative to ruminant animals in utilizing high fibrous forages.

Because geese produce relatively few eggs, their nutrient requirements for egg production are not greatly increased over maintenance - or at least not increased for any sustained period. In order to control body weight, breeder candidates should be offered holding diets soon after selection, and this feed offered on a restricted basis up to time of maturity. Specialized breeder diets can be introduced 2-3 weeks prior to anticipated first egg, or alternatively the birds fed increasing quantities of the holding diet together with 3 - 4 g calcium/d as large particle limestone or oyster shell. If breeders are retained for subsequent breeding seasons, then holding diets and/or grains with minerals and vitamins, should be allocated according to maintenance of body weight.

CHAPTER 8

Selected References

A) DUCKS

Adams, R.L., P.Y. Hester and W.J. Stadelman. 1983. *The effect of dietary lysine levels on performance and meat yields of White Pekin ducks.* Poultry Sci. 62:616-620.

Elkin, R.G. 1986. *Methionine requirement of male White Pekin ducklings.* Poultry Sci. 65:1771-1776.

Elkin, R.G. 1987. *A review of duck nutrition research.* World Poultry Sci. J. 43:84-106.

Farrell, D.J. and P. Stapleton. 1985. *Duck Production-Science and World Practice.* Publ. Univ. of New England, Armidale, Australia.

Jeschke, N., and P.E. Nelson. 1987. *Toxicity to ducklings of Fusarium moniliforme isolated from corn intended for use in poultry feed.* Poultry Sci. 66:1619-1623.

Leeson, S., J.D. Summers and J. Proulx. 1982. *Production and carcass characteristics of the duck.* Poultry Sci. 61:2456-2464

Olver, M.D. 1995. *Effect of restricted feeding during the rearing period a "forced moult" at 40 weeks of production on productivity of Pekin breeder ducks.* Br. Poult. Sci. 36:737-746.

Plavnik, I. 1988. *Protein requirements of Muscovy male ducklings.* Nut. Rep. Int. 39:13-17.

Stadelman, W.J. and C.F. Meinart. 1977. *Some factors affecting meat yield from young ducks.* Poultry Sci. 56:1145-1147.

Veltmann, J.R. and J.S. Sharlin. 1981. *Influence of water deprivation on water consumption, growth and carcass characteristics of ducks.* Poultry Sci. 60:637-642.

Vernam, J. 1995. *Assessing the mule duck as a meat producer.* Wld. Poult. 11: No. 5, p. 44.

Wilson, B.J. 1975. *The performance of male ducklings given starter diets with different concentrations of energy and protein.* Br. Poultry Sci. 16:617-625.

Yalda, A.Y. and J.M. Forbes 1995. *Effect of wet feeding on the growth of ducks.* Br. Poult. Sci. 36:878-879.

B) GEESE

Hollister, A.G., H.S. Nakaue and G.H. Arscott. 1984. *Studies with confinement reared goslings. 1. Effects of feeding high levels of dehydrated alfalfa and Kentucky Bluegrass to growing goslings.* Poultry Sci. 63:532-537.

Nitsan, Z. and I. Nir. 1977. *A comparative study of the nutritional and physiological significance of raw and heated soya beans in chicken and goslings.* Br. J. Nutr. 37:81-87.

Serafin, J.A. 1981. *Studies on the riboflavin, pantothenate, nicotinic acid and choline requirements of growing Embden geese.* Poultry Sci. 60:1910-1915.

Storey, M.L. and N.K. Allen. 1982. *Apparent and true metabolizable energy of feedstuffs for mature non-laying female Embden geese.* Poultry Sci. 61:739-745.

Summers, J.D., G. Hurnik and S. Leeson. 1987. *Carcass composition and protein utilization of Embden geese fed varying levels of dietary protein supplemented with lysine and methionine.* Can. J. Anim. Sci. 67:159-164.

CHAPTER 9. FEEDING PROGRAMS FOR GAME BIRDS, RATITES AND PET BIRDS

9.1 GAME BIRDS

Nutrient requirements of game birds such as the pheasant and quail have not been well established. In the past, diets formulated for turkeys have been fed to these classes of bird with reasonably good results. Another problem associated with the feeding of game birds is that they are often produced for various types of markets. For example, quail are produced mainly for meat and eggs, with a few being released for hunting while pheasant are raised both for release and for meat. For birds designed for release, fast growth or maximum body weight are not major considerations. This of course would not be true for birds raised for meat or egg production. Consequently, bird management is a major consideration in selecting a feeding program for game birds.

Pheasants

Table 9.1 outlines starter, grower and breeder diet specifications for pheasants. The pheasant starter diet should be fed to 6 weeks of age, followed by first grower diet to 12 weeks of age, then the second grower diet up to the time that the birds are sold for meat or until they are selected for breeding. Diets shown in Table 9.2 are complete diets and need not be supplemented with grain. However, the feeding of 5 to 10% cracked grain can be utilized after 12 weeks of age for birds that are to be released for hunting. The grain portion should be switched to whole-grain at 16 weeks of age at which time one half of the feed allotment can be grain. Such a feeding program results in a stronger, hardier bird and one more able to forage for itself when released.

The pheasant breeder diet should be fed to the birds starting at least 2 weeks before eggs are expected. Again this is a complete diet and no supplements should be added to it. Table 9.3 indicates weight gain and feed intake data for male and female pheasants to 18 weeks of age.

TABLE 9.1 Diet specifications for pheasants

	Starter	Grower		Breeder
		1	2	
Approximate protein level (%)	28.0	20.0	15.0	17.0
Amino acids (% of diet)				
Arginine	1.6	1.2	.80	.72
Lysine	1.6	1.2	.82	.73
Methionine	.60	.45	.34	.41
Methionine + cystine	1.1	.80	.61	.62
Tryptophan	.27	.22	.15	.15
Histidine	.60	.45	.35	.35
Leucine	1.9	1.4	1.1	.70
Isoleucine	1.2	1.7	1.3	.60
Phenylalanine	1.2	.93	.74	.61
Phenylalanine + tryosine	1.9	1.4	1.1	1.0
Threoine	1.1	.81	.62	.63
Valine	1.3	1.0	.75	.66
Metabolizable energy (kcal/kg)	2900	2900	2700	2800
Calcium (%)	1.2	1.0	0.8	2.5
Available phosphorus (%)	0.60	0.50	0.40	0.40
Sodium (%)	0.18	0.18	0.18	0.18
Vitamins (per kg of diet)				
Vitamin A (I.U.)	5000	4000		5000
Vitamin D_3 (I.U.)	2500	2000		2500
Choline equivalents (mg)	1000	750		1000
Riboflavin (mg)	4.0	3.0		4.0
Pantothenic acid (mg)	12.0	9.0		12.0
Vitamin B_{12} (mg)	.012	.010		.012
Folic acid (mg)	1.0	0.8		1.0
Biotin (mg)	0.25	0.20		0.25
Niacin (mg)	60.0	40.0		60.0
Vitamin K (mg)	2.0	1.5		2.0
Vitamin E (I.U.)	25	15		25
Thiamin (mg)	2.5	2.0		2.5
Pyridoxine (mg)	5.0	4.0		5.0
Trace minerals (per kg of diet)				
Manganese (mg)	70	55		70
Iron (mg)	80	70		80
Copper (mg)	10	8		10
Zinc (mg)	80	60		80
Selenium (mg)	0.3	0.2		0.3
Iodine (mg)	.40	.40		.40

TABLE 9.2 Examples of diets for pheasants (kg)				
	Starter	Grower1	Grower2	Breeder
Ingredients				
Corn	388	518	307	483
Soybean meal (48%)	520	295	120	220
Barley	---	129	400	200
Wheat bran	---	---	120	---
Limestone	15	14	12.5	60
Dicalcium phosphate (20% P)	22.5	19	15	12.5
Salt (Iodized)	3.5	3.5	3.5	3.5
Fat	40	10	10	10
Methionine	1.5	1.5	1.0	1.0
Vitamin-Mineral Premix[1]	10	10	10	10
L-Lysine HC1	---	---	1.0	---
Calculated analysis				
Crude protein (%)	28.0	20.3	15.1	17.1
Digestible protein (%)	27.4	18.2	12.8	15.3
Crude fat (%)	5.6	3.4	3.6	3.4
Crude fiber (%)	2.6	3.2	5.6	3.5
Metabolizable energy (kcal/kg)	2910	2920	2720	2800
Calcium (%)	1.16	1.0	0.9	2.6
Av. Phosphorus (%)	0.60	0.50	0.45	0.36
Sodium (%)	0.19	0.18	0.18	0.18
Methionine (%)	0.61	0.49	0.34	0.39
Methionine + cystine (%)	1.09	0.80	0.56	0.65
Lysine (%)	1.86	1.18	0.76	0.88

[1]Use additional choline if premix does not contain this vitamin

TABLE 9.3 Growth rate, feed consumption and feed efficiency of ring-necked pheasants						
	Average weight (g)		Cumulative feed consumption (g)		Feed efficiency	
Age (weeks)	Male ♂	Female ♀	Male ♂	Female ♀	Male ♂	Female ♂
2	85	85	144	144	1.71	1.71
4	220	200	430	416	1.98	2.07
6	380	350	866	794	2.25	2.25
8	620	520	1496	1352	2.43	2.61
10	830	660	2161	1915	2.61	2.88
12	1050	820	3136	2747	2.97	3.33
14	1300	960	4092	3640	3.15	3.78
16	1475	1025	5163	4709	3.51	4.59
18	1530	1080	6338	5827	4.14	5.40

Quail

Quail diet specifications are shown in Table 9.4. The quail starter diet (Table 9.5) should be fed as a complete feed up to 6 weeks of age. At this time, the birds should be placed on the grower diet either until they are marketed as meat or until one week before table or hatching eggs are expected from the females. As mentioned above, a small percentage of scratch grain may be employed. Table 9.6 shows body weight and feed intake data for both male and female quail to 10 weeks of age. Table 9.7 shows potential reproductive performance data that can be expected from a flock of laying quail.

TABLE 9.4 Diet specifications for quail and guinea fowl

	Quail			Guinea Fowl	
	Starter	**Grower**	**Layer Breeder**	**Starter**	**Breeder**
Approximate protein level (%)	28	18	20	25	18
Amino acids (% of diet)					
Arginine	1.7	.90	.90	1.5	.91
Lysine	1.5	.84	.80	1.3	.83
Methionine	.60	.50	.53	.52	.55
Methionine + cystine	1.1	.71	.81	.91	.74
Tryptophan	.25	.20	.22	.22	.17
Histidine	.55	.40	.45	.54	.41
Leucine	1.7	.82	1.4	1.5	.80
Isoleucine	1.1	.72	1.3	1.0	.73
Phenylalanine	1.1	.71	.90	1.0	.74
Phenylalanine + tyrosine	1.7	1.1	1.3	1.5	1.0
Threonine	1.0	.70	.82	.93	.71
Valine	1.1	.71	1.0	1.1	.72
Metabolizable energy (kcal/kg)	2900	2900	2950	2900	2900
Calcium (%)	1.3	1.0	3.0	1.2	3.0
Available phosphorus (%)	.60	.45	.40	.50	.40
Sodium (%)	0.18	0.18	0.18	0.18	0.18

Vitamin and mineral levels as shown for pheasants (Table 9.1).

TABLE 9.5 Examples of diets for quail and guinea fowl (kg)

Ingredient	Quail Starter	Quail Grower	Quail Breeder	Guinea Fowl Starter	Guinea fowl Breeder
Corn	370	554	513	466	622
Barley	---	150	---	---	---
Soybean meal (48%)	515	240	320	435	260
Limestone	17.5	14	70	20	70
Dicalcium phosphate (20% P)	22.5	17.5	12.5	20	12.5
Fat	60	10	70	45	20
Salt	3.5	3.5	3.5	3.5	3.5
DL-methionine	1.5	1.5	1.5	1.0	2.0
Vitamin-Mineral Premix[1]	10	10	10	10	10
Calculated analysis					
Crude protein (%)	28.0	18.1	19.9	24.9	18.0
Digestible protein (%)	25.6	16.3	18.1	22.8	16.3
Crude fat (%)	7.7	3.5	9.1	6.5	4.5
Crude fiber (%)	2.5	3.3	2.2	2.5	2.3
Metabolizable energy (kcal/kg)	3015	2950	3070	3010	2895
Calcium (%)	1.2	0.95	3.0	1.3	3.0
Available phosphorus (%)	0.6	0.47	0.38	0.5	0.38
Sodium (%)	0.18	0.18	0.18	0.19	0.18
Methionine (%)	0.58	0.46	0.48	0.5	0.51
Methionine + cystine (%)	1.03	0.74	0.71	0.89	0.78
Lysine (%)	1.72	0.94	1.13	1.50	0.96

[1]Use additional choline if premix does not contain this vitamin

TABLE 9.6 Mean body weight and feed intake of male and female quail to 10 weeks of age

	Male		Female	
Age (weeks)	Body wt (g)	Cumulative feed intake (g)	Body wt (g)	Cumulative feed intake (g)
2	40	50	40	50
4	90	180	100	190
6	120	300	130	330
8	130	350	160	450
10	140	400	170	510

TABLE 9.7 Reproductive performance of quail from 6 to 20 weeks of age				
50% Production		Eggs per hen	Fertility	Hatch of set
Age days	Body wt (g)		(%)	(%)
50	150	80	94	81

Guinea fowl

Guinea fowl production has increased in North America during the past few years. It is already a well established industry in some parts of Europe and especially in France. Rearing and feeding of guinea fowl does not differ significantly from that of commercial poultry. While commercial poultry diets have been employed in rearing guineas, the specifications shown in Tables 9.4 and 9.5 may more closely meet the requirements of the guinea at various times in the life cycle. The starter diet shown in Table 9.5 should be fed to 4 weeks of age, and this is followed by the 2 grower diets as shown for pheasants. Grower diet 1 should be fed from 4 to 10 weeks of age and grower diet 2 from 10 weeks until the birds are marketed. Breeding females are switched to the guinea breeding diet shown in Table 9.5 approximately 2 weeks before eggs are expected. This is a complete diet and no grain or grit supplements need to be fed. Table 9.8 indicates body weight and feed consumption for a mixed sex flock of guinea fowl up to 12 weeks of age.

TABLE 9.8. Body weight and feed intake of guinea fowl (mixed sexes) to 12 weeks of age			
Age wks	Body wt (g)	Cumulative feed intake (g)	Feed/gain
2	150	240	1.60
4	370	650	1.76
6	650	1400	2.15
8	930	2300	2.47
10	1200	3200	2.67
12	1400	4400	3.14

9.2. RATITES

Ratites refer to flightless birds such as the ostrich and emu, and in recent years, an industry has developed in many areas of the world to commercialize these large birds. Besides being a potential source of lean, red meat, the feathers and also the oil from these birds is in demand. These birds have no keel and no crop. In the ostrich and emu the food storage function is provided, in part, by an elongated proventriculus, which can be prone to impaction with large intakes of fibrous material.

Ostriches are by far the largest birds of this group with adults reaching in excess of 150 kg, and hens can produce 20 to 40 eggs per year, each averaging around 1.25 kg. Emus are about half the size of ostriches and while they produce a similar number of eggs as do ostriches, they are only about half the size. Emus are a more docile than the ostrich and thus easier to handle in a farm type environment. The large intestine of the ostrich is approximately 3 times the size of the small intestine, with food transit time of about 40 hours. The ostrich's digestive system is therefore adapted to handle large quantities of roughage material allowing it to obtain a significant portion of its energy from hind gut fermentation. The emu's digestive system, unlike that of the ostrich, is more like that of poultry, with rate of food passage being around 5 to 6 hours. Despite the apparent limitation in fermentation sites, there are reports suggesting that emus can digest more fibrous material than do other types of commercial poultry.

Ostrich diets usually contain 6 to 15% fiber, depending on the age of the bird, while diets similarly vary in protein from 23 to 15%. It has been reported that the ostrich will metabolize 30 to 40% more energy from a diet than will poultry. This could account for some of the obesity problems seen in breeders fed "poultry diets". Both the ostrich and the emu are prone to leg problems as well as apparent vitamin E and selenium deficiencies. Consequently, mineral fortification should be closely monitored, especially if forage or dehydrated alfalfa make up a significant proportion of their diet. Vitamin E fortification of up to 80 IU/kg of diet is usually recommended, while selenium supplementation must remain within approved levels.

Young ostrich and emu chicks have the tendency to search for and pick up any appropriate size material in their surroundings and so litter eating can be a problem in commercial units. To avoid this occurrence, chicks can be started on some type of rough paper or burlap. However care must be taken to avoid anything with a smooth surface because ratite chicks are prone to leg problems. During the first week, the young chicks have a relatively slow rate of growth and growth rate is not constant as occurs with other meat birds. Growth in the first 7 - 10 d is quite slow and then this period is followed by maximum growth rate up to about 6 months of age. Subsequent slower growth continues up to maturity at around 30 months of age.

There have been few reports of research involving ostrich chicks, although Gandini *et al.* (1986) cite convincing evidence for protein requirements of around 18 - 20%. Table 9.9 outlines estimates of nutrient requirements of ostrich chicks and breeders. There does not seem to be any major restriction on ingredient use, although most reports of nutritional studies involve complete diets together with free-choice forage such as alfalfa for chicks and hay for breeders. With this intake of forage, it seems important to provide insoluble grit - pebbles are often referred to in feeding of adults.

Ostrich chicks show fast early growth rate, as described earlier and estimates of feed efficiency suggest a value of around 3:1 to 8 weeks of age (Table 9.10). However such feed intake data are approximate due to problems of feed wastage especially with forage material.

Ostrich chicks suffer from leg problems similar to those seen with broiler chicks, although due to the relative size of the leg bones, such disorders are more easily noticed. Supplemental calcium feeding, as calcium borogluconate solution, seems useful for severely affected individuals, suggesting perhaps that attention should be given to calcium levels in the diet, and especially calcium availability of forages.

TABLE 9.9 Diet specifications for ostriches[1]

	Starter (0-8 weeks)	Breeder
Crude protein (%)	18.0	15.0
Metabolizable energy (kcal/kg)	2750	2650
Methionine (%)	0.36	0.30
Methionine+cystine (%)	0.70	0.62
Lysine (%)	0.90	0.72
Calcium (%)	1.40	1.80
Av Phosphorus (%)	0.70	0.45
Sodium (%)	0.18	0.17

[1]Plus free choice forage - actual energy level therefore reduced to equivalent of around (2300 kcal/kg)

TABLE 9.10 Body weight and feed intake of ostrich chicks

	Weeks of age							
	1	2	3	4	5	6	7	8
Body wt (kg)	1.1	1.7	2.5	3.0	4.0	5.0	7.0	10.0
Cumulative Feed intake (kg)	1.6	3.1	4.8	6.0	9.0	12.5	18.5	31.0

After 4 to 6 weeks of age, liberal quantities of green forage can be offered to the birds either as a supplement or from pasture grazing. At 6 months dietary protein level can be reduced to 13 to 15% with energy level remaining at 2700 - 2800 kcal ME/kg. Since ostriches make relatively good use of high fibrous feedstuffs their diets can contain up to 20% fiber. Thus liberal quantities of forage or alfalfa hay can be fed along with the prepared diet. Emus cannot handle as much fiber as do ostriches, and so their fiber intake should not make up more than 10 to 15% of their diet. Because of this high fibre intake, it is common practice to supply grit to the birds at least once a week.

During the breeding season, the females markedly reduce their feed intake somewhat like a turkey breeder. Thus it is important that they be in good condition carrying sufficient nutrient reserves, so that quality eggs are produced, leading to potentially healthy offspring. Table 9.11 shows gross composition of eggs from ostriches and emus.

TABLE 9.11 Ratite egg components

	Weight	Albumen	Yolk	Shell
	(g)	(%)	(%)	(%)
Ostrich	1200	54	32	14
Emu	600	53	34	13

There is relatively little information available on carcass composition of ratites, because in N. America at least, there have been relatively few animals killed for meat production to date. Table 9.12 shows some carcass and body composition data of ostriches killed at 100 kg live weight.

TABLE 9.12 Carcass composition of 100 kg ostrich

	% Yield of live weight
Feathers	1.8
Blood	3.0
Hide	7.0
Feet	2.5
Carcass	60.0
Heart	1.0
Liver	1.5
Abdominal fat	4.0
Viscera	8.5

9.3. PET BIRDS

Most pet birds are fed diets based on whole seeds rather than complete feeds as pellets. Feeding birds along the guidelines outlined in Table 9.13 would seem more logical from a nutritional viewpoint, although nutrition *per se* does not always seem to be the major factor involved in diet selection by owners.

Use of complete pelleted feeds has been met with resistance by owners, and not all species of bird readily accept conventional pellets. However by using extruded pellets and incorporation of color/taste/smell additives, both the owner and bird can be coaxed into using such complete feeds. With increased demand for hand-reared birds due to subsequent behaviourial disposition and the potential for a ban in trade in exotic species, there is increased demand for information on diets for hand-feeding of newly-hatched birds. While many such hatchlings are fed on baby food/peanut butter mixes, a gruel formulated to the specifications shown in Table 9.13, and composed of more conventional ingredients, should prove more economical for large-scale operations. However due to the monetary value of many of these pet-birds, economics of nutrition is not always a major factor, especially when one considers the actual feed intake of these very small birds.

For adults and young birds that are self-feeding, then conventional ingredients as detailed in Chapter 2 can be considered. Fruit flavours may be an advantage with some of the more exotic species. Feed wastage can be a major problem with many pet birds, especially when pelleted feeds are first introduced to birds that are more accustomed to whole grain diets. Under these conditions, feed wastage can approach 10x that of the actual feed intake. Due to the difficulty of measuring feed wastage, and the obvious problem of equating feed intake with feed disappearance from the feeder, monitoring of body weight becomes the most rigorous criteria to be used in assessing any new feeding program. Regardless of the feeding program used, behaviourial problems are often reduced if birds have access to other appropriate nutrient sources, such as cuttle fish bones and/or fresh fruit etc.

TABLE 9.13 Diet specifications for pet birds

	Budgie		Parrots		Hand-feeding
	Young	Adult	Young	Adult	
Crude protein(%)	23.0	15.0	21.0	14.0	26.0
Metabolizable energy (kcal/kg)	3000	2900	2900	2800	3200
Crude fat (%)	5.0	5.0	5.0	4.0	10.0
Crude fiber (%)	3.0	3.0	3.0	3.0	3.0
Calcium (%)	1.0	1.0	1.0	0.9	1.4
Available Phosphorus (%)	0.45	0.45	0.45	0.4	0.7
Sodium (%)	0.17	0.17	0.16	0.14	0.18
Methionine (%)	0.50	0.30	0.43	0.25	0.60
Methionine+cystine (%)	1.32	0.61	1.00	0.52	1.20
Lysine (%)	1.30	0.75	1.20	0.68	1.40

[1]**Mineral-vitamin premix as per turkey with 200 mg/kg Vit C**

CHAPTER 9

Selected References

Chah, J.C. 1982. *Scientific formulation of diets for captive birds.* in Proc. 2nd Dr. Scholl Conf. on Nutr. of Captive Wild Animals. Lincoln Park Zoo, Chicago, Illinois.

Flegal, C.J. 1993. *Diets for growing and breeding ostrich.* Proc. Multi-state Big Bird Conf., April 24 - 25.

Gandini, G.C.M., R.E.J. Burroughs and H. Ebedes. 1986. *Preliminary investigation into the nutrition of ostrich chicks under intensive conditions.* J. South African Vet. Sci. March 1986. 39-42.

Muirhead, S. 1995. *Ratite gastrointestinal physiology, nutrition principles explored.* Feedstuffs Oct. 2., P. 12.

Roudybush, T. 1986. *The nutrition of altricial birds.* in Proc. 6-7th Dr. Scholl Conf. on Nutr. of Captive Wild Animals. Lincoln Park Zoo, Chicago, Illinois.

Tables AEC. 1987. *Nutrient requirement tables for various game bird species.* A.E.C. Rhone-Poulenc. Paris, France.

Vohra, P. 1992. *Information on ostrich nutritional needs still limited.* Feedstuffs July 13, P 16.